HAZARDOUS CHEMICAL SPILL CLEANUP

HAZARDOUS CHEMICAL SPILL CLEANUP

Edited by J.S. Robinson

NOYES DATA CORPORATION
Park Ridge, New Jersey, U.S.A.
1979

Library of Congress Catalog Card Number: 79-16362
ISBN: 0-8155-0767-4
Printed in the United States

Published in the United States of America by
Noyes Data Corporation
Noyes Building, Park Ridge, New Jersey 07656

Library of Congress Cataloging in Publication Data

Main entry under title:

Hazardous chemical spill cleanup.

(Pollution technology review ; no. 59)
Bibliography: p.
Includes index.
1. Hazardous substances. I. Robinson, J. S., 1936– II. Title: Chemical spill cleanup. III. Series.

TD811.5.H39 604'.7 79-16362
ISBN 0-8155-0767-4

FOREWORD

This book is intended to serve as a practical guide and manual for the handling and control of hazardous material spills. Oil spills, their prevention and removal techniques, are described in other volumes published by Noyes Data Corporation.

Spill control and cleanup methods have been emphasized. The introductory overview deals with the legislative aspects as they now stand, while the five chapters that follow give practical advice and response techniques for individual types of materials. This includes the preferred method of disposal for the recovered substances, and spill damage restoration.

Cost figures provided are those given in the particular report cited, the date of which is always given. When the dates of the cost figures themselves are given, we have included them.

Advanced composition and production methods developed by Noyes Data have been employed to bring this durably bound book to you in a minimum of time. Special techniques are used to close the gap between manuscript and completed book. Industrial technology is progressing so rapidly that time-honored, conventional typesetting, binding and shipping methods are no longer suitable. We have bypassed the delays in the conventional book publishing cycle and provide the user with an effective and convenient means of reviewing up-to-date information in depth.

The expanded table of contents is organized in such a way as to serve as a subject index. It provides easy access to the information contained in this book which is based on various studies produced by and for diverse governmental agencies under grants and contracts. These primary sources are listed at the end of the volume under the heading Sources Utilized. The titles of additional publications pertaining to topics in this book are found in the text.

CONTENTS AND SUBJECT INDEX

INTRODUCTION

THE UPWARD TREND IN SPILLS OF HAZARDOUS MATERIALS

The information in this section is based on *Evaluation of MTF for Testing Hazardous Material Spill Control Equipment*, EPA-670/2-74-073 prepared by C.R. Thomas and G.M.L. Robinson of Environment Control Corporation and E.J. Martin of Environmental Quality Systems, Inc. as subcontracted by Hancock County Port and Harbor Commission for the Environmental Protection Agency.

American industry, the major producer of hazardous materials, handles these materials at virtually every industrial location throughout the country and regularly transports them by various modes such as truck, barge, ship, train, pipeline and aircraft. During the ten-year period between 1958 and 1968, production of organic and inorganic chemicals, respectively, increased 130% and 90%. In comparison, during the 1958-1968 period, the rate of increase in population growth was a relatively small 15% (1).

The increased production and transport of chemicals have, in turn, increased the incidence rate of accidental spills in the United States. In 1971, the U.S. Coast Guard received reports of 8,496 spills, as compared to 3,711 reports in 1970 (2). This represents approximately a 229% increase in incidence of spills in that one-year period; however, a large part of the increase was due to improved reporting requirements.

In addition to the U.S. Coast Guard data, the U.S. Department of Transportation (USDOT) reported 12,664 landspills associated with overland shipments of various hazardous materials between January 1, 1971 and December 31, 1973 (3).

Although the USDOT figures do not include pipeline transport or storage container data, the 12,664 land-related spills indicate: (1) a large and significant number of non-water-related spills, and (2) the probability of a greater number of spills which are not reported under present reporting techniques.

The USDOT figures also indicate a progressive increase in the number of accidental spills reported over the three-year period of operation by the Office of Hazardous Materials of the Accident Analysis Office. The number of reports for 1973 represents an increase of 37% over the reports of 1971 and represents 48% of all known reports since the USDOT Accidental Spills Reporting Program began in 1971.

HAZARDOUS MATERIAL SPILLS–THE NATIONAL PROBLEM

> The material in this section is based on a paper given at Conference 1972 by L.D. Attaway of the Environmental Protection Agency. The title of the paper was "Hazardous Material Spills–The National Problem."

In a statement which very succinctly describes the general feeling of the nation, President Nixon said: "The 1970s absolutely must be the years when America pays its debt to the past, by reclaiming the purity of its air, its waters, and our living environment. It is literally now or never."

The mandate leading to this statement comes in part from a long-smoldering reaction by the people to the gross pollution which has been destroying the natural beauty and the quality of life in the United States, but it is also based upon hard scientific data. The biological sciences have provided alarms which quite rightly have given the solution of this problem such a high priority.

Through an increased understanding of the effects of industrialization, people have become very much aware that a natural balanced environment is a tightly integrated combination of chemical, physical and biological elements. A change of any one of these three may greatly affect the other two.

A large number of trace elements and chemicals are quite valuable, in low concentrations, for maintaining the biological stability of a community. But although an organism may require the presence of these elements within a certain concentration range, in many cases an excess of these same elements can be toxic. For example, it is well known that zinc is an essential micronutrient. It has been found, however, that concentrations as low as 0.1 ppm will exhibit toxic effects on some fish.

In nature, the probability of gross changes of chemical properties, while present, is usually very small. These changes, as they occur, normally would be painstakingly slow compared to the rates of change we see today. The chances for excess concentrations to occur before the era of production of industrial chemicals were very limited.

Now, through man's activities, this natural balance system has been subjected to an estimated 50 billion gallons per year of polluted water, 142 million tons of air pollutants, 360 million tons of garbage, and 1.5 billion pounds of industrial solid waste. The equilibrium between chemical, physical and biological elements has undergone a severe shift, very much to the detriment of the biological element. Quite rightly the President, Congress, and State Legislatures have spoken through increasingly meaningful legislation. It is clear to everyone that the waters, air and land are to be restored.

Estimates of the cost vary with the level of restoration and the approach, but they range from $75-125 billion in the next five to ten years to obtain a quality of water environment essentially equivalent to 1915. As a percentage of GNP, however, which is one crude method for measuring economic impact, these costs are certainly well within the capacity of the nation. Annualized costs would run to 1.5% of the GNP.

It is the industrial revolution that produced this ecological imbalance. It now remains for an industrial resolution to correct it.

Added to this industrial resolution is a further need to prevent unnecessary spills or minimize accidental spills.

It is in terms of an environment far improved over the present situation that we must evaluate the impact of hazardous material spills. In view of the tremendous national effort and the important historical significance of this task, what would be the public reaction, to an industry which releases, over a period of one year, in concentrated slugs, at unexpected times and places: 1.7 million lb of sulfuric acid, 3.2 million lb of various industrial alcohols, 3.4 million lb of fertilizer solution, 44,000 lb of phenol, 2 to 3 million gallons of industrial wastes, 360,000 lb of acetone, 600,000 lb of various monomers, 3 million lb of other organic chemicals, 19,000 lb of especially toxic materials, such as pesticides, or 2,000 million gallons of mining wastes.

These are undoubtedly only a fraction of the spills actually occurring, which include both in-plant and transportation spills. Many of these spills occurred in streams now biologically sterile from chronic pollution, and have escaped the widespread publicity which could occur. This will not be the case a few years hence.

Does the cost for prevention of spills compare to the environmental and human safety benefits received? The major characteristics of a hazardous material spill, either through a transportation or in-plant accident, are these: (1) Not only are spills unpredictable as to time and location and type of waterway, but, in addition they vary widely as to the properties of the material. This makes response difficult, and requires a considerable amount of specific technical knowledge. (2) Spills often present immediate severe threats to human health, and in many cases the prevention of adverse environmental effects becomes secondary. (3) This one is going to assume increasing importance. A spill may in a few hours cancel the results of years of environmental improvement efforts. (4) From sad experience, we have found that very small spills of certain materials can have far-reaching effects.

So, with these special properties of a hazardous material spill, it is apparent that we have the potential for undermining a large portion of our restoration effort from the aforementioned tremendous outlay of funds. A major improvement program is essential and now is the time to begin.

The following discussion describes, in very general terms, first the legislative background relating to spills and then the effort that the Federal Government is making to provide its share of the solution.

Legislative Background

The Torrey Canyon incident in 1967 directed attention to the problem of immediate and long-term damages to the environment from spills of oil. This, and subsequent major oil spills, resulted in correcting legislation as well as focusing attention on problems associated with accidents involving chemicals. The Water Quality Improvement Act of 1970, therefore, contained a specific section, entitled "Control of Hazardous Polluting Substances." This section required the promulgation of regulations designating hazardous materials.

It also gave the President the authority to establish means for the removal of such substances. Another provision of the section requires reporting of spills.

Also, when the owner or operator of a facility where a spill has occurred refuses to take appropriate action, the President is to have the authority to arrange for removal, at the owner's expense.

Under this legislation, EPA has directed an effort to the problem of hazardous material spill control. A national contingency plan has been established, and teams have been established at both headquarters and in the regions to meet this problem.

For general discussion, the problem of hazardous materials spill control may be separated into three lines of defense: (1) prevention; (2) countermeasures; and (3) restoration.

Prevention, the first and most important line of defense against spills, is outside the scope of this book; however, assuming that very great improvements can and will be made on the prevention of spills within the next few years, there is still the unfortunate probability that some spills will occur. Under the authority given by the Water Quality Act of 1970, EPA has undertaken research and development work aimed at giving better tools to the people who face the control of these spills in the field.

The first need for applying countermeasures is an improved reporting and information system. Quick response to a spill is essential if any of the treatment systems are to have a chance of working. Keeping in mind the vast numbers of chemicals which can be spilled, a large amount of technical information must be available for the choosing of the proper countermeasure. Computerized information systems are being developed which will place this information in the hands of the people in the field.

For spills which are not reported, or for locating and following reported spills, there is a need for portable instruments which can be used for the detection and identification of an array of chemicals.

An effort is also being made to put together technology which will mitigate the effects of spills. The necessity for delivery on very short notice has introduced certain engineering constraints. There is little doubt that this second line of defense will be expensive to call in, and it will merely mitigate, not eliminate, the damage to the environment. Keeping in mind the variety of properties which one can expect to face, it is apparent that there will not be one cure-all countermeasure for all spills.

The development of countermeasures is an area where the private sector has apparently had no economic incentive to enter, and one where a large void exists.

Restoration, returning of the damaged environment to its prespill condition, can be a sad and expensive task faced by the spiller. The effects of a shock exposure to extremely high concentrations has an effect which can be different than the same material slowly metered into the environment. Restoration encompasses a survey of the failure of our prevention program and our countermeasures program and an attempt to rebuild to the natural environment. In most cases this requires the disposal of any remaining pollutants and setting the stage for nature to cure its wounds. Restoration is an aspect that can be minimized by an aggressive prevention plan.

Since a spill, when it does occur, must be handled under crisis conditions, it is extremely important that the procedure for handling of spills be previously thought out and that cooperation between the state agencies, industry, and the Federal Government be obtained. Because of the many overlapping areas of responsibility this is no small task.

The Federal Government, through an interagency agreement, has attempted to put its own house in order. Under this agreement DOT has prime responsibility for the prevention of accidents while in transit. EPA has the lead responsibility for prevention of in-plant spills. Once a spill threatens a watercourse, EPA has lead responsibility. Thus, the EPA leads in the countermeasure programs.

As a further effort to provide preplanning for the handling of spills, the Council for Environmental Quality has, as directed by the Water Quality Act of 1970, established a National Oil and Hazardous Substances Contingency Plan. This plan establishes an organization for handling spills which are beyond the capability of regional teams.

The cooperation between industry, the states, and the Federal Government has been rapidly improving as everyone realizes the importance of attacking this problem. There is no doubt that the increased effort to prevent and mitigate the effects of spills will intensify the need for still closer cooperation.

To summarize, a significant reduction in the amount of hazardous materials being spilled into the environment is required to conform to the National Effort for a Restored Environment. Prevention should be the main thrust. It must be the industrial sector which applies its technical and managerial expertise to lead the way.

Efforts for the second line of defense, containment and treatment, will be dictated by the success of the prevention program, and hopefully will get smaller.

Restoration will be required, but it, too, should become minimal and insignificant with time, depending upon the degree of success of the other two more important aspects.

EPA and other participating agencies are anxious to have industry take the leadership in these areas, and to have the Federal Government participate and cooperate to help resolve spill problems.

GOVERNMENT ROLES IN CONTROLLING HAZARDOUS CHEMICALS

The material in this section is based on a paper given at Conference 1978 by Lt. Gregory N. Yaroch of U.S. Coast Guard Headquarters. The title of the paper was "Government Roles in Controlling Hazardous Chemicals."

Introduction

Public regulation of industry is but one phase of the relationship of government to economic life. Normally, regulation develops by the more or less gradual process of adjustment to new situations as they arise. As long as this change is not too rapid, serious disruptions do not emerge as there is time to gain a reasonable grasp of the implications of the change.

Today, however, all phases of human activity are developing rapidly. An increasingly progressive industrial society has seen a rapid proliferation of oil and petrochemical production and usage. The economic advantages derived from these products are significant but steps must continually be taken to insure that these benefits are not reaped at the expense of public safety and welfare. A striking potential for harm to the environment arising from accidental release of hazardous chemicals, as well as the increasing public awareness of this problem, has led to the comprehensive regulation of chemicals in production, utilization and transportation. Viewing the progression of legislation since the turn of the century, one can readily observe that the seemingly concentrated environmental legislative activity within just the last six years is long overdue. Congress moved to establish a national objective, beginning in the early 1970s, with far-reaching legislation aimed at restoring and maintaining the chemical, physical and biological integrity of the environment.

Repealing the Oil Pollution Act of 1924 the Water Quality Improvement Act of 1970 was the first comprehensive effort toward environmental protection declaring a national policy that there should be no discharges of oil into or upon the navigable waters of the United States, the adjoining shorelines, or from vessels operating in the Contiguous Zone. The Water Quality Improvement Act, as amended in 1972, is commonly called the Federal Water Pollution Control Act or the FWPCA. Within the FWPCA there is another expansion of the nation's attention toward hazards to man and the environment by addressing, the heretofore ignored, "hazardous polluting substances."

New levels of environmental consciousness have evolved in the passage of the Clean Water Act of 1977. Amending the FWPCA, this Act extends the prohibition against discharges of oil and hazardous substances to include not only discharge into the waters of the United States and Contiguous Zone, but also discharges in connection with activities under the Outer Continental Shelf Lands Act and the Deepwater Port Act or which may affect natural resources under the authority of the United States, including resources under the Fishing Conservation and Management Act.

Historically, Federal regulatory authority in the area of hazardous chemicals has been directed toward controlling some particular hazard associated with the use of, or establishing specific conditions of exposure to hazardous chemicals.

To more broadly control the wide spectrum of hazardous chemical substances, the Toxic Substances Control Act was enacted in 1976. Implemented on January 1, 1977, TSCA acts to regulate chemicals in commerce, without regard to their particular use or application. If one establishes whether a commodity presents "an unreasonable risk of injury to health or the environment," it becomes possible to control that chemical at its source before human or environmental exposure has occurred.

Whereas TSCA clearly reflects the increased sense of environmental responsibility in the seventies by authorizing testing and review of chemical products prior to their dissemination into commerce, the Resource Conservation and Recovery Act of 1976 goes in the opposite direction to monitor and control the ultimate disposal of hazardous waste following the use of such commodities. Amending the Solid Waste Disposal Act of 1965, this new legislation provides a program to eliminate open dumping and supply assistance in the improvement of solid waste management programs. It further expands current provisions of the Solid Waste Disposal Act requiring the promulgation of guidelines on alternatives in solid waste management intended to aid state and regional authorities in their development or enhancement of solid waste management plans.

To more clearly understand governmental direction towards controlling the hazards associated with the chemical industry, a brief summary is provided of the policy and intent of Congress in their most recent legislative efforts. Concentrating on trend setting Federal environmental and safety statutes of the 1970s, one may see the steady formulation of a comprehensive move toward governmental and industrial cooperation in overseeing the public health and welfare.

Toxic Substances Control Act of 1976

Confronting the reality that the advancement of an industrialized society also brings with it an array of unhealthy risks for man and environment, the Toxic Substances Control Act was signed into law on October 11, 1976 to regulate the production and distribution of toxic substances which were not controlled by previous legislation. An understanding of the intent of the Act is facilitated in a review of the Congressional statements of policy within the Act itself (4). "It is the policy of the United States that:

(1) adequate data should be developed with respect to the effect of chemical substances and mixtures on health and the environment and that the development of such data should be the responsibility of those who manufacture and those who process such chemical substances and mixtures;

(2) adequate authority should exist to regulate chemical substances and mixtures which present unreasonable risk of injury to health or the environment, and to take action with respect to chemical substances and mixtures which are imminent hazards; and

(3) authority over chemical substances and mixtures should be exercised in such a manner as not to impede unduly or create unnecessary economic barriers to technological innovation while fulfilling the primary purpose of this Act to assure that such innovation and commerce in such chemical substances and mixtures do not present an unreasonable risk of injury to health or the environment."

The structure of this legislation authorizes the Environmental Protection Agency, as administrator of the Act, to require industry to provide data on chemical substances and mixtures. Through this mechanism, the EPA may regulate the production, processing, dissemination in commerce, utilization and ultimate disposal of a particular chemical substance or mixture, thus providing an "early stage" of control under Federal regulatory authority which did not generally exist in previous legislation. Additionally, TSCA, through the unprecedented comprehensive collection of chemical information provides the basis for examining the totality of health hazards as a result of environmental exposures.

Resource Conservation and Recovery Act of 1976

Signed into law on October 21, 1976 as an amendment to the Solid Waste Disposal Act, the Resource Conservation and Recovery Act evolved with an intent to provide technical and financial assistance for the development of solid waste management plans. These plans are intended to provide the mechanism for the recovery of energy and other resources from discarded materials and for the safe disposal of discarded materials and to regulate the management of hazardous waste.

The substance of the Act is embodied in the Congressional presentation of objectives (5): "The objectives of this Act are to promote the protection of health and the environment and to conserve valuable material and energy resources by:

(1) providing technical and financial assistance to state and local governments and interstate agencies for the development of solid waste management plans which will promote improved solid waste management techniques, new and improved methods of collection, separation, and recovery of solid waste, and the environmentally safe disposal of nonrecoverable residues;
(2) providing training grants in occupations involving the design, operation, and maintenance of solid waste disposal systems;
(3) prohibiting future open dumping on the land and requiring the conversion of existing open dumps to facilities which do not pose a danger to the environment or to health;
(4) regulating the treatment, storage, transportation, and disposal of hazardous wastes which have adverse effects on health and the environment;
(5) providing the promulgation of guidelines for solid waste collection, transport, separation, recovery, and disposal practices and systems;
(6) promoting a national research and development program for improved solid waste management and resource conservation techniques, more effective organizational arrangements, and new and improved methods of collection, separation, and recovery, and recycling of solid wastes and environmentally safe disposal of nonrecoverable residues;
(7) promoting the demonstration, construction, and application of solid waste management, resource recovery and resource conservation systems which preserve and enhance the quality of air, water and land resources; and
(8) establishing a cooperative effort among the Federal, state and local governments and private enterprise in order to recover valuable materials and energy from solid waste."

Clean Water Act of 1977

The Clean Water Act of 1977, amending the Federal Water Pollution Control Act, was signed into law on December 28, 1977. This Act provides the nation's most comprehensive legislation toward the preservation of the waters of the nation, dramatically expanding not only the Federal role in water pollution control, but also the jurisdictional coverage of the provisions of the Act (6).

"As stated by Congress, the objective of this Act is to restore and maintain the chemical, physical, and biological integrity of the nation's waters. In order to achieve this objective it is declared that, consistent with the provisions of this Act–

(1) it is the national goal that the discharge of pollutants into the navigable waters be eliminated by 1985;
(2) it is the national goal that wherever attainable, an interim goal of water quality which provides for the protection and propagation of fish, shellfish, and wildlife and provides for recreation in and on the water be achieved by July 1, 1983;
(3) it is national policy that the discharge of toxic pollutants in toxic amounts be prohibited;
(4) it is the national policy that Federal financial assistance be provided to construct publicly owned waste treatment works;
(5) it is the national policy that areawide waste treatment management planning processes be developed and implemented to assure adequate control of sources of pollutants in each state; and
(6) it is the national policy that a major research and demonstration effort be made to develop technology necessary to eliminate the discharge of pollutants into the navigable waters, waters of the Contiguous Zone, and the oceans."

The Clean Water Act now extends U.S. prohibition against discharges of oil and hazardous substances beyond the Contiguous Zone to those areas or activities under the Outer Continental Shelf Lands Act and the Deepwater Port Act or which may affect natural resources under the management authority of the United States, including resources falling under the purview of the Fishing Conservation and Management Act. The Act also provides Federal authority to initiate containment and removal activities in response to any discharge occurring within the expanded coverage of jurisdiction, imposing an unprecedented challenge in environmental protection.

A look at the objectives of these recent legislative controls lends perspective to the expanding interplay between governmental authorities and the chemical industry. Control in the industrial community necessitates compliance with established rules of conduct; and the emphasis, naturally, is on the compulsory aspects. However, it should be recognized that control also involves consent, and that successful control can only be expected when the established rules give expression to generally accepted standards of conduct with industry. In summary, control may very well involve compulsion, but it cannot be maintained under significant resistance to the stated objectives of the law. Thus, in a broader sense, governmental control actually relies upon voluntary cooperation.

EPA REGULATIONS

The material in this section is based on a paper presented at Conference 1978 by Joseph I. Lewis and Alexandre R. Tarsey of the U.S. Environmental Protection Agency. The title of the paper was "EPA's Hazardous Spill Control Regulations."

In December 1977 the President signed the Clean Water Act of 1977 (6). This law amended the Federal Water Pollution Control Act (7) which previously had been extensively rewritten by Public Law 92-500 (8).

This section will first address briefly those parts of these laws that concern the discharge of hazardous substances. It will then discuss the regulations promulgated by EPA in compliance with these laws.

The objective of the law is to eliminate the discharge of pollutants into the navigable waters of the U.S. by 1985. Section 311 deals with the discharge of hazardous substances. It provides for the Executive Branch:

(1) to designate elements and compounds that are hazardous when discharged into the waters of the U.S.;
(2) to make a determination of which of them are removable;
(3) to designate harmful quantities;
(4) to exercise a variety of powers to regulate the discharge of hazardous substances, and mitigate their effects;
(5) to collect the costs of such removal from the owners or operators of responsible facility, which costs can run very high;
(6) to assess fines for discharging substances that are not removable. These fines either may be between $500 and $5,000, or alternatively may range up to $500,000 for discharges from shore facilities and $5 million for discharges from vessels, in accordance with a schedule to be established, based on the quantity and the identity of the discharge. The application of this alternate schedule of fines is discretionary; and
(7) to assess an additional civil fine of up to $5,000 administered by the Coast Guard for any spill of oil or hazardous material.

Section 311 also provides for a variety of other items including the establishment of a revolving cleanup fund and a trained strike force, and a $10,000 fine or one year imprisonment for failure to report, as required, the discharge of oils or hazardous materials.

The Clean Water Act of 1977 amended Section 311 in several ways. First, it clarified the status of mitigating action when nonremovable substances are discharged. Up to that time it had been clear that the Coast Guard could charge a civil penalty of up to $5,000 for discharges of oils and all types of hazardous materials, and that spills of oil or removable hazardous substances could be cleaned up at the expense of the owner or operator; and that EPA could collect one of the two alternate additional penalties if the substance discharged was not removable. The law was silent on whether EPA could act to mitigate effects of nonremovable substances by means of the contingency fund. The 1977 Act specifically adds the authorization for EPA to act to mitigate damages to public health and welfare resulting from discharges of nonremovable substances, and to assess the costs to the dischargers. These are examples of such costs: (1) containment; (2) measures to warn and protect the public; (3) monitoring of temporary water supplies; (4) monitoring the spread of pollution; (5) efforts to raise sunken vessels; (6) emergency treatment facilities; and (7) dredging, such as occurred in one area of a harbor in the case of spillage of PCB's, a substance determined to be not removable.

The Clean Water Act also revises liability charges. (1) Cleanup liability for vessels is now generally limited to $150 per gross ton unless they carry oil or hazardous substances, in which case the cleanup liability is at least $250,000. Special limits have been set for inland oil barges. (2) Cleanup liabilities for onshore facilities are now subject to a $50,000,000 maximum in lieu of the old $8,000,000 maximum. The President may lower the maximum for specific categories of facilities. (3) There is no maximum cleanup liability in case of unlawful misconduct or negligence.

The Clean Water Act also extends the area of applicability far out into the ocean beyond the contiguous zone to include activities under the Continental Shelf Land Act, or the Deepwater Port Act of 1974. However, on the outer continental shelf the Coast Guard $5,000 penalty is limited to vessels under U.S. jurisdiction, as is the $10,000 penalty or up to one year imprisonment for failure to report a discharge.

On March 3, 1978, in conformance with an Executive Order (9), EPA published four regulations that have been determined by the Agency to be key to the establishment of a hazardous substances regulatory program. The program is based on the dual concepts of encouraging proper handling and cleanup through civil penalties for unauthorized discharges and assessment of costs incurred in cleanup and removal.

Prior to promulgating these rules, EPA had published proposed rules for public comment (10). More than 160 comments were received and carefully considered.

The first of the rules, 40 CFR 116, designates as hazardous 271 chemical substances together with any hydrates, isomers, and solutions or mixtures containing these substances. That does not imply that among the millions of known chemical compounds there are not many others that are also hazardous; it is just a start. To make the list manageable, priority has been given to those substances meeting EPA toxicological criteria, which are produced in large quantities and are sold at low unit prices or have past histories of spillage.

The classification scheme originally proposed for defining harmful quantities was somewhat similar to that of IMCO (11). It included aquatic animal toxicity, oral mammalian toxicity, dermal mammalian toxicity, inhalation toxicity, and phytotoxicity. As it turned out the more than 300 substances on the original list exhibited aquatic animal toxicity at or below the upper aquatic toxicity limit of 500 ppm and only a very few met any of the other toxicological criteria. In response to comments by the public, a total of 36 substances that had been proposed are not listed in the final rules, mostly because of the limited usage or marginal toxicity. The final 271 substances all qualify on the basis of aquatic toxicity. Expanding on the criteria to include carcinogenic, mutagenic, teratogenic, or radioactive substances is being planned for the future.

Section 311 also requires EPA to determine whether any such designated substance "can actually be removed." Rule 40 CFR 117 makes that determination, and based upon public comments, physical data and existing oil removal technology, designates ten of the hazardous substances as removable. All of them resemble petroleum oils in their behavior when discharged to water. They are cohesive, they float, and they dissolve in water at a ratio of less than one part to a thousand parts of water. The remaining 261 designated hazardous substances were not determined to be removable and their discharges are subject to the two schedules of civil penalties of $500 to $5,000 or up to $500,000 for shore facilities and $5 million for vessels.

Rule 40 CFR 118 determines harmful quantities. This regulation is a fine example of the extent to which constructive public comments can contribute to improve significantly both concept and formulation. Because subjective judgment is required in using IMCO type guidelines, they were not used in the defining of harmful quantities in the final rulemaking. When first proposed, part 118 had divided all designated materials into four categories, A, B, C, and D, covering toxicity ranges 1 ppm and below, greater than 1 to 10 ppm, greater than 10 to 100 ppm, and greater than 100 to 500 ppm, respectively. Harmful quantities for these categories were then defined as 1, 10, 100, and to 500 lb, so that the least toxic substance in each category theoretically could do equivalent damage to equal volumes of water. It was pointed out by several commentators that this scheme did not give the desired results.

These public comments noted that Category A, by covering all designated substances which kill fish at concentrations of less than 1 ppm, in fact lumped together a much wider toxicity range than the other categories. As a result the designation of harmful quantities was seen to be unnecessarily biased against the less toxic substances. For example, it was stated that 500 lb of a typical D substance could be expected to do less harm than 1 lb of a typical A substance.

A new category, Category X, has therefore been introduced. It covers substances with median lethal concentrations below one part in 10 million. Harmful quantities for Categories X, A, B, C, and D are now 1, 10, 100, 1,000 and 5,000 lb, respectively. This will probably reduce the number of discharges that will have to be reported, but will not affect those large discharges that most seriously endanger the public welfare.

Public comment also resulted in changing the proposed regulation with respect to the effective date. The waterborne shipping industry commented that it would be difficult for members to obtain certification for proof of financial responsibility in a 90-day period, because first the insurance industry will have to establish rates, and then documents will have to be processed for thousands of ships. The regulations therefore become effective 90 days after publication, except for vessels for which the period is 180 days after publication.

EPA has received many questions about the relationship between Section 311 and the NPDES permit system. Some persons think that Section 311 applies only to accidental transportation spills. That is not the case. Discharges of a given material can be equally harmful whether they come out of a pipe or out of an overturned railroad car.

The rules now make that quite clear. Discharges of designated hazardous substances which are in compliance with an NPDES permit are exempted. Note the word "compliance." An NPDES permit is not a license for indiscriminate discharges. A discharger is in compliance when the discharge does not exceed the maximum daily amount expressly allowed, nor exceed the average daily discharge not limited expressly in the permit, but is as disclosed in the permit application.

The fourth of the rules, 40 CFR 119, covers the penalty rates. The penalty either may be in the $500 to $5,000 bracket or alternatively may be based on the quantity of the discharge, in which case it is based on the identity and the quantity of the substance discharged. Penalties in the latter case range between $100 and $1,000 per harmful quantity, depending on factors such as solubility, volatility, and whether the substance sinks or floats. This alternate penalty for nonremovable substances may range up to half a million dollars for on- or offshore facilities and $5 million for vessels. The choice of the penalty, according to the law, is at the "Discretion of the Administrator." The proposed rules used "gross negligence" as the discriminator for determining whether or not to impose penalties from the high penalty schedule. Proving or disproving gross negligence would have resulted in much litigation and has taken away the Administrator's "discretion." Therefore the rules now state that he will make the determination based on "the gravity of the offense."

Finally, the rules now contain administrative procedures. They provide for notices of violation; for opportunities for interested third parties, including the State, to submit written comments; for a hearing, including a prehearing conference; and for appeal procedures.

We must now see how well the rules will work in practice. If the rules are found to have excessive shortcomings, consideration may be given to amending them. In any case, the list of designated materials will be expanded. An additional list of 28 substances has already been prepared for public comments to be published simultaneously with the final rulemaking for the 271 substances. Included on it are several substances such as Kepone, Mirex, and carbon tetrachloride, the discharge of which has occurred since publication of the initial proposed lists and has attracted nationwide attention.

One fringe benefit we will derive from the rules published in March is data on spill frequencies. Since spills of hazardous materials have not been reportable, the magnitude of the problem is unknown. We may all be surprised, one way or the other.

Future designations of hazardous substances may be expected to include additional materials on the sole basis of aquatic toxicity. But you may also expect the designation criteria to be expanded so as to include other types of hazardous materials. No firm decisions have as yet been made, but additional emphasis on nondegradable substances does not seem unreasonable, while the Agency defines more clearly the hazards of carcinogenic, mutagenic and teratogenic substances.

One thing is clear: the extent of amendments to these regulations and the quality of future regulations will likely depend on the inputs EPA obtains from the public, industry, and the scientific community.

THE NATIONAL CONTINGENCY PLAN–PRESENT AND FUTURE

The material in this section is based on a paper given at Conference 1972 by R.E. Hess of the Environmental Protection Agency. The title of the paper was "The National Contingency Plan–Present and Future."

The key to protecting the environment from spills of hazardous substances is multifaceted in that it involves design, engineering and construction of facilities and equipment to prevent spills; stockpiles of equipment and available trained manpower to clean up spills; and contingency plans to insure prompt and effective implementation of countermeasures to mitigate the effects of spills.

This discussion deals with the National Contingency Plan, its history, its basis, its functions, and how its provisions relate specifically to spills of hazardous substances.

The National Contingency Plan has its roots in oil, having been developed initially in 1968 after the Torrey Canyon and Ocean Eagle tanker accidents. Additional impetus for establishing a National basis for responding to spills stemmed from the following incident:

> On January 1, 1968, at about 9:30 PM, two trains including five tank cars carrying chemicals, crashed near Dunreith, Indiana (12). There were two cars carrying acetone cyanohydrin, and one filled with each of the following: vinyl chloride, ethylene oxide, and methyl methacrylate. Each car contained 20,000 gallons. Explosion and fire, with the evolution of toxic gases, followed immediately after the accident, and townspeople were evacuated. About noon, on January 2, motorists reported noticing dead cattle and fish along Buck Creek, south of and downstream from Dunreith. This confirmed fears that, although there appeared to be no outlet from the wreck site to the stream, chemicals from the tank cars entered Buck Creek. During the investigations of the stream in the vicinity of the fish and animal kill, the characteristic odor of cyanide in the water gave a clue to at least one cause of the mortalities. Analysis showed a maximum concentration of over 400 mg/l cyanide into the creek.
>
> After locating the slug of pollutants containing approximately 1,600 lb cyanide, and following its progress, the decision was made by Indiana State Board of Health to treat the east fork of the White River upstream from the city of Seymore, which is the first municipality to take its water from the river. Accordingly, 6,200 lb of calcium hypochlorite were added to the Big Blue River between the evening of January 5 and early morning on January 6. By the time the slug reached Seymore, the cyanide concentration had been reduced to 0.015 mg/l, well within the upper limit of 0.020 mg/l specified in drinking water standards.
>
> In addition to the impact of materials on surface waters, concentrations of cyanide were also found in groundwater in the immed-

> diate vicinity of the accident. Recovery wells coupled with the calcium hypochlorite treatments were employed until May 23, when the operation was declared completed.

The report from which the foregoing example was abstracted states: "This incident demonstrates that, with a growing population and advanced civilization, the hazards to public health from hazardous chemicals are increasing. All pollution control agencies should and must prepare themselves to handle emergency problems of this nature." That is what contingency planning is all about: preparing our agencies or companies to handle emergency problems.

A Brief History of the National Contingency Plan

As stated above, a contingency plan called the National Multiagency Oil and Hazardous Materials Pollution Contingency Plan, was prepared by five Federal agencies at the direction of the President. This was essentially an interagency agreement among the Departments of Interior; Transportation; Defense; and Health, Education and Welfare; and the Office of Emergency Preparedness. Each of the agencies, through statute, executive order or Presidential directive, was assigned primary roles because of its resources or expertise that would be particularly useful in cleanup and countermeasure actions.

The Federal Government's efforts to deal with spills of hazardous substances were placed on a statutory base with the President signing on April 3, 1970, the Water Quality Improvement Act of 1970. This new Act included, in Section 12, important provisions for coping with spills of hazardous substances and directed, in Section 11, that within 60 days after the effective date of the Act there should be published a National Contingency Plan to provide for efficient, coordinated and effective action to minimize damage from spills.

Section 12, mentioned above, included a requirement that the discharger of a hazardous substance should immediately notify the appropriate agency of the United States in order to facilitate removal of the discharged material; authorization for the Federal Government to act to remove a hazardous substance discharged to the navigable waters of the U.S. when removal actions are not immediately undertaken by the discharger; and authority to utilize the cleanup fund provided for by Section 11 to effect such removal by the Federal Government.

Pursuant to the provisions of Section 11, the Council on Environmental Quality published on June 3, 1970, a National Oil and Hazardous Materials Pollution Contingency Plan. This Plan was essentially the same as the 1968 Plan mentioned previously because assignments of responsibility for carrying out the new Act and regulations implementing its specific elements had not yet been issued. The same five agencies also were still involved.

Regulations required to fully implement the new Act coupled with Reorganization Plans Nos. 3 and 4, 1970, (establishment of EPA and NOAA, respectively) made it imperative that the National Plan be again revised.

National Contingency Plan—1971

The reorganizations mentioned above caused a restructuring of the response organization. Specifically, there are now eight Federal agencies involved, four

Primary and four Advisory. The Primary Agencies are: Environmental Protection Agency, Department of Transportation, Department of Defense and Department of the Interior. The Advisory Agencies, the Departments of Commerce; Treasury; Health, Education and Welfare; and the Office of Emergency Preparedness, are those departments or agencies that could have significant inputs during specific pollution spills, e.g., in a spill in the international boundary waters, the State Department would be deeply involved.

The National Plan establishes a planning and response organization called the National Response Team (NRT) which includes representatives of the Primary Agencies and representatives from the Advisory Agencies, as appropriate. The NRT has two major functions; these are:

(1) it serves as the national body for preparedness and preplanning before a pollution spill occurs, and
(2) it acts as an emergency response team to be activated when a spill:
  (a) exceeds the response capability of the region in which it occurs;
  (b) involves National security or,
  (c) presents a major hazard to substantial numbers of persons or nationally significant amounts of property (13).

Each region has its Regional Response Team (RRT) with functions comparable to those of the NRT, but at the regional level. Federal response activities at the site of a spill are under coordination and direction of the On-Scene Coordinator (OSC), whose duties are specified in the National Contingency Plan (14). The OSC is the single executive agent, predesignated by regional plans, to coordinate and direct the Federal pollution control activities. The OSC must be a Federal officer or employee and must be predesignated in appropriate regional plans if Federal resources are to be employed. The Plan establishes that the Environmental Protection Agency shall furnish or provide for the OSCs for all inland navigable waters; the U.S. Coast Guard shall furnish or provide for the OSCs for the high seas, coastal and contiguous zone waters and for Great Lakes coastal waters, ports and harbors. It is interesting to note that, without regard to the Agency supplying the OSC, any environmental pollution control activity must receive the concurrence of the NRT.

The National Plan contains general information on response actions and separates the response into five relatively distinct classes or phases; these are:

Phase I: Discovery and Notification – Reports of spills may be the result of deliberate discovery, through the actions of a discharger or by random, incidental observations by private citizens or others. As mentioned earlier, Federal law requires that any person in charge of a vessel or of an on-shore facility shall, as soon as he has knowledge of any discharge that violates Federal law or regulation, immediately notify the appropriate agency of the United States Government of such discharge.

Phase II: Containment and Countermeasures – These are actions to be initiated as soon as the existence of a spill is known. Common sense actions, like closing valves and activating pumps, would be first order countermeasures. Deployment of containment booms to trap material floating on water, buttressing dikes to prevent escape of liquid from a storage area to a water course and notification of users of the affected waters are also included under this phase. With specific reference to hazardous substances that could not be removed from the water environment, the addition of a neutralizer as in the example cited earlier would be considered to be a countermeasure if accomplished within several hours of the spill's occurrence.

Phase III: Cleanup and Disposal — This includes physical removal of the contaminant from the water environment; safe, noncontaminating disposal of any collected polluting material; and in situ neutralization of the effects of the pollutant.

Phase IV: Restoration — This means actions taken to return the water environment to its prespill conditions. Actions that might be taken under this phase are reseeding shellfish beds decimated by a spill, restocking finfish in the affected water body, removal and replacement of contaminated beach sand, or filtering contaminated water through material that removes the pollutant.

Phase V: Recovery of Damages and Enforcement — This can include a variety of activities, depending upon the location of and the circumstances surrounding a particular spill. Enforcement activities under appropriate Federal authority such as the Refuse Act of 1899, the Federal Water Pollution Control Act, as amended, and under appropriate State statutes and local ordinances are included. Recovery of damages done to Federal, State or local property is included; recovery of third party damages, however, is not dealt with in the National Contingency Plan.

Inasmuch as a fundamental precept of most Federal emergency plans is that of committing Federal resources when needed to assist State and local forces, it is important that Federal and State contingency plans be interrelated. To achieve this goal, regional contingency plans developed under authority of the statute and the National Contingency Plan are structured so as to include subregional plans. Development of subregional plans is generally based on State boundaries, and, in the inland water areas, most subregional plans have been or are being developed cooperatively by Federal and State agencies.

This has been a brief discussion of the National Contingency Plan, its history and some of its important provisions. This plan or any other plan cannot or should not be expected to take the place of top-quality design and construction of facilities, adequate training of operating personnel, and liberal applications of good operating practices. What a contingency plan can and should be designed to do is insure that resources are identified and the operating mechanism is available to move into action when a spill threatens or occurs.

SPILL DAMAGE RESTORATION

> The material in this section is based on a paper given at Conference 1972 by Allen L. Jennings of the Office of Water Programs, Environmental Protection Agency. The title of the paper was "Spill Damage Restoration."

Section 12(f)(2)(A) of the 1970 Federal Water Pollution Control Act (P.L. 91-224) defines the removal of spilled hazardous polluting substances to include

". . .the taking of such action as may be necessary to minimize or mitigate damage to the public health or welfare, including, but not limited to, fish, shellfish, wildlife, and public and private property, shorelines and beaches."

The purpose of this paper is to explore the need for mitigating environmental damages and some possible means of implementing an effective program.

Prevention of spills is the first line of defense against environmental damage, and the development of effective prevention plans and a code of clean practice can effectively reduce spill incidents, but the unforeseen and sometimes unpreventable spills will continue to pollute our waters.

Rapid response resulting in containment and physical cleanup or neutralization of toxicity are, of course, the primary means to minimize the areal extent and severity of environmental damages when these spills occur. The containment of a soluble hazardous material spill in open waters is, at present, practically impossible. In these cases, most often the only practical response action is a warning to downstream water users. Environmental damages are usually inflicted until dilution decreases the level of pollutant below the toxic or damaging concentration. The resulting environmental damages range from subtle alterations of population balances by the elimination of sensitive or marginal species to obvious and massive fish kills. An example of the severity of environmental damage that can result is the recent phosphate mining waste discharge on the Peace River that was reported as severely depressing approximately 75 river miles of aquatic life (15).

The time required for natural recovery from spill insults is extremely variable. Factors influencing this time include the general health of the upstream environment, flow rate of the waterway or flushing rate of impoundments and estuaries, persistence of the material, and the seasonal time of the spill. In a controlled experiment, Cairns showed that the benthic community of a fast-flowing mountain stream required a time period of about one month to recover from a sulfuric-acid-induced pH stress (16). The "spill" conditions of the experiment can be described as mild, when compared to more toxic chemicals, such as insecticides, and of relatively short duration. The conditions for recovery in this situation were nearly ideal with a nonstressed upstream environment, rapid flushing, no damage to habitat or littoral life forms, and a less than total kill.

In 1967, a dike failure released approximately 130 million gallons of caustic solution from a fly ash holding pond into the Clinch River. As a result, some 200,000 fish were killed in a total of 90 river miles in both Virginia and Tennessee. Follow-up studies of benthic and fish communities indicated that full biological recovery was not achieved after two years (17).

Blumer (18)(19) has documented the long-term effects of oil spills on shellfish and estimated the recovery of shellfish populations from a diesel oil fuel spill in Buzzards Bay, Massachusetts, to be at least two years. Additionally, wide-range and long-term biological damages, resulting from oil spills in the marine environment, have been indicated from recent EPA investigations of spills in the Gulf of Mexico and Puget Sound (20)(21).

Preliminary data now exist which indicate a possible correlation between oil contaminated substrates and gonadal tumors in clams (22). As more detailed studies become available, previously unknown long-term effects will probably be brought to light. The need for an accelerated restoration program to ease the long-term impact of spills is becoming increasingly important. In the case of persistent, slowly degraded materials (such as oil or chlorinated hydrocarbons), thorough and effective cleanup must be an integral part of aquatic restoration.

Lakes, impoundments, and estuaries have slow exchange or flushing rates when compared to free-flowing rivers and streams. A spill in such contained waters, may therefore, result in a longer exposure time for aquatic organisms and the probability of a higher percentage of mortality. As a higher percentage of any aquatic community is destroyed, the time required to establish prespill population densities is increased. The slow exchange rate also leads to significantly decreased benthic organism drift. In contained waters, slow drift repopulation will probably be the limiting factor in the recovery of the benthic community, and one would expect an overall decrease in productivity for at least one seasonal cycle.

Since most of our waterways are continually subjected to man-induced stresses such as continuous industrial and municipal discharges, agricultural chemical runoff, mine drainage, and silt loadings from construction, natural restoration may be significantly delayed and require many years to reach prespill quality. During the natural recovery processes, a succession of population balances can occur and exhibit varying degrees of stability. Because of a complex phenomenon that may most easily be addressed as competition, certain species become predominant, and undesirable floral and faunal species may become established.

The minimization or mitigation of biological damages and the resultant denial of a beneficial water use such as recreational or decreased esthetic value can only take the form of accelerated restoration. A successful restoration program could significantly lessen the long-term impact of hazardous materials spills by returning a water body to its prespill condition in the shortest possible time. For example, the restoration of sport fishing within months rather than years would significantly minimize damages to the public welfare.

The impact of a major spill on aquatic life could be significantly lessened by the timely enlargement of the potential breeding stock. Organisms from experimental aquaculture facilities at universities and State hatcheries as well as selective harvesting from unaffected areas are potential sources of new breeding stock. Selective harvesting would distribute the impact of the spill over a larger area so that the local severity would be lessened.

There is, however, the potential for creating even larger scale ecosystem imbalances if due care is not exercised with the harvesting method. The use of cultured breeding stock does not carry this potential for harm, and would seem to be the method of choice when cultured organisms are available. A notable example of culture technology is that of shrimp. Postlarval shrimp are currently available on a limited scale at about 0.25 cent each. Future considerations should, therefore, be given to enlarging both the numbers and species of reserve organisms. Emphasis should be given to those species which are both commercially and recreationally valuable as well as to those organisms that occupy important niches in the food web.

Most states, today, levy a charge for fish killed in a pollution incident and efforts toward restoring the fish population are made. Restocking of game fish has sometimes improved sport fishing, but the effective period may be relatively short. For example, placing fish in a river or impoundment without restoring the necessary food sources, habitat, or water quality has little long-term beneficial effect. If possible, the fish will undoubtedly migrate to areas of the stream most agreeable to them. In impoundments where migration to suitable areas is not always possible, inadequate food supplies and spawning habitat will be manifested in depressed populations. Increasing knowledge of the delicate ecosystem balances and the interdependence of the many varied species within the aquatic environment indicates that a broader approach to restoring damaged waterways is needed.

To be effective, aquatic restoration must consider the physical habitat, food web organisms, fish, shellfish, vegetation, and other aquatic life as well as littoral life forms. The Environmental Protection Agency, through programs such as the Division of Oil and Hazardous Materials, is in the process of developing a comprehensive restoration and mitigation program concerned with the orderly rebuilding of all trophic levels within the aquatic ecosystem. The eventual goal of this or any restoration-mitigation effort should be the maximum diversity of all aquatic communities yielding a more stable ecosystem.

New techniques and knowledge must be developed in this area before a large-scale program is functional. Typically, the tools required for such a program may be summarized as: (1) baseline data, (2) characterization of typical ecosystems and the food web interrelationships, (3) long-term chronic effects of spills, (4) aquatic population dynamics and the factors that influence them, (5) effects of physical alterations of habitat on ecosystems, and (6) effects of natural stresses.

Baseline data are critical for the accurate assessment of spill damages and for providing a target for restoration activities. Because of the great variety of ecosystems, as well as the effect of local and seasonal conditions, baseline information should be gathered from an unaffected area near the spill and at a corresponding season.

This highly specific data should be backed up by a more general characterization of typical ecosystems. A great deal of information is currently available and this task could quite conceivably be accomplished through a literature review and compilation of existing data. The generated profiles should highlight the predicted species and food web interrelations that one would expect to find in each typical ecosystem. Such data would be valuable in assessing the impact of previous man-induced or natural stresses on the aquatic environment.

Research aimed at established chronic effects of hazardous material on aquatic organisms must be expanded. For example, a sublethal effect may, in the next spawning season, be expressed at impaired reproductive capability. Mutagenic and teratogenic effects may not be observed for at least one generation. The eventual impact of these potential sublethal effects can be as severe as immediate mortality. The success of a restoration effort may, in some cases, hinge on the degree of accuracy with which chronic effects are assessed.

Attempts to repopulate a waterway by species introduction will require a firm understanding of population dynamics and the factors that influence them. A familiar form of population dynamics management is that of using large predator fish to control the population of smaller, more prolific species. This concept of predator pressure needs to be expanded and applied to all communities within the ecosystem.

In addition to chemical and biological factors, consideration must be given to the effects of physical alterations of the habitat. Cleanup operations to remove dense, insoluble materials such as creosote wastes from the water may necessitate dredging of a stream or lake bed. Since the detrimental effects of stream channelization and dredging on aquatic life have been documented (23)(24)(25), a restoration effort must evaluate these effects.

Finally, natural stresses and normal seasonal fluctuations in community balances must be evaluated in terms of their effect on damages and restoration. These factors will also be important considerations in assessing the liability for environmental damages.

One approach to a viable restoration effort could be through the use of response teams assembled in key areas. Following a spill, such a team could be activated and attempt a variety of techniques in real situations and environments, with follow-up studies gauging the degree of success. Circulation of a newsletter to all participating teams describing all attempts both successful and nonsuccessful would decrease duplication of effort and eventually evolve supportive data and proven techniques.

Through the Federal Excise Tax on firearms and fishing tackle, Americans paid $40 million to expand, develop and preserve our wildlife resources. An effort should be made to protect this investment by the implementation of a restoration program that effectively reverses spill damages.

REFERENCES

(1) McConnaughey, et al, *Chemical Engineering Progress*, p. 58 (February 1970).

(2) U.S. Coast Guard, *The Third Annual Report of the Council on Environmental Quality*, p. 120, Government Stock No. 4111-0011, Washington, D.C. (August 1972).

(3) U.S. Department of Transportation, telephone communication with personnel in the Accident Analysis Office of the Office of Hazardous Materials, Washington, D.C.

(4) Toxic Substances Control Act of 1976, Public Law 94-469; 90 STAT 2003 (1976).

(5) Resource Conservation and Recovery Act of 1976, Public Law 94-580; 90 STAT 2795 (1976).

(6) Clean Water Act of 1977, Public Law 95-217 (Dec. 27, 1977).

(7) Water Quality Improvement Act of 1970, Public Law 21-224 (April 3, 1970).

(8) Federal Water Pollution Control Act Amendments of 1972, Public Law 92-500 (Oct. 18, 1972).

(9) Executive Order 11735, *Assignment of Functions under Section 311 of the Federal Water Pollution Control Act, as Amended* (August 7, 1973).

(10) Environmental Protection Agency, *Hazardous Substances*, 40 CFR 116-119, Federal Register 59959-60017 (Dec. 30, 1975).

(11) U.S. Senate Committee on Commerce, *1973 Intergovernmental Maritime Consultative Organization Conference*. Hearings, 93rd Congress, First Session, Series 93-52, GPO, Washington, D.C. (1973).

(12) Indiana State Board of Health, *Cyanide Pollution and Emergency Duty, Train Wreck, Dunreith, Indiana* (January 1968).
(13) *National Oil and Hazardous Substances Pollution Contingency Plan*, Section 303.3 (August 1971).
(14) *Ibid*, Section 504.
(15) EPA, Region IV, Atlanta, Georgia, Enforcement Division report (1971).
(16) Cairns, J., Jr., et al, *26th Purdue Indiana Waste Conference* (May 4-6, 1971).
(17) Cairns, J., Jr., Crossman, J.S., Dickson, K.L., and Herricks, E.E., *The Association of Southeastern Biologists Bulletin,* 18, 79 (1971).
(18) Blumer, M., et al, *Environment,* 13(2), 2 (1971).
(19) Blumer, M., et al, *Marine Biology,* 5, 195 (1970).
(20) *Biological Assessment of Diesel Spill*, Anacortes, Washington (1971).
(21) *Fate and Effect Study, Shelf Platform B Incident* (1971).
(22) Tarzwell, C.M., Unpublished communication.
(23) Martin, E.C., *Proceedings of 23rd Annual Conference*, Southeastern Association of Game and Fish Commission, Mobile, Alabama (1969).
(24) Peters, J.C., and Alvord, W., *Trans. North Am. Wildlife Nat. Resources Conf.,* 29, 93 (1964).
(25) Gebharts, S., *Idaho Wildlife Review,* 22, (5) (1970).

PROCEDURES LEADING TO CLEANUP

NOTIFICATION

The information in this section is based on *Manual for the Control of Hazardous Material Spills. Volume 1: Spill Assessment and Water Treatment Techniques,* EPA-600/2-77-227 prepared by K.R. Huibregtse, R.C. Scholz, R.E. Wullschleger, J.H. Moser, E.R. Bollinger and C.A. Hansen of Envirex for the Environmental Protection Agency.

The objective of this section is to assist and encourage the user of this manual to determine the proper initial notification procedure for the immediate reporting of a hazardous material spill in his area. This procedure should be determined before any actual spill occurs, and proper contacts and telephone numbers should be recorded. In this regard the National, State, and Regional Contingency Plans should serve as reference documents.

It should be pointed out that all reporting requirements must be met and that state reporting requirements are often more encompassing than federal requirements. For example, a spill on land only may be covered by state but not by federal regulation. Also, this discussion covers initial and immediate reporting only; there may also be requirements for later detailed reports to the Regional or State Response Centers and the U.S. Department of Transportation.

Pertinent Federal Regulations

The National Oil and Hazardous Substances Pollution Contingency Plan was developed in compliance with the Federal Water Pollution Control Act (Public Law 92-500). The Plan provides for a pattern of coordinated and integrated response by departments and agencies of the federal government to protect the environment from the damaging effects of pollution discharges. The Plan as published in the *Federal Register,* Vol 40, No 28, outlines the notification requirements. In this regard Annex V of the Plan states:

"1503.2 The initial reporting of a pollution discharge by agencies participating in this plan shall be in accordance with the information and format as described in the regional plans. Reports of medium or major discharge received by the National Response Center (NRC) shall be relayed by telephone to predesignated On-Scene Coordinator (OSC)."

The Plan also specified the NRC, located at Headquarters, U.S. Coast Guard, Washington, D.C., as the headquarters site for activities relative to pollution emergencies. The National Response Team, consisting of representatives from various federal agencies, serves planning and response functions and is to work closely with the National Response Center.

Regional Response Centers

The Plan also provides for Regional Response Centers and Regional Response Teams. The Regional Response Center is the regional site for pollution emergency response activities. Each region has prepared a contingency plan to deal with oil and hazardous material spills in its region. The Regional Response Team performs response and advisory functions in its specific region. EPA regional offices are listed in Table 1.1, and U.S. Coast Guard Districts are given in Table 1.2.

Table 1.1: EPA Regional Offices

Region	States Included
Environmental Protection Agency Region I, Room 2303 John F. Kennedy Federal Building Boston, Massachusetts 02203 Tel: (617) 223-7265	Maine, Vermont, New Hampshire, Massachusetts, Rhode Island and Connecticut
Environmental Protection Agency Region II, Room 908 26 Federal Plaza New York, New York 10007 Tel: (201) 548-8730	New York, New Jersey and Puerto Rico
Environmental Protection Agency Region III Curtis Building 6th and Walnut Streets Philadelphia, Pennsylvania 19106 Tel: (215) 597-9898	Pennsylvania, Maryland, Delaware, West Virginia and Virginia
Environmental Protection Agency Region IV 1421 Peachtree Street NE Atlanta, Georgia 30309 Tel: (404) 526-5062	Kentucky, Tennessee, North Carolina, South Carolina, Georgia, Alabama, Mississippi and Florida
Environmental Protection Agency Region V 230 South Dearborn Street Chicago, Illinois 60604 Tel: (312) 896-7591	Ohio, Michigan, Indiana, Illinois, Wisconsin and Minnesota

(continued)

Table 1.1: (continued)

Region	States Included
Environmental Protection Agency Region VI, Suite 1600 1600 Patterson Street Dallas, Texas 75201 Tel: (214) 749-3840	Arkansas, Louisiana, Oklahoma, Texas and New Mexico
Environmental Protection Agency Region VII 1735 Baltimore Avenue Kansas City, Missouri 64108 Tel: (816) 374-3778	Iowa, Missouri, Nebraska and Kansas
Environmental Protection Agency Region VIII, Suite 900 1860 Lincoln Street Denver, Colorado 80203 Tel: (303) 837-3880	North Dakota, South Dakota, Montana, Wyoming, Utah and Colorado
Environmental Protection Agency Region IX 100 California Street San Francisco, California 94111 Tel: (415) 556-6254	Nevada, Arizona, California, Hawaii and Guam
Environmental Protection Agency Region X 1200 Sixth Avenue Seattle, Washington 98101 Tel: (206) 442-4343	Idaho, Oregon, Washington and Alaska

Source: EPA-600/2-77-227

Table 1.2: U.S. Coast Guard Districts

1st Coast Guard District
150 Causeway Street
Boston, Mass. 02114
Duty Officer: (617) 223-6650

2nd Coast Guard District
Federal Building
1520 Market Street
St. Louis, Mo. 63101
Duty Officer: (314) 622-4614

3rd Coast Guard District
Governors Island
New York, N.Y. 10004
Duty Officer: (212) 264-4800

5th Coast Guard District
Federal Building
431 Crawford Street
Portsmouth, Va. 23705
Duty Officer: (703) 393-9611

7th Coast Guard District
Room 1018, Federal Building
51 SW 1st Avenue
Miami, Fla. 33130
Duty Officer: (305) 350-5611

8th Coast Guard District
Customhouse
New Orleans, La. 70130
Duty Officer: (504) 527-6225

9th Coast Guard District
1240 East 9th Street
Cleveland, Ohio 44199
Duty Officer: (216) 522-3984

11th Coast Guard District
Hearwell Building
19 Pine Avenue
Long Beach, Calif. 90802
Duty Officer: (213) 590-2311

(continued)

Table 1.2: (continued)

12th Coast Guard District 630 Sansome Street San Francisco, Calif. 94126 Duty Officer: (415) 556-5500	14th Coast Guard District 677 Ala Moana Blvd. Honolulu, Hawaii 96813 Duty Officer: (808) 546-7109 Commercial Only AUTOVON 421-4845
13th Coast Guard District 618 2nd Avenue Seattle, Washington 95104 Duty Officer: (206) 524-2902	17th Coast Guard District P.O. Box 3-5000 Juneau, Alaska 99801 Duty Officer: (907) 586-7340 Commercial Only AUTOVON 388-1121

Source: EPA-600/2-77-227

INFORMATION SOURCES

The information in this and the next eight sections is based on *Manual for the Control of Hazardous Material Spills. Volume 1: Spill Assessment and Water Treatment Techniques,* EPA-600/2-77-227 prepared by K.R. Huibregtse, R.C. Scholz, R.E. Wullschleger, J.H. Moser, E.R. Bollinger and C.A. Hansen of Envirex for the Environmental Protection Agency.

There are a number of information systems whose main function is to provide assistance during hazardous materials spills. There are also information retrieval services, both computerized and manual, which provide information or a list of titles or abstracts of articles dealing with a specific subject. The availability of an on-line computer usually indicates a short turn-around time for responses. This is often important in an emergency situation.

There are also available numerous reference texts and handbooks which contain information on the properties of hazardous chemicals. Those likely to be faced with a hazardous material spill may find it helpful to obtain one or more of these books for future reference. In this section, each system designed specifically to provide information on hazardous materials is discussed in terms of what it is, what information it contains, how it operates, and how it can be accessed by responsible people at the emergency scene.

EPA Oil and Hazardous Materials Technical Assistance Data System (OHM-TADS)

The OHM-TADS is a computerized information retrieval file on more than 850 oil and hazardous substances. The system is presently on-line and available to assist in identification of a spilled material from certain observations (color, smell, etc.) made at the site.

The OHM-TADS has a random access provision which enables the user to solve problems involving unidentified pollutants by inputting color, odor or other physical/chemical characteristics as observed on-scene. The system responds with a list of the materials meeting the input characteristics. The output is displayed on the user's terminal. The user can then refine the search if necessary to narrow the list of possible materials. Access to OHM-TADS is through the oil and hazardous material spill coordinator at the EPA Regional Office (Regional Response Center). A listing may be found in Table 1.1.

U.S. Coast Guard CHRIS

This system consists of four manuals, a regional contingency plan, a hazard assessment computer system (HACS), and an organizational entity at the Coast Guard Station (4). The four manuals are as follows:

> Vol 1 (CG-446-1), *Condensed Guide to Chemical Hazards,* contains essential information on those hazardous chemicals that are shipped in large volumes by marine transportation.
>
> Vol 2 (CG-446-2), *Hazardous Chemical Data Manual*, contains detailed information on the chemical, physical, and toxicological properties of hazardous chemicals, in addition to all the information in Vol 1.
>
> Vol 3 (CG-446-3), *Hazard Assessment Handbook*, contains methods of estimating the rate and quantity of hazardous chemicals that may be released and methods for predicting the potential toxic, fire, and explosive hazards.
>
> Vol 4 (CG-446-4), *Response Methods Handbook*, contains information on existing methodology for handling spills; the Appendix to this volume contains a list of manufacturers of equipment which may be useful in a spill situation.

The contingency plan is part of the National Contingency Plan. The Hazard-Assessment Computer System is the computerized counterpart of Vol 3 and makes it possible to obtain detailed hazard evaluations. Although calculations can be performed by hand using Vol 3, the HACS permits one to make a more complex and usually more accurate assessment of the spill situation.

Vol 1, *Condensed Guide to Chemical Hazards*, is intended for use by port security personnel and others who may be first to arrive at the scene of the accident. It contains easily understood information about the hazardous nature of the chemical, assuming the chemical is identified. It is intended to assist those present in quickly determining the actions that must be taken immediately to safeguard life, property and the environment. Vol 1 contains a list of the information needed to assess potential hazardous effects through the use of Volume 3.

Volumes 2,3 and 4 are intended for use by the OSC office and the Regional and National Response Centers. Coast Guard stations, especially those in major ports, will usually also have these manuals. The computer system (HACS) is also designed for use by OSC personnel.

Volumes 2 and 3 are designed to be used together. For example, Vol 2, *The Hazardous Chemical Data Manual*, contains a hazard-assessment code for each chemical. This code is used in Vol 3, *The Hazard-Assessment Handbook*, to select the appropriate calculation procedures for the hazard assessment, enabling the user to estimate the rate and quantity of hazardous chemicals that may be released under different situations. For example, procedures are provided for estimating the concentration of hazardous chemicals (both in water and in air) as a function of time and distance from the spill. The Hazard-Assessment Computer System (HACS) is the computerized counterpart of Vol 3 and makes it possible to obtain detailed hazard evaluations quickly. The HACS system is intended primarily for use by OSC personnel through Coast Guard headquarters. While the input needed for evaluation will depend on the specific accident situation and that part of the system which is to be used, the following information should be supplied to Coast Guard headquarters as applicable.

Material discharged	Hole diameter
Quantity spilled	River depth
Quantity originally in tank	River width
Location of spill	Stream velocity
Time of occurrence	Temperature (air)
Tank dimensions	Temperature (water)
Other cargos or nearby chemicals	Cloud cover (percent)

Depending on which model it is decided to use, other information may be needed by Coast Guard headquarters. In this case a call-back number should be given so that headquarters personnel can request additional information if necessary.

Vol 4, *The Response Methods Handbook*, contains descriptive and technical information on methods of spill (primarily oil) containment. This manual is intended for use by Coast Guard OSC personnel who have had some training or experience in hazard response.

Access to the CHRIS manuals can be obtained through the Coast Guard District office (see Table 1.2). The HACS can be assessed on an emergency basis through the Regional Response Center, the Coast Guard District Office, or directly through the Department of Transportation National Response Center at Coast Guard headquarters by telephoning 800-424-8802.

U.S. Coast Guard National Strike Force

The Coast Guard's National Strike Force (NSF) is part of the National Contingency Plan established under authority of the Federal Water Pollution Control Act Amendments of 1972, Section 311 (1). It consists of high-seas equipment and trained personnel available to assist the OSC upon request during Phase III (Containment and Countermeasures), Phase IV (Cleanup, Mitigation and Disposal), and Phase V (Documentation and Cost Recovery), as defined in the National Contingency Plan.

There are three Coast Guard Strike Teams, located on the East, West, and Gulf coasts. Each strike team consists of 18 or 19 men, including 3 or 4 officers. Each strike team is capable of responding to a pollution incident in its area with four or more men within 2 hours and at full strength in 12 hours. The Strike Team can provide communications support and assistance and advice on ship salvage, diving and removal techniques. Available equipment, primarily designed for air transport, consists of the following: Air Deliverable Antipollution Transfer System (ADAPTS), consists of a pumping system to off-load stricken cargo vessels (details are given in the chapter on mobile units); Yokohama fenders, used for side protection during vessel-to-vessel cargo transfer; high-seas containment barrier; and high-seas skimmer.

The services of the National Strike Force are available to any OSC anywhere in the country. Requests for assistance can be made through the NRC through its 24-hour emergency telephone number, 800-424-8802. The specific details of the emergency situation should be given.

U.S. Army Technical Escort Center Chemical Emergency Response Team

The U.S. Army Technical Escort Center maintains, on standby, a 14-man alert team at Aberdeen Proving Ground, Maryland, ready to respond to chemical emergencies within 2 hours. If necessary, additional personnel are available for mobilization. The team is trained and experienced in handling chemical emergencies and has available to it special equipment such as decontamination trucks, detection devices, and protective clothing.

The U.S. Army Technical Escort Center's team responds to a chemical emergency when directed by the higher command. While the team was formed mainly to respond to emergencies involving Department of the Army chemicals, it has assisted other agencies such as the U.S. Coast Guard.

To obtain the assistance of the escort team, initial contact should be made with the regional EPA office. EPA personnel will then contact the Dept. of the Army Operations Center. Upon receipt of each request, the Army Operations Center determines if the specific services of the Technical Escort Center are needed. If the determination is made for the emergency team to respond, intermediate commands are notified and the team dispatched.

Chemical Transportation Emergency Center (CHEMTREC)

CHEMTREC serves a clearing house function by providing a single emergency 24-hour telephone number for chemical transportation emergencies. Upon receiving notification of a spill, CHEMTREC immediately contacts the shipper of the chemicals involved for assistance and follow-up. CHEMTREC also provides warning and limited guidance to those at the scene of the emergency if the product can be identified either by the chemical or trade name. The CHEMTREC system covers over 3,600 items which have been submitted by manufacturers as their primary items of shipment. CHEMTREC is sponsored by the Manufacturing Chemists Association although nonmembers are also served.

The CHEMTREC emergency telephone number is widely distributed to emergency service personnel, carriers, and throughout the chemical industry. The number is usually given on the bill of lading. When an emergency call is received by CHEMTREC, the person on duty records the essential information in writing. He tries to obtain as much information as possible from the caller. The person on duty will give out information as furnished by the chemical producers on the chemical reported to be involved. This would include information on hazards of spills, fire, or exposure. After advising the caller, the person on duty immediately notifies the shipper of the chemical by phone, giving him the details of the situation. At this point, responsibility for further guidance passes to the shipper.

CHEMTREC'S function is basically to serve as the liaison between the person with the problem and the chemical shipper and/or manufacturer, the theory being that the manufacturer of the chemical or material will know the most about his product and its properties. CHEMTREC also serves as a contact point for the Chlorine Institute, the National Agricultural Chemicals Association (pesticides), and the Energy Research and Development Administration (radioactive materials).

CHEMTREC can be accessed through its emergency telephone number, 800-424-9300. As much of the following information should be provided by the caller as possible:

Name of caller and call-back number;

Location of problem;

Shipper or manufacturer;

Container type;

Rail car or truck number;

Carrier name;

Consignee; and

Local conditions.

NACA Pesticides Safety Team Network

The National Agricultural Chemicals Association through its members operates a national pesticide information and response network. Its function is to provide advice and on-site assistance when the spill situation warrants it. Access to the network is through CHEMTREC. Upon receiving notification of an emergency involving a pesticide, the manufacturer is contacted by CHEMTREC. The manufacturer will provide specific advice regarding the handling of the spill. If necessary, spill response teams are available on a geographical basis to assist at the emergency scene.

Transportation Emergency Assistance Plan (TEAP)

TEAP serves a function in Canada similar to that of CHEMTREC in the United States. Canada is divided into eight geographic areas, each served by a regional control center. Depending on the location of the spill, one of these control centers is called and notified of the emergency. The functions of TEAP are to provide emergency advice, to get knowledgeable personnel (usually the manufacturer) in touch with responsible people at the emergency scene, and to see that on-the-scene assistance is provided if needed.

When a call is received at a regional control center, the attendant records basic information on a record sheet and obtains a call-back number. He may also give preliminary information from standard references if the name of the product is known. The attendant will then call one of the center's technical advisers with the preliminary information. The technical adviser will then call the accident scene to obtain as much detail as possible. At this time he may also be able to provide additional advice on coping with the emergency. The adviser will then try to contact the producer from the listed references.

If the producer can be contacted, the adviser will turn the problem over to him as the most knowledgeable contact. If the producer cannot be reached, or if distances are great, the regional control centers will contact a company familiar with the product. The center is also prepared to send men and equipment to the scene if necessary. Once contact has been established between the producer and the local authorities on the scene, the technical adviser assumes a follow-up role and notifies the Canadian Chemical Producers' Association.

Access to TEAP is through the regional control centers given in Table 1.3. Each has a 24-hour number. Essential information that should be provided includes: Exact name of the product spilled, name of the producer and name of the carrier.

Table 1.3: Transportation Emergency Assistance Plan (TEAP)

Hooker Chemicals Division Vancouver, British Columbia 604-929-3441; geographic location: British Columbia	Cyanamid of Canada, Ltd., Niagara Falls, Ontario; 416-356-8310 geographic location: Eastern Ontario
Celanese Canada Ltd., Edmonton, Alberta 403-477-8339; geographic location: Prairie Provinces	DuPont of Canada, Ltd., Maitland, Ontario; 613-348-3616; geographic location: Western Ontario
Canadian Industries Ltd., Copper Cliff, Ontario 705-682-2881 geographic location: Northern Ontario	Allied Chemical Canada Ltd., Valleyfield, Quebec, 514-373-8330 geographic location: Quebec, south of St. Lawrence
Dow Chemical of Canada, Ltd., Sarnia, Ontario 519-339-3711 geographic location: Central Ontario	Gulf Oil Canada Ltd., Shawinigan, Quebec, 819-537-1123; geographic location: Quebec, north of St. Lawrence

Source: EPA-600/2-77-227

Chlorine Emergency Plan (CHLOREP)

Chlorine manufacturers in the U.S. and Canada through the Chlorine Institute have established the Chlorine Emergency Plan to handle chlorine emergencies. This is essentially a mutual aid program whereby the manufacturer closest to the emergency will provide technical assistance even if it involves another manufacturer's product.

Upon receiving an emergency call, CHEMTREC notifies the appropriate party in accord with the mutual aid plan. This party then contacts the emergency scene to determine if it is necessary to send a technical team to provide assistance. Each participating manufacturer has trained personnel and equipment available for emergencies. Access to CHLOREP on a 24 hour basis is through CHEMTREC.

Spill Cleanup Equipment Inventory System (SKIM)

> The material in this section is based on a paper presented at Conference 1978 by R.J. Imbrie, K.R. Karwan and C.M. Stone of the U.S. Coast Guard. The title of the paper was "The Spill Cleanup Equipment Inventory System (SKIM)."

Information on the availability of spill cleanup equipment is addressed in varying formats and degrees of detail in regional and local contingency plans. Review of the information available from this source indicated a lack of central organization, difficulty in retrieval of specific information and a lack of timely updating of the information. In considering alternative methods of establishing an inventory system that would improve these conditions, the National Emergency Equipment Locator System (NEELS)(see next section) developed by Environmental Canada was found to alleviate many of the difficulties in the equipment listings contained in the regional and local contingency plans. Environmental Canada developed a computerized inventory system to provide their emergency managers with information on equipment available for combating oil spills. During the last few years NEELS has demonstrated the general success of this inventory concept.

The Coast Guard Office of Marine Environment and Systems, Pollution Response Branch undertook the task of developing the Spill Cleanup Equipment Inventory System (SKIM), a computer based inventory of equipment available for pollution response in the United States (including Puerto Rico and Guam). The inventory includes public equipment, such as owned by the U.S. Coast Guard, U.S. Navy and other agencies, as well as cleanup and containment equipment maintained by contractors, cooperatives and private companies. With the cooperation of industry and all involved federal agencies, it is anticipated that the inventory will become the most complete source of its kind in the world.

SKIM has been implemented as a computerized system in order to enable effective and timely retrieval and updating of information as it is required. The information is readily available to interested parties through a network of computer terminals that are allowed to access the system.

In response to the Presidential Initiative, SKIM was designed with a variety of potential uses in mind. The system has the capabilities to:

Provide current and up to date information for the OSC. When a spill occurs and the predesignated OSC in that area is notified of the specific details, he is responsible for summoning the best combination of response resources to bear on the problem. SKIM will serve as an aid for his making of such decisions.

Provide updates to equipment listings in local contingency plans. Whereas currently these plans are often out of date, SKIM will give local and regional response teams the capability to obtain up to date listings.

Serve as an informational aid for Coast Guard Marine Safety Offices, Captain of the Ports and District and Headquarters managers. This will enhance the decision making capability of those concerned with budgeting, resource allocation and planning for pollution response.

Be used by all federal agencies involved with national or regional response teams.

Provide a means for international cooperation for spills on or near the borders of the United States. This would include the possibility of an interface with the Canadian NEELS system.

Information input to SKIM covers equipment available in both the public and private sectors. Twenty-six individual types of equipment were identified as appropriate for compiling the data base.

The data gathering effort commenced in January of 1978. Coast Guard Predesignated On-Scene Coordinators collected data on equipment from the private sector and on Coast Guard owned equipment for their areas of responsibility from telephone inquiries and facility visits after utilizing information in existing Regional and Local Contingency Plans. In a like manner, EPA Predesignated On-Scene Coordinators began collecting information within their jurisdiction for inclusion in SKIM.

Other members of the National Response team have advised their representatives on the Regional Response Team to insure that data on equipment maintained by each federal agency are entered into the inventory system. To assist with the data collection from federal resources Coast Guard District Marine Environmental Protection Branches have made an initial distribution of inventory forms to members of the Regional Response teams and will coordinate collection at the regional level if participating agencies desire.

Participation in SKIM by contractors, private companies and cooperatives is voluntary. However, strong support for the inventory system has been received from the American Petroleum Institute and the Oil Spill Control Association. Information received from members of these organizations and others have enabled the establishment of a data base in excess of 10,000 records. The value of the system will be constantly increased in the future as new information is added to the data base by field units using dataphone terminals.

The SKIM has been implemented on the TYMCOM-370 computer owned by Tymshare. Using Tymshare's own data communications network called Tymnet, users are permitted to access the SKIM data base through dataphone terminals. Local telephone access numbers are available in a number of major and intermediate cities across the country. Other users can also tie into the system on a long distance basis through one of these numbers.

System 2000, developed by MRI, Incorporated, was chosen as the data base management system to be used with SKIM. This software package was designed to operate on a variety of hardware systems and has already been tested in conjunction with a number of other Coast Guard data bases. Very specific retrievals and changes in the data base can be made most efficiently. Complex management reports can also be generated through a special System 2000 module called Report Writer or through Fortran or Cobol programs and the knowledge of Host Language Interface.

Using System 2000 on the Tymshare network, it has been possible to write special purpose computerized programs which can be accessed by users of SKIM. At this time, users of the system are permitted to obtain listings of equipment from the data base by type and area, e.g., all booms located in Coast Guard District 1 or all skimmers in EPA Region IV.

Information can also be obtained by distance from a specific location input at the terminal by the user. This location, specified in terms of latitude and longitude, would generally represent a spill site or staging site for equipment to be used for a specific spill. For example, the user might want to know a listing of all vacuum/pumper trucks available within 10 miles of a given location.

Other Information Retrieval Systems

The information in this section is based on *Manual for the Control of Hazardous Material Spills. Volume 1: Spill Assessment and Water Treatment Techniques,* EPA-600/2-77-227 prepared by K.R. Huibregtse, R.C. Scholz, R.E. Wullschleger, J.H. Moser, E.R. Bollinger and C.A. Hansen of Envirex for the Environmental Protection Agency.

Organizations providing information retrieval should be considered as secondary sources of information because information is from the published literature or past events, and because interaction is limited since the contact usually has no special expertise in spills technology or hazardous chemicals. A list of information retrieval sources is given in Table 1.4. Because of the emergency nature of most spills, a short response time is important. In this regard an on-line computerized system is desirable, although a manual search could also have a short turnaround time.

Table 1.4: Information Retrieval Systems

Information Source	On-Line Computer System	Contact
Lockheed Information Systems	yes	415/493-4411 ext. 45635
Editec Inc.	yes	312/427-6760
Illinois Institute for Environmental Quality Library	yes	312/793-3870
Institute for Scientific Information	yes	215/923-3300
NIOSH Technical Information Center	yes	513/684-8328
National Technical Information Service	yes	202/967-4349
National Emergency Equipment Locator System (NEELS-Canadian)	yes	819/997-3742
National Analysis of Trends in Emergencies System (NATES-Canadian)	yes	819/997-3742
NASA–Scientific and Technical Information Office	yes	202/755-3548
NASA–Industrial Application Centers		
University of Connecticut, Storrs, CT	yes	203/486-4533
Research Triangle Park, NC	yes	919/549-8291
University of Pittsburgh, PA	yes	412/624-5211
Indiana University, Bloomington, IN	yes	812/337-8884
University of New Mexico, Albuquerque NM	yes	505/277-3622
University of Southern California, Los Angeles, CA	yes	213/746-6132
Global Engineering Documentation Services	no	714/540-9870 213/624-1216
U.S. Department of Commerce Maritime Administrator	no	212/967-5136
National Bureau of Standards Fire Technology Library	no	301/921-3246
NASA/Aerospace Safety Research and Data Institute	no	216/443-4000 ext. 285
Chemical Abstract Service, Ohio State University	no	614/421-6940
Computer Search Center, Illinois Institute of Technology Research Institute	no	312/225-9630

(continued)

Table 1.4: (continued)

Information Source	On-Line Computer System	Contact
Fire Research Section Southwest Research Institute	no	512/684-5111 ext. 2415
Environmental Engineering Division, Texas A&M University	no	713/845-3011
Toxicology Data Bank, National Library of Medicine	no	301/496-1131

Source: EPA-600/2-77-227

IDENTIFICATION AND ASSESSMENT

> The information in this section is based on *Manual for the Control of Hazardous Material Spills. Volume 1: Spill Assessment and Water Treatment Techniques,* EPA-600/2-77-227 prepared by K.R. Huibregtse, R.C. Scholz, R.E. Wullschleger, J.H. Moser, E.R. Bollinger and C.A. Hansen of Envirex for the Environmental Protection Agency.

The information in this section deals with stepwise identification and assessment of the danger potential in a spill situation. The material presented has been divided into two parts which are directed at two people associated with the spill control effort. One person is called the Man on the Scene (MOS) who is the OSC for the region or his designated representative at the spill site. The MOS is to provide information to a remote OSC who has available to him the additional information critical to accurate spill identification and assessment.

Directions for the Man on the Scene

There are various assumptions inherent in this presentation of the duties of the MOS. These assumptions include:

> The MOS is the OSC or his designated representative at the site of a spill situation. The MOS is aware of the risks involved in spill control.
>
> The responsibility of the MOS is to provide information to a remote OSC which will aid in spill identification and assessment of the human danger potential.
>
> The responsibility for evacuation, fire fighting, or crowd control is not the duty of the appointed MOS but is assigned to other agencies, divisions, or personnel.
>
> More than one person is available to perform the functions of the MOS, if necessary. Therefore, notification can be done concurrently with information gathering for identification.

The safety of the people who arrive on the scene of a hazardous spill is critical to consider before any action is taken. All spills are considered extremely hazardous. If the identity of the spill is not known, then the MOS should not approach the spill and should await the arrival of more highly trained, experienced personnel or a volunteer who is more aware of the risks involved with hazardous materials. In all cases, fully protective clothing should be worn by personnel at the spill site. In addition, the following general safety precautions should be followed:

Always approach a spill from upwind;

Don't touch the material and avoid any indirect or direct contact with it;

Remove all possible ignition sources, do not smoke;

Restrict access to the area;

Do not touch any container unless full knowledge of the hazards involved is available; and

If unidentified fuming liquids or gases are present, do not approach.

The MOS is the eyes and ears of a remote OSC. There are a series of steps to follow which will allow the MOS to gather sufficient information for transfer to the OSC so an identification can be made. The MOS should note the source of the spill and any visible label and observe any easily visible physical properties. Once this information has been assembled, the OSC should be contacted. The information can be transmitted to the OSC who will then direct the MOS to perform additional steps to aid in identification or will direct him to begin to assess the danger potential of the situation. If so directed by the OSC the MOS can take one or more of the following steps to aid in specific identification of the spilled material:

Make on-site inspections to determine which cargo is damaged and undamaged. This procedure requires the MOS to be very cautious and board the vehicle to check the cargo. The procedure is required only for mixed loads and will allow elimination of various cargoes which have not been damaged from the list of possible materials involved in the spill.

Take samples for chemical testing: The OSC may direct the MOS to collect samples and then ship or deliver these samples to a specified laboratory. Care must be taken at all times to protect the MOS. The spill should not be approached without fully protective clothing.

Once the spill has been identified as to its specific constituents, the magnitude of the spill and its potential danger to human safety must be established. To determine this, the MOS must answer a series of questions as directed by the OSC. The OSC can then establish the effect of the spill using various references. Information regarding environmental conditions and location of the spill is necessary to determine the danger which exists from the spill itself. There are again various questions which must be answered by the MOS, and the information must be transferred to the OSC.

Functions of the On-Scene Coordinator

The OSC will use the information relayed to him from the MOS and apply his experience and resources to establish a positive identification of the material involved. This identification process may require only one or two steps plus confirming identification by physical properties or it may be rather lengthy.

Identification may be made from shipping papers or placards and labels. In addition to the shipping papers and placards, other information is also available. Among the most helpful are the Chemcards recommended and produced by the Manufacturing Chemists Association (2). The information on the Chemcards includes:

- Identification of the cargo;
- Appearance and odor;
- Statement of hazards involved and instructions for safe handling and, as applicable, the need for special cargo environments;
- Emergency procedures and precautions and
- Fire fighting procedures and precautions.

However, if mixed loads of hazardous materials are involved, further identification steps are necessary.

Another aid to the OSC is a requirement by the Coast Guard that all foreign vessels carrying hazardous materials register the intended route of the ship with the nearest Captain of the Port Coast Guard Office. Also, all vessels carrying any of the 40 specified dangerous cargoes (see Table 1.5) must notify the Captain of the Port at least 24 hours prior to arrival. Therefore, when ships are the source of a spill, more detailed information can be obtained from the Coast Guard.

Table 1.5: Hazardous Chemicals Designated by the Coast Guard

Acetaldehyde	Hydrochloric acid
Acetone cyanohydrin	Methane
Acetonitrile	Methyl acrylate
Acrylonitrile	Methyl bromide
Allyl alcohol	Methyl chloride
Allyl chloride	Methyl methacrylate (Monomer)
Ammonia, anhydrous	Nonylphenol
Aniline	Oleum
Butadiene	Phenol
Carbolic acid	Phosphorus, elemental
Carbon disulfide	Propane
Chlorine	Propylene
Chlorohydrins, crude	Propylene oxide
Crotonaldehyde	Sulfuric acid
1,2-Dichloropropane	Sulfuric acid, spent
Dichloropropane	Tetraethyl lead
Epichlorohydrin	Tetraethyl lead mixtures
Ethylene	Vinyl acetate
Ethyl ether	Vinyl chloride
Ethylene oxide	Vinylidene chloride

Source: EPA-600/2-77-227

If there is no direct information available from shipping papers or other sources, the physical properties can aid in establishing the identity of the material involved. These properties are also useful in confirmation of the identity of the spill.

OHMTADS can then be contacted to run a computer search for chemical identity. Another way to utilize the information is through the EPA publication *Field Detection and Damage Assessment Manual for Oil and Hazardous Material Spills* (3). The physical properties of many chemicals are listed in this manual and eliminations can be done to establish the identity of the chemicals.

The final type of identification is done by sampling the spill in the contaminated area and then analyzing to determine the specific chemical involved. The simplest approach is to indicate to the laboratory performing the analysis what the possible contaminants are. This information can be established from shipping papers, warning labels or physical property identification.

The OSC must contact a qualified analytical laboratory and then relay their specifications to the MOS for sampling. It is important that the sample be delivered to the laboratory as soon as possible to hasten identification of the chemical. Therefore, a laboratory in close proximity to the spill is desirable.

Assessment of Spill Magnitude and Human Danger Potential

Once the identity of the chemical spilled is known, the assessment of spill magnitude and human danger potential must be established. This process requires utilization of the input from the MOS into prepared programs to assess the danger involved. The OSC will need to contact the CHRIS system and access the HACS computer. This procedure, however, does require both knowledge of the CHRIS system and gathering of necessary input data.

The most comprehensive resource available for assessment of human danger is the *CHRIS Hazard Assessment Handbook,* CG-446-3, the third volume in the four volume set of CHRIS manuals. This handbook and HACS, the associated computer program, can provide extensive information for many types of spill situations. Determination of the appropriate program to use is directly related to the identity of the chemical spill. The second CHRIS manual lists a hazard assessment code for which various calculation procedures can be followed to determine the extent of the hazard involved.

CHRIS requires inputs of primary and secondary information to directly assess the effect of the hazard. This information is gathered by the MOS and must be transformed into an acceptable form for the HACS or CHRIS input.

CLASSIFICATION SYSTEMS

Categorization of Representative CHRIS Chemicals

> The information in this section is based on *Agents, Methods and Devices for Amelioration of Discharges of Hazardous Chemicals on Water,* CG-D-38-76 prepared by W.H. Bauer, D.N. Borton and J.J. Bulloff of Rensselaer Polytechnic Institute for the U.S. Coast Guard.

The work reported here was performed to identify, test and evaluate methods and devices advantageous for amelioration of spill situations of selected representative hazardous chemicals. Spills of chemicals on and into water are possibly more dangerous to the health and welfare of the public and to the environment than oil spills. Much work has been done by governmental agencies and private industry to develop methods for control, cleanup and disposal of oil spills; less has been done to reduce the danger and damage threatened by spills of hazardous chemicals. Discharge of these tasks faces the obstacle of the vast number of hazardous chemicals which exist. It is not expeditious or even practicable to evaluate every potential ameliorant method against each hazardous chemical.

One approach to this complex problem is to group hazardous chemicals into categories that can be ameliorated by a particular method, and to then select a representative hazardous chemical for each such amelioration response category. Study of selected representative chemicals contributes to the development of an amelioration system that may possibly deal with the entire category satisfactorily. If, as is possible for some hazardous chemicals, an effective system may not be attainable, such chemicals would remain classified as unrecoverable. Presumably, preventive regulation would be applied to their handling.

In accordance with this approach, the U.S. Coast Guard CHRIS list of selected hazardous chemicals has been classified into four behavior response categories and thirty amelioration response categories, and for each amelioration response category a representative hazardous chemical has been identified. This information is given in Table 1.6.

Table 1.6: Categories Representative of the CHRIS 400 Chemicals

RHC*	Behavior Response Category	ARC**
	.... I—Vaporizes	
Benzene	Heavier than air, flammable	I A
Chlorine	Heavier than air, nonflammable	I B
Hydrogen fluoride	Lighter than air, nonflammable	I C
Liquefied natural gas, LNG	Heavier than air when spilled, then lighter	I D
Ammonia, anhydrous	Water-soluble	I E
	.... II—Floats	
Pentadecanol	Solids	II A
Ethyl ether	Liquids, high flammability	II B
Kerosene	Liquids, low flammability	II C
n-Amyl alcohol	Liquids that spread	II D
Naphtha: solvent	Liquids that do not spread	II E
Ethyl acetate	Liquids, oxygen-containing: esters	II F
Hexane	Liquids, simple hydrocarbons	II G
n-Butyraldehyde	Liquids, oxygen-containing: others	II H
	.... III—Sinks	
Aldrin	Solids	III A
Bromine	Inorganic liquids	III B
Chloroform	Liquids, halogenated organics	III C
Nitrobenzene	Liquids, aromatic organics	III D
	.... IV—Dissolves	
Sulfuric acid, concentrated	Inorganic acids	IV A
Sodium hydroxide	Inorganic bases	IV B
Potassium cyanide	Inorganic, ionic	IV C
Cadmium chloride	Inorganic containing heavy metal	IV D
Bromine	Halogens	IV E

(continued)

Table 1.6: (continued)

RHC*	Behavior Response Category	ARC**
	.... IV—Dissolves.	
Oxalic acid	Organic acids	IV F
Methyl alcohol	Oxygen containing: alcohols	IV G
Ethyl acetate	Organic, oxygen containing: esters	IV H
Ethylenediamine	Organic, nitrogen containing: amines	IV J
Dimethyl sulfoxide	Organic, sulfur-containing	IV K
Epichlorohydrin	Organic, chlorine containing	IV L
Sodium alkylbenzene-sulfonates	Organic with ionically bound metal; nonbiodegradable	IV M
Phenol	Organic, biodegradable	IV N

*Representative Hazardous Chemical.
**Amelioration Response Category.

Source: CG-D-38-76

The amelioration response categories were defined in a report to the U.S. Coast Guard prepared by Lyman, et al of Arthur D. Little, Inc. (5). Schemes were developed for categorizing hazardous chemicals according to their physical and chemical nature. The categorization schemes were developed for 400 chemicals judged to present the greatest risks in water transport (6).

Four principal behavioral categories were based on whether the hazardous chemicals vaporize, float, sink or mix with water. The primary categories were further subdivided according to key physical and chemical properties potentially important to the amelioration of spills. Using the criteria established for each of the categories, it is possible to list those of the 400 hazardous chemicals which fit the requirements of the category. Such lists are given in the following compilation, Table 1.7. In the following lists representative hazardous chemicals are designated RHC.

Table 1.7: Categorization of CHRIS Chemicals

Category I A—Vapors: Heavier Than Air, Flammable

1. Acetaldehyde
2. Acetone
3. Acetonitrile
4. Allyl chloride
5. Benzene (RHC)
6. Butadiene, inhibited
7. Butane
8. n-Butyl alcohol
9. sec-Butyl alcohol
10. tert-Butyl alcohol
11. Butylene
12. n-Butyraldehyde
13. Cyclohexane
14. Diethylamine
15. Diisobutylene
16. Dimethylamine
17. 1,1-Dimethylhydrazine
18. Distillates: flashed feed stock
19. Distillates: straight run
20. Ethyl acetate
21. Ethyl alcohol
22. Ethylbenzene
23. Ethyl chloride
24. Ethyleneimine
25. Ethylene oxide
26. Ethyl ether
27. Gasoline blending stocks: alkylates
28. Gasoline blending stocks: reformates
29. Gasolines: automotive (<4.23 g lead/gal)
30. Gasolines: aviation (<4.86 g lead/gal)
31. Gasolines: casinghead
32. Gasolines: polymer

(continued)

Table 1.7: (continued)

Category I A—Vapors: Heavier than Air, Flammable

33. Gasolines: straight run
34. Heptane
35. 1-Heptene
36. Hexane
37. 1-Hexene
38. Hydrogen sulfide
39. Isobutane
40. Isobutylene
41. Isobutyraldehyde
42. Isohexane
43. Isopentane
44. Isoprene
45. Isopropyl alcohol
46. Liquefied petroleum gas
47. Methyl acrylate
48. Methyl chloride
49. Methyl ethyl ketone
50. Methyl methacrylate
51. Naphtha: VM&P (75% naphtha)
52. Nitromethane
53. Pentane
54. 1-Pentene
55. Phosphorus pentasulfide
56. Propane
57. Propionaldehyde
58. Propylene
59. Toluene
60. Trimethylamine
61. Vinyl acetate
62. Vinyl chloride

Category I B—Vapors: Heavier than Air, Nonflammable

1. Aluminum chloride
2. Benzoyl chloride
3. Chlorine (RHC)
4. Chlorosulfonic acid
5. Cyanogen chloride
6. 1-Decene
7. Dichlorodifluoromethane
8. Ethylene glycol dimethyl ether
9. Fluorine
10. Hydrogen chloride
11. Isoamyl alcohol
12. Methyl bromide
13. Mineral spirits
14. Monochlorodifluoromethane
15. Naphtha: coal tar
16. Naphtha: solvent
17. Naphtha: Stoddard solvent
18. Nitrogen tetroxide
19. Nitrosyl chloride
20. Oleum
21. Oils, miscellaneous: penetrating
22. Phenol
23. Phosgene
24. Phosphorus oxychloride
25. Phosphorus trichloride
26. Sulfur dioxide
27. Sulfur monochloride
28. Sulfuryl chloride
29. Toluene 1,4-diisocyanate
30. Trichlorofluoromethane

Category I C—Vapors: Lighter than Air, Nonflammable

1. Ammonia, anhydrous
2. Hydrogen fluoride (RHC)

Category I D—Vapors: Heavier than Air when Spilled, Then Lighter

1. Ammonia, anhydrous
2. Ethane
3. Ethylene
4. Liquified natural gas (RHC)
5. Methane

Category I E—Vapors: Water-Soluble

1. Acetaldehyde
2. Acetone
3. Acetonitrile
4. Aluminum chloride
5. Ammonia, anhydrous (RHC)
6. Ammonium hydroxide (<28% aqueous ammonia)
7. Benzoyl chloride
8. n-Butyl alcohol
9. sec-Butyl alcohol
10. tert-Butyl alcohol
11. n-Butyraldehyde
12. Chlorosulfonic acid
13. Cyanogen chloride
14. Diethylamine
15. Dimethylamine
16. 1,1-Dimethylhydrazine
17. Ethyl acetate

(continued)

Table 1.7: (continued)

18. Ethyl alcohol
19. Ethylene glycol dimethyl ether
20. Ethyleneimine
21. Ethylene oxide
22. Ethyl ether
23. Formaldehyde solution
24. Fluorine
25. Hydrogen chloride
26. Hydrogen cyanide
27. Hydrogen fluoride
28. Isoamyl alcohol
29. Isobutyraldehyde
30. Isopropyl alcohol
31. Methyl alcohol
32. Methyl acrylate
33. Methyl chloride
34. Methyl ethyl ketone
35. Methyl methacrylate
36. Nitromethane
37. Nitrosyl chloride
38. Oleum
39. Paraformaldehyde
40. Phenol
41. Phosgene
42. Phosphorus oxychloride
43. Propionaldehyde
44. Sodium amide
45. Sulfur dioxide
46. Sulfuryl chloride
47. Trimethylamine
48. Vinyl acetate

Category II A–Floats: Solids

1. Asphalt
2. Asphalt blending stocks: roofers flux
3. Asphalt blending stocks: straight run residue
4. Camphor oil
5. Waxes: carnauba
6. Ethylenediaminetetraacetic acid
7. Isodecyl alcohol
8. Isooctyl alcohol
9. Linear alcohols (12-15 carbons)
10. Paraffin
11. Pentadecanol (RHC)
12. Petrolatum
13. Phenol

Category II B–Floats: Liquids, High Flammability

1. Acrylonitrile
2. Allyl chloride
3. Amyl acetate
4. n-Amyl alcohol
5. n-Butyl alcohol
6. Butadiene, inhibited
7. n-Butyl acetate
8. sec-Butyl acetate
9. n-Butyraldehyde
10. Diethyl carbonate
11. Diisobutylene
12. Distillates: flashed feed stocks
13. Distillates: straight run
14. Ethyl acetate
15. Ethyl acrylate
16. Ethyl benzene
17. Ethyl ether (RHC)
18. Gasoline blending stocks: alkylates
19. Gasoline blending stocks: reformates
20. Gasolines: automotive (<4.23 g lead/gal)
21. Gasolines: aviation (<4.86 g lead/gal)
22. Gasolines: casinghead
23. Gasolines: polymer
24. Gasolines: straight run
25. Heptane
26. 1-Heptene
27. Hexane
28. 1-Hexene
29. Isobutyl alcohol
30. Isobutyl acetate
31. Isobutyraldehyde
32. Isohexane
33. Isopentane
34. Isoprene
35. Isopropyl acetate
36. Jet fuel: JP-3
37. Jet fuel: JP-4
38. Methyl acrylate
39. Methyl isobutyl ketone
40. Methyl methacrylate
41. Naphtha: VM&P (75% naphtha)
42. Nonene
43. 1-Nonene
44. 1-Octene

(continued)

Table 1.7: (continued)

45. Oils, miscellaneous:
    coal tar
46. Oils: crude
47. Pentane
48. 1-Pentene
49. Petroleum naphtha
50. n-Propyl acetate
51. Styrene
52. Toluene
53. Triethylamine
54. Turpentine
55. Valeraldehyde
56. Vinyl acetate
57. Vinyl chloride
58. m-Xylene
59. o-Xylene

Category II C–Floats: Liquids, Low Flammability

1. Aldrin
2. Aniline
3. Benzaldehyde
4. tert-Butyl hydroperoxide
5. Cumene
6. Cyclohexanone
7. 1-Decene
8. n-Decyl alcohol
9. Dibutyl phthalate
10. Dicyclopentadiene
11. Diethylbenzene
12. Diisobutylcarbinol
13. Dioctyl adipate
14. Dioctyl phthalate
15. Dodecanol
16. Dodecene
17. 1-Dodecene
18. Ethyl butanol
19. 2-Ethyl hexanol
20. Ethyl hexyl tallate
21. 2-Ethyl-3-propylacrolein
22. Gas oil: cracked
23. Glycidyl methacrylate
24. Heptanol
25. Hexanol
26. Isoamyl alcohol
27. Isodecaldehyde
28. Isooctaldehyde
29. Jet fuels: JP-1 (kerosene)
30. Jet fuels: JP-5 (kerosene, heavy)
31. Kerosene (RHC)
32. Lauryl mercaptan
33. Methyl amyl acetate
34. Methyl amyl alcohol
35. Methylethylpyridine
36. Methyl isobutyl carbinol
37. Mineral spirits
38. Nonanol
39. Nonylphenol
40. Naphtha: coal tar
41. Naphtha: solvent
42. Naphtha: Stoddard solvent
43. Octanol
44. Oils: clarified
45. Oils: crude
46. Oils: diesel
47. Oils, edible: castor
48. Oils, edible: fish
49. Oils, edible: olive
50. Oils, edible: soya bean
51. Oils, edible: vegetable
52. Oils, fuel: No. 1 (kerosene)
53. Oils, fuel: No. 1-D
54. Oils, fuel: No. 2
55. Oils, fuel: No. 2-D
56. Oils, fuel: No. 4
57. Oils, fuel: No. 5
58. Oils, miscellaneous: absorption
59. Oils, miscellaneous: lubricating
60. Oils, miscellaneous: mineral
61. Oils, miscellaneous: mineral seal
62. Oils, miscellaneous: motor
63. Oils, miscellaneous: neatsfoot
64. Oils, miscellaneous: penetrating
65. Oils, miscellaneous: range
66. Oils, miscellaneous: resin
67. Oils, miscellaneous: road
68. Oils, miscellaneous: rosin
69. Oils, miscellaneous: sperm
70. Oils, miscellaneous: spindle
71. Oils, miscellaneous: spray
72. Oils, miscellaneous: tall
73. Oils, miscellaneous: tanner
74. Oils, miscellaneous: transformer
75. Polybutene
76. Propylene butylene polymer
77. Propylene tetramer
78. 1-Tetradecene
79. Tetrahydronaphthalene
80. Triethylbenzene
81. 1-Tridecene
82. 1-Undecene
83. Vinyltoluene

Category II D–Floats: Liquids That Spread

1. Acrylonitrile
2. n-Amyl Alcohol (RHC)

(continued)

Table 1.7: (continued)

3. n-Butyl acetate
4. n-Butyl alcohol
5. tert-Butyl hydroperoxide
6. n-Butyraldehyde
7. Cresol
8. Diethyl carbonate
9. Ethyl acetate
10. Ethyl acrylate
11. Ethyl ether
12. Heptanol
13. Hexanol
14. Isoamyl alcohol
15. Isobutyl alcohol
16. Isobutyraldehyde
17. Isopropyl acetate
18. Methyl acrylate
19. Methyl amyl alcohol
20. Methyl isobutyl carbinol
21. Methyl isobutyl ketone
22. Methyl methacrylate
23. Nonanol
24. Octanol
25. n-Propyl acetate
26. Tetradecanol
27. Tridecanol
28. Triethylamine
29. Undecanol
30. Valeraldehyde
31. Vinyl acetate

Category II E–Floats: Liquids That Do Not Spread

1. Adiponitrile
2. Asphalt blending stocks: roofers flux
3. Asphalt blending stocks: straight run residue
4. Benzene
5. Camphor oil
6. Cumene
7. Decaldehyde
8. 1-Decene
9. Dibutyl phthalate
10. Diethylbenzene
11. Diisobutylcarbinol
12. Dioctyl adipate
13. Dioctyl phthalate
14. Distillates: flashed feed stocks
15. Distillates: straight run
16. Dodecene
17. 1-Dodecene
18. Dowtherm
19. Epoxidized vegetable oils
20. Ethylbenzene
21. Ethylhexyl tallate
22. 2-Ethyl-3-propylacrolein
23. Gas oil: cracked
24. Gasoline blending stocks: alkylates
25. Gasoline blending stocks: reformates
26. Gasolines: automotive (<4.23 g lead/gal)
27. Gasolines: aviation (<4.86 g lead/gal)
28. Gasolines: casinghead
29. Gasolines: polymer
30. Gasolines: straight run
31. Heptane
32. 1-Heptene
33. Hexane
34. 1-Hexene
35. Isohexane
36. Isopentane
37. Isoprene
38. Jet fuels: JP-1 (kerosene)
39. Jet fuels: JP-3
40. Jet fuels: JP-4
41. Jet fuels: JP-5 (kerosene, heavy)
42. Kerosene
43. Lauryl mercaptan
44. Mineral spirits
45. Naphtha: coal tar
46 Naphtha: solvent (RHC)
47. Naphtha: Stoddard solvent
48. Naphtha: VM&P (75% naphtha)
49. Nonene
50. 1-Nonene
51. Nonylphenol
52. 1-Octene
53. Oils: clarified
54. Oils: crude
55. Oils: diesel
56. Oils, edible: castor
57. Oils, edible: cottonseed
58. Oils, edible: fish
59. Oils, edible: olive
60. Oils, edible: peanut
61. Oils, edible: soya bean
62. Oils, edible: vegetable
63. Oils, fuel: No. 1 (kerosene)
64. Oils, fuel: No. 1-D
65. Oils, fuel: No. 2
66. Oils, fuel: No. 2-D
67. Oils, fuel: No. 4
68. Oils, fuel: No. 5

(continued)

Table 1.7: (continued)

69. Oils, fuel: No. 6
70. Oils, miscellaneous: absorption
71. Oils, miscellaneous: coal tar
72. Oils, miscellaneous: lubricating
73. Oils, miscellaneous: mineral
74. Oils, miscellaneous: mineral seal
75. Oils, miscellaneous: motor
76. Oils, miscellaneous: neatsfoot
77. Oils, miscellaneous: penetrating
78. Oils, miscellaneous: range
79. Oils, miscellaneous: resin
80. Oils, miscellaneous: road
81. Oils, miscellaneous: rosin
82. Oils, miscellaneous: sperm
83. Oils, miscellaneous: spindle
84. Oils, miscellaneous: spray
85. Oils, miscellaneous: tall
86. Oils, miscellaneous: tanner's
87. Oils, miscellaneous: transformer
88. Pentane
89. 1-Pentene
90. Petroleum naphtha
91. Polybutene
92. Propylene butylene polymer
93. Propylene tetramer
94. Styrene
95. Tallow
96. 1-Tetradecene
97. Tetrahydronaphthalene
98. Toluene
99. Toluene 1,4-diisocyanate
100. 1-Tridecene
101. Triethylbenzene
102. Turpentine
103. 1-Undecene
104. Vinyltoluene
105. m-Xylene
106. o-Xylene
107. p-Xylene

Category II F–Floats: Liquids, Oxygen-Containing: Esters

1. Amyl acetate
2. n-Butyl acetate
3. sec-Butyl acetate
4. n-Butyl acrylate
5. Dibutyl phthalate
6. Dioctyl adipate
7. Dioctyl phthalate
8. Ethyl acetate (RHC)
9. Ethyl acrylate
10. Ethylhexyl tallate
11. Isobutyl acetate
12. Isobutyl acrylate
13. Isopropyl acetate
14. Methyl acrylate
15. Methyl amyl acetate
16. Methyl methacrylate
17. Oils, edible: castor
18. Oils, edible: fish
19. Oils, edible: olive
20. Oils, edible: soya bean
21. Oils, edible: vegetable
22. Oils, miscellaneous: neatsfoot
23. Oils, miscellaneous: sperm
24. n-Propyl acetate
25. Vinyl acetate

Category II G–Floats: Liquids, Simple Hydrocarbons

1. Heptane
2. Hexane (RHC)
3. Isohexane
4. Isopentane
5. Pentane

Category II H–Floats: Liquids, Oxygen Containing: Others

1. Benzaldehyde
2. tert-Butyl hydroperoxide
3. n-Butyraldehyde (RHC)
4. Cyclohexanone
5. Diethyl carbonate
6. Ethyl ether
7. 2-Ethyl-3-propylacrolein
8. Isobutyraldehyde
9. Isooctaldehyde
10. Methyl isobutyl ketone
11. Oils, miscellaneous: resin
12. Oils, miscellaneous: rosin
13. Oils, miscellaneous: tall
14. Oils, miscellaneous: tanner's
15. Valeraldehyde

Category III A–Sinks: Solids

1. Aldrin (RHC)
2. Aluminum chloride

(continued)

Table 1.7: (continued)

3.	Aluminum fluoride	19.	Dibenzoyl peroxide
4.	Asphalt	20.	p-Dichlorobenzene
5.	Asphalt blending stocks: roofers flux	21.	2,4-Dichlorophenol
		22.	Dichloropropene
6.	Asphalt blending stocks: straight run	23.	DDT
		24.	2,4-Dinitroaniline
7.	Barium carbonate	25.	2,4-Dinitrophenol
8.	Benzene hexachloride	26.	Diphenylmethanediisocyanate
9.	Benzoic acid	27.	Lead arsenate
10.	Bisphenol A	28.	Naphthalene, molten
11.	Calcium carbide	29.	Pentachlorophenol
12.	Calcium fluoride	30.	Phosphorus, red
13.	Calcium hydroxide	31.	Phosphorus, white
14.	Calcium oxide	32.	Phthalic anhydride
15.	Camphor oil	33.	Polychlorinated biphenyl
16.	Carbaryl	34.	Sulfur (liquid)
17.	Carbolic oil	35.	Toxaphene
18.	Cresols	36.	Trichlorophenol

Category III B–Sinks: Inorganic Liquids

1.	Bromine (RHC)	2.	Mercury

Category III C–Sinks: Liquids, Halogenated Organics

1.	Benzoyl chloride	10.	Ethylene dichloride
2.	Carbon tetrachloride	11.	Motor fuel antiknock compounds containing lead alkyls
3.	Chlorobenzene		
4.	Chloroform (RHC)	12.	Tetrachloroethylene
5.	o-Dichlorobenzene	13.	Trichloroethane
6.	Dichloromethane	14.	Trichloroethylene
7.	Dichloropropane	15.	Trichlorofluoromethane
8.	Epichlorohydrin	16.	Vinylidene chloride, inhibited
9.	Ethylene dibromide		

Category III D–Sinks: Liquids, Aromatic Organics

1.	Acetophenone	6.	Dibutyl phthalate
2.	Aniline	7.	o-Dichlorobenzene
3.	Benzaldehyde	8.	Methylparathion
4.	Benzoyl chloride	9.	Nitrobenzene (RHC)
5.	Chlorobenzene	10.	Tricresyl phosphate

Category IV A–Dissolves: Inorganic Acids

1.	Aluminum chloride	12.	Oleum
2.	Benzoyl chloride	13.	Phosphoric acid
3.	Chlorosulfonic acid	14.	Phosphorus oxychloride
4.	Hydrochloric acid	15.	Phosphorus pentasulfide
5.	Hydrofluoric acid	16.	Phosphorus trichloride
6.	Hydrogen chloride	17.	Polyphosphoric acid
7.	Hydrogen cyanide	18.	Sulfur monochloride
8,	Hydrogen fluoride	19.	Sulfuric acid, spent
9.	Nitric acid	20.	Sulfuric acid (RHC)
10.	Nitrogen tetroxide	21.	Sulfuryl chloride
11.	Nitrosyl chloride	22.	Titanium tetrachloride

Table 1.7: (continued)

Category IV B—Dissolves: Inorganic Bases

1. Ammonium hydroxide (28% aqueous ammonia)
2. Caustic potash solution
3. Caustic soda solution
4. Lithium aluminum hydride
5. Potassium hydroxide
6. Sodium
7. Sodium amide
8. Sodium borohydride
9. Sodium hydride
10. Sodium hydroxide (RHC)

Category IV C—Dissolves: Inorganic, Ionic

1. Ammonium nitrate
2. Ammonium perchlorate
3. Ammonium sulfate
4. Antimony trifluoride
5. Cadmium chloride
6. Calcium hypochlorite
7. Chromic anhydride
8. Copper sulfate
9. Ferrous sulfate
10. Hydrazine
11. Hydrogen peroxide
12. Nickel sulfate
13. Phosphorus pentasulfide
14. Potassium cyanide (RHC)
15. Potassium dichromate
16. Potassium permanganate
17. Potassium iodide
18. Silver nitrate
19. Sodium bisulfite
20. Sodium chlorate
21. Sodium cyanide
22. Sodium dichromate
23. Sodium ferrocyanide
24. Sodium fluoride
25. Sodium hypochlorite
26. Sodium silicate
27. Sodium sulfide
28. Sodium sulfite
29. Suifur dioxide
30. Zinc chloride

Category IV D—Dissolves: Inorganic Containing Heavy Metal

1. Antimony trifluoride
2. Cadmium chloride (RHC)
3. Chromic anhydride
4. Copper sulfate
5. Ferrous sulfate
6. Nickel sulfate
7. Potassium dichromate
8. Potassium permanganate
9. Silver nitrate
10. Sodium dichromate
11. Sodium ferrocyanide
12. Titanium tetrachloride

Category IV E—Dissolves: Halogens

1. Bromine (RHC)
2. Chlorine
3. Fluorine

Category IV F—Dissolves: Organic Acids

1. Acetic acid
2. Acetic anhydride
3. Acrylic acid
4. Formic acid
5. Maleic anhydride
6. Oxalic Acid (RHC)
7. Propionic acid

Category IV G—Dissolves: Oxygen-Containing, Alcohols

1. Acetone cyanohydrin
2. Allyl alcohol
3. n-Amyl alcohol
4. 1,4-Butanediol
5. 1,4-Butenediol
6. n-Butyl alcohol
7. sec-Butyl alcohol
8. tert-Butyl alcohol
9. 1,4-Butynediol
10. Corn syrup
11. Cyclohexanol
12. Dextrose solution
13. Diacetone alcohol
14. Diethylene glycol
15. Diethylene glycol monoethyl ether
16. Diethylene glycol monomethyl ether

(continued)

Table 1.7: (continued)

17. Dipropylene glycol
18. Ethoxylated dodecanol
19. Ethoxylated pentadecanol
20. Ethoxylated tetradecanol
21. Ethoxylated tridecanol
22. Ethoxytriglycol
23. Ethyl alcohol
24. Ethylene cyanohydrin
25. Ethylene glycol
26. Ethylene glycol monobutyl ether
27. Ethylene glycol monoethyl ether
28. Ethylene glycol monomethyl ether
29. Hexylene glycol
30. Glycerin
31. Isoamyl alcohol
32. Isobutyl alcohol
33. Isopropyl alcohol
34. Methyl alcohol (RHC)
35. Methyl amyl alcohol
36. Methyl isobutyl carbinol
37. Polypropylene glycol methyl ether
38. n-Propyl alcohol
39. Propylene glycol
40. Propylene glycol methyl ether
41. Sorbitol
42. Triethylene glycol

Category IV H—Dissolves: Organic, Oxygen-Containing (Esters)

1. n-Butyl acetate
2. Ethyl acetate (RHC)
3. Ethyl acrylate
4. Ethylene glycol monoethyl ether acetate
5. Isopropyl acetate
6. Methyl acrylate
7. Methyl methacrylate
8. n-Propyl acetate
9. Vinyl acetate

Category IV J—Dissolves: Organic, Nitrogen-Containing (Amines)

1. Aminoethanolamine
2. Aniline
3. Cyclohexylamine
4. Diethanolamine
5. Diethylamine
6. Diethylenetriamine
7. Diisopropanolamine
8. Dimethylformamide
9. 1,1-Dimethylhydrazine
10. Ethylenediamine (RHC)
11. Ethyleneimine
12. Hexamethylenediamine
13. Hexamethylenetetramine
14. Methylethylpyridine
15. Monoethanolamine
16. Monoisopropanolamine
17. Morpholine
18. Pyridine
19. Triethanolamine
20. Triethylamine
21. Triethylenetetramine
22. Trimethylamine
23. Urea

Category IV K—Dissolves: Organic, Sulfur-Containing

1. Dimethyl sulfate
2. Dimethyl sulfoxide (RHC)
3. Sodium alkyl benzenesulfonates
4. Sodium alkyl sulfates
5. Sulfolane

Category IV L—Dissolves: Organic, Chlorine-Containing

1. Chloroform
2. Chlorohydrins (crude)
3. Dichloromethane
4. Epichlorohydrin (RHC)

Category IV M—Dissolves: Organic with Ionically Bound Metal, Nonbiodegradable

1. Sodium alkylbenzene sulfonate (RHC)

Category IV N—Dissolves: Organic, Biodegradable

1. Acetic acid
2. Acetic anhydride

(continued)

Table 1.7: (continued)

3.	Acetone	25.	Isoamyl alcohol
4.	Acrylic acid	26.	Isobutyl alcohol
5.	n-Amyl alcohol	27.	Isopropyl acetate
6.	Aniline	28.	Isopropyl alcohol
7.	Benzaldehyde	29.	Maleic anhydride
8.	Benzoic acid	30.	Methyl alcohol
9.	n-Butyl acetate	31.	Methyl ethyl ketone
10.	n-Butyl alcohol	32.	Methylethylpyridine
11.	n-Butyraldehyde	33.	Methyl methacrylate
12.	Carbolic oil	34.	Monoethanolamine
13.	Corn syrup	35.	Oxalic acid
14.	Cresols	36.	Paraformaldehyde
15.	Crotonaldehyde	37.	Phenol (RHC)
16.	Dextrose solution	38.	Phthalic anhydride
17.	Diethylene glycol	39.	Propionaldehyde
18.	Dimethyl formamide	40.	Propionic acid
19.	Ethyl acetate	41.	n-Propyl alcohol
20.	Ethylene diamine	42.	Pyridine
21.	Ethylene glycol monoethyl ether	43.	Sodium alkyl sulfates
		44.	Sorbitol
22.	Formaldehyde solution	45.	Triethylene glycol
23.	Furfural	46.	Urea
24.	Glycerin	47.	Vinyl acetate

Source: CG-D-38-76

In Table 1.17 in the section on Response Techniques are listed the most advantageous agents for the representative hazardous chemical (RHC) in each of thirty amelioration categories, and the hazardous chemicals in each category to which the agents may apply.

Chemicals That Sink

The information in this section is based on *A Feasibility Study of Response Techniques for Discharges of Hazardous Chemicals That Sink,* CG-D-56-78 prepared by T.D. Hand, A.W. Ford, P.G. Malone, D.W. Thompson and R.B. Mercer of the U.S. Army Engineer Waterways Experiment Station for the U.S. Coast Guard.

The Federal Water Pollution Control Act Amendments of 1972 require the United States Coast Guard to respond to hazardous chemical spills in U.S. waters. Because of the disparity between the Coast Guard's statutory obligation to respond to such spills and the current undeveloped state of spill response and pollution control technology, the Coast Guard has undertaken a research program to bring its capabilities in line with its responsibilities.

In the early stages of this research program, the Coast Guard compiled a list of 900 hazardous chemicals and identified from these a list of 400 which are transported in significant quantities on water and are therefore of immediate concern (4). These chemicals were classified as floaters, sinkers, vapors, and those that mix or dissolve. Subsequent research (5)(7)(8)(9)(10) has generally focused on one or more of these categories, recognizing that greatly differing responses are appropriate to each.

Sinking hazardous chemicals, the subject of this study, have been defined by the Coast Guard to be those chemicals (or their reaction products) that are more dense than water and relatively insoluble in water (those for which a saturated solution is 10% or less by weight at 20°C). Of the 400 chemicals listed and subsequently described in the CHRIS, 70 have been designated sinkers. They include inorganic and organic compounds that are solid or liquid at normal temperatures, have a diverse range of physical and chemical properties, and present a variety of associated hazards.

To simplify the discussion, the 70 chemical substances are arranged in groups according to similarities in physical and chemical characteristics. A major division is made according to the physical state of the substances as they are normally shipped, i.e., solid or liquid. Further subdivisions are made based on their chemical composition as follows:

I. Inorganics—substances that do not contain carbon atoms as the major constituent of the molecule.

II. Organics—substances that do contain carbon atoms as the major constituent of the molecule.

(a) Halogenated—organic compounds that also contain any of the halogens, i.e., fluorine, chlorine, iodine, or bromine (especially chlorine or bromine).

(b) Nonhalogenated—all other organic compounds.

It will be noted that this categorization of the 70 chemicals differs from the system proposed by Lyman et al (5) and further developed by Bauer et al (8) in previous studies for the Coast Guard. Though these two references represent valuable pioneering work, it was felt that the four categories of sinkers they propose (III A, Solids; III B, Inorganic Liquids; III C, Halogenated Organic Liquids; and III D, Aromatic Organic Liquids) and the representative hazardous chemicals (RHC) they select were not suited to the purposes of this study in which an attempt is made to consider each chemical and a number of specific responses.

Each substance is listed in Table 1.18 or 1.19, along with a qualitative indication of its toxicity, flammability, reactivity with water, solubility in water, persistence in the aquatic environment, potential for bioaccumulation, and a very brief comment on any specific hazards relating to its recovery or burial and a recommended response. These tables are found in the next section, Response Techniques.

Chemicals That Float

The information in this section is based on *Feasibility Study of Response Techniques for Discharges of Hazardous Chemicals That Float on Water,* CG-D-56-77 prepared by J.S. Greer of MSA Research Corporation for the U.S. Coast Guard.

Organic compounds are the primary source for hazardous chemicals which float on water. Toxicity hazards for this class of chemicals range from common foodstuffs through carcinogens and lethal poisons. Most are flammable, with the rate of decomposition varying between self-propagation and detonation. Approximately one hundred sixty-seven of the four hundred chemical species contained in the CHRIS, are categorized as floating, hazardous chemicals, and were of primary interest in this study. The chemicals considered in this program are listed in Table 1.8.

Table 1.8: Hazardous Chemicals That Float

Acrylonitrile
Adiponitrile
Aldrin
Allyl alcohol
Amyl acetate
n-Amyl alcohol
Aniline
Asphalt
Asphalt blending stocks:
  roofers flux
  straight run residue
Benzaldehyde
Benzene
Butadiene, inhibited
n-Butyl acetate
sec-Butyl acetate
n-Butyl acrylate
Isobutyl acrylate
n-Butyl alcohol
tert-Butyl hydroperoxide
n-Butyraldehyde
Camphor oil
Cresols
Cumene
Cyclohexane
Cyclohexanone
Decaldehyde
1-Decene
n-Decyl alcohol
Dibutyl phthalate
Dicyclopentadiene
Diethylbenzene
Diethyl carbonate
Diisobutyl carbinol
Diisobutylene
Dioctyl adipate
Dioctyl phthalate
Distillates:
  straight run
  flashed feed stocks
Dodecanol
Dodecene
Dowtherm
Epoxidized vegetable oils
Ethyl acetate
Ethyl acrylate
Ethylbenzene
Ethyl butanol
Ethyl ether
2-Ethyl hexanol
Ethyl hexyl tallate
2-Ethyl-3-propylacrolein
Gas oil: cracked
Gasolines:
  automotive (<4.23 g lead/gal)
  aviation (<4.86 g lead/gal)
  casinghead
  polymer
  straight run
Gasoline blending stocks:
  alkylates
  reformates
Heptane
Heptanol
1-Heptene
Hexane
Hexanol
1-Hexene
Isoamyl alcohol
Isobutyl acetate
Isobutyl alcohol
Isodecaldehyde
Isodecyl alcohol
Isohexane
Isooctyl alcohol
Isooctaldehyde
Isopentane
Isoprene
Isopropyl acetate
Jet fuels:
  JP-1 (Kerosene)
  JP-3
  JP-4
  JP-5 (Kerosene, heavy)
Kerosene
Lauryl mercaptan
Linear alcohols
Methyl acrylate
Methyl amyl acetate
Methyl amyl alcohol
Methyl ethyl ketone
Methylethylpyridine
Methyl isobutyl carbinol
Methyl isobutyl ketone
Methyl methacrylate

(continued)

Table 1.8: (continued)

Mineral spirits	rosin
Naphtha:	sperm
coal tar	spindle
solvent	spray
Stoddard solvent	tall
VM&P (75% naphtha)	tanner's
Nonanol	transformer
Nonene	Pentadecanol
1-Nonene	Pentane
Nonylphenol	1-Pentene
Octanol	Petrolatum
1-Octene	Petroleum naphtha
Oils:	Phenol
clarified	Polybutene
crude	n-Propyl acetate
diesel	Propylene-butylene polymer
Oils, edible:	Propylene tetramer
castor	Styrene
cottonseed	Tallow
fish	Tetradecanol
olive	1-Tetradecene
peanut	Tetrahydronaphthalene
soya bean	Toluene
vegetable	Toluene-2,4-diisocyanate
Oils, fuel:	Tridecanol
No. 1 (Kerosene)	1-Tridecene
No. 1-D	Triethylamine
No. 2	Triethylbenzene
No. 2-D	Turpentine
No. 4	Undecanol
No. 5	1-Undecene
No. 6	Valeraldehyde
Oils, miscellaneous:	Vinyl acetate
absorption	Vinyl chloride
coal tar	Vinyltoluene
lubricating	Waxes:
mineral	carnauba
motor	paraffin
neatsfoot	o-Xylene
penetrating	m-Xylene
range	p-Xylene
resin	

Source: CG-D-56-77

Evaluating the hazards surrounding the accidental discharge of these chemicals involves determining the potential impact upon human health from initial exposure, detrimental environmental reactions or slow erosion from residual activity. The hazardous situations created by such diverse chemicals present such a variety that reasonable comparisons are difficult. The toxicity of vapors, both from compounds and decomposition products, and their flammability were the primary hazards used in ranking floating, hazardous chemicals. Recent experience with pesticides, insecticides and carcinogenic chemicals has accentuated the necessity of determining benefit/risk potentials for hazardous chemicals. Similar evaluations have been made concerning the transport of chemicals, where spillage and accidents are responsible for the generation of potential hazards.

Lyman et al (5) divided the major category of floating spills into eight subdivisions, each with a representative compound. These representatives and a brief description are presented in Table 1.9.

All candidate response techniques evaluated in this program were assessed in terms of their reported, extrapolated or estimated capabilities in treating spills of these eight chemicals. Benzene was added to this list since it is lighter than water and its vapors are both toxic and flammable. Benzene is also transported in very large volumes.

Comparison and correlation of potential hazards for the purpose of ranking required some estimation and extrapolation. The basic assumption used in this assessment was that toxic vapors, no matter how they were generated, present a greater risk to response personnel than flammable vapors.

Table 1.9: Floating Spill Types and Representative Chemicals

Amelioration Category	Representative Chemical
A Floats: solids	Pentadecanol
B Floats: liquids, high flammability	Ethyl ether
C Floats: liquids, low flammability	Kerosene
D Floats: liquids that spread	n-Amyl alcohol
E Floats: liquids that do not spread	Naphtha: solvent
F Floats: liquids, esters	Ethyl acetate
G Floats: liquids, simple hydrocarbons	Hexane
H Floats: liquids, oxygen containing: other	n-Butyraldehyde

Source: CG-D-56-77

Chemical vapor hazards can be measured by several criteria, including threshold limits, 50% lethal dosage, short-term inhalation limits, nuisance concentrations, flash points and flammability limits.

The environmental conditions and chemical and physical properties of the hazardous chemicals, at the time of the spill, combine to determine the feasibility of a candidate response technique. Assessing the feasibility of individual response techniques often involved the extrapolation or estimation of some properties, due to the differences in the response techniques. Correlating and ranking these techniques involves the evaluation and assessment of data from several different areas.

The results of this review of chemical properties are presented in Table 1.10. This tabulation provided the basic data for correlating the potential hazards of individual chemicals with chemical and physical properties which could be used for their containment and removal.

Table 1.10: Ranking Chemical Spill Hazards*

A.D. Little Data (5)	Evaporation Rates**	Threshold Limit Value	Vapor Pressure**	Flash Point
Acrylonitrile	Pentane	Isoprene	Butadiene	Butadiene
Benzene	Ethyl ether	Toluene-2,4-diisocyanate	Vinyl chloride	Isoprene
Ethyl benzene	Hexane	Vinyl chloride	Hexane	Ethyl ether
2-Ethyl hexanol	Methyl acrylate	Allyl alcohol	Isopentane	Gasoline
Hexane	Triethyl amine	Camphor oil	1-Pentene	Isohexane
Isopropanol	Heptane	Cresols	Ethyl ether	JP-4
Methyl ethyl ketone	Ethyl acetate	Aniline	Isoprene	Hexane
Phenol	Benzene	Dicyclopentadiene	Isohexane	Petroleum naphtha
Styrene	1-Propyl acetate	Phenol	1-Hexene	1-Pentene
Toluene	Vinyl acetate	Dioctylphthalate	Pentane	Benzene
Vinyl acetate	Methyl amyl acetate	Dibutylphthalate	Vinyl acetate	1-Hexene
Vinyl chloride	n-Propyl acetate	Methyl acrylate	Acrylonitrile	n-Butyraldehyde
Xylenes	Toluene	Vinyl acetate	Ethyl acetate	Ethyl acetate
Butadiene	Acrylonitrile	Benzene	n-Butyraldehyde	Heptane
Pentadecanol	Methyl isobutyl ketone	Acrylonitrile	Methyl acrylate	Methyl acrylate
Ethyl ethers	Isobutyl acetate	Ethyl acrylate	Isopropyl acetate	Acrylonitrile
Kerosene	n-Butyl acetate	Triethylamine	Heptane	Methyl methacrylate
n-Amyl alcohol	Allyl alcohol	Methyl amyl alcohol	Methyl methacrylate	Pentane
Solvent naphtha	Xylenes	Isobutanol	1-Heptene	n-Propyl acetate
Ethyl acetate	Ethyl benzene	Methyl amyl acetate	Toluene	Ethyl benzene
n-Butyraldehyde	Amyl acetate	Adiponitrile	n-Propyl acetate	Ethyl acrylate
Valeraldehyde	Styrene	Amyl acetate	Allyl alcohol	Isobutyl acetate
Butyl acetates	Cumene	Amyl alcohols	Gasolines	Allyl alcohol
Hexanol	Diethyl carbonate	Ethyl benzene	Valeraldehyde	1-Octene
Octanol	n-Butyl alcohol	Hexane	Isobutyl acetate	n-Butyl acetate
Turpentine	Cyclohexanone	Methyl isobutyl ketone	Diethyl carbonate	Methyl isobutyl ketone
Gasoline	Coal tar naphthas	Methyl methacrylate	Butyl acetate	Amyl acetate
	n-Butyraldehyde	Hexane	Xylenes	Diethyl carbonate
	Isoamyl alcohol	Coal tar naphthas	Methyl isobutyl ketone	Asphalt
	Methyl amyl alcohol	Solvent naphthas	Butyl acrylate	Isobutyl alcohol
	n-Amyl alcohol	Styrene	Amyl acetate	Isobutyl acetate
	Benzaldehyde	Toluene	Amyl alcohol	Styrene
	Tetrahydronaphthalene	Turpentine	Cumene	Dicyclopentadiene
		Vinyl toluene	1-Decene	n-Amyl alcohol
		Xylenes	Methyl amyl alcohol	Kerosene
		Butyl acetates		

*Listing in order of decreasing risk assessment (top to bottom) of floating, hazardous chemicals only.
**Ambient temperature.

Source: CG-D-56-77

Hazardous chemicals were ranked according to several individual and composite sets of data. Five of these compilations are presented in Table 1.10.

Chemicals That Disperse Through the Water Column

> The information in this section is based on *A Feasibility Study of Response Techniques for Discharges of Hazardous Chemicals That Disperse Through the Water Column, CG-D-16-77* prepared by E. Drake, D. Shooter, W. Lyman and L. Davidson of Arthur D. Little, Inc., for the U.S. Coast Guard.

Chemicals which mix (dissolve) in water present one of the major current challenges to spill amelioration. Until recently, almost all research and development in spill clean-up has been devoted to those (insoluble) chemicals which are present in the water as separate phases; either floating (as oils) on the surface of the water, or deposited on the bottom. The deployment problems in removal of these chemicals are often difficult, particularly in rough waters, but the ability to physically separate the two phases significantly aids the detection, containment and removal of these spilled chemicals.

In contrast, soluble chemicals, which comprise a significant fraction of hazardous chemicals transported by water or likely to find their way into waterways, produce spills which are usually invisible, do not remain localized and cannot be removed by simple physical means such as booming, skimming or settling. Since the dissolved chemicals are present in the water body as separate molecules, rather than discrete phases, they disperse much more rapidly, and their treatment is made more difficult as they become more dilute and dispersed. Finally, since they are dispersed, the bulk properties of the chemicals (such as density, viscosity, freezing point, etc.) are largely irrelevant, and their removal depends upon the specific chemical or physico-chemical behavior of the particular chemical. Because of these severe difficulties of rapid dispersion and chemical specificity amelioration techniques for soluble chemicals have been largely neglected.

The Coast Guard has now identified four treatment techniques which appear, at least on the basis of theoretical and laboratory data, to be particularly promising for the amelioration of soluble chemical spills. These are neutralization of acids and bases, solvent extraction, precipitation and chelation of metal ions. Three classes of chemicals have been specified for treatment by each of these techniques. (A single class of chemicals has been assigned for potential treatment by either precipitation and/or chelation.) These chemicals are listed in Table 1.11.

Table 1.11: Chemicals Categorized by Treatment Techniques

A. Chemicals Subject to Neutralization

1. Acetic acid	8. Anhydrous ammonia
2. Acetic anhydride	9. Aniline
3. Acetophenone	10. Benzoyl chloride
4. Acrylic acid	11. Bromine
5. Aluminum chloride	12. Caustic potash solution
6. Aminoethanolamine	13. Caustic soda solution
7. Ammonium hydroxide	14. Chlorosulfonic acid

(continued)

Table 1.11: (continued)

15. Cyclohexylamine
16. Diethanolamine
17. Diethylamine
18. Diethylenetriamine
19. Diisopropanolamine
20. Dimethylformamide
21. 1,1-Dimethylhydrazine
22. Ethyleneimine
23. Ethylenediamine
24. Formic acid
25. Hexamethylenediamine
26. Hexamethylenetetramine
27. Hydrazine
28. Hydrochloric acid
29. Hydrofluoric acid
30. Hydrogen chloride
31. Hydrogen cyanide
32. Hydrogen fluoride
33. Lithium aluminum hydride
34. Maleic anhydride
35. Methylethylpyridine
36. Monoethanolamine
37. Monoisopropanolamine
38. Morpholine
39. Nitrogen tetroxide
40. Nitric acid
41. Nitrosyl chloride
42. Oleum
43. Oxalic acid
44. Phosphoric acid
45. Phosphorus oxychloride
46. Phosphorus pentasulfide
47. Phosphorus trichloride
48. Polyphosphoric acid
49. Potassium hydroxide
50. Propionic acid
51. Pyridine
52. Sodium
53. Sodium amide
54. Sodium hydride
55. Sodium hydroxide
56. Sulfur monochloride
57. Sulfuric acid
58. Sulfuric acid (spent)
59. Sulfuryl chloride
60. Titanium tetrachloride
61. Triethanolamine
62. Triethylamine
63. Triethylenetetramine
64. Trimethylamine
65. Urea

B. Chemicals Subject to Solvent Extraction

1. Acetone cyanohydrin
2. Allyl alcohol
3. Aminoethanolamine
4. n-Amyl alcohol
5. 1,4-Butanediol
6. 1,4-Butenediol
7. n-Butyl acetate
8. n-Butyl alcohol
9. sec-Butyl alcohol
10. tert-Butyl alcohol
11. 1,4-Butynediol
12. Carbon bisulfide
13. Chloroform
14. Chlorohydrins (crude)
15. Corn syrup
16. Dextrose solution
17. Diacetone alcohol
18. Dichloromethane
19. Diethanolamine
20. Diethylene glycol
21. Diethylene glycol dimethyl ether
22. Diethylene glycol monoethyl ether

(continued)

Table 1.11: (continued)

23. Diethylene glycol monomethyl ether
24. Diisopropanolamine
25. Dimethyl sulfate
26. Dimethyl sulfoxide
27. 1,4-Dioxane
28. Dipropylene glycol
29. Epichlorohydrin
30. Ethoxytriglycol
31. Ethoxylated dodecanol
32. Ethoxylated pentadecanol
33. Ethoxylated tetradecanol
34. Ethoxylated tridecanol
35. Ethyl acetate
36. Ethyl acrylate
37. Ethyl alcohol
38. Ethylene cyanohydrin
39. Ethylene glycol
40. Ethylene glycol monobutyl ether
41. Ethylene glycol monoethyl ether
42. Ethylene glycol monomethyl ether acetate
43. Ethylene glycol monomethyl ether
44. Ethyleneimine
45. Formaldehyde solution
46. Glycerin
47. Hexylene glycol
48. Isoamyl alcohol
49. Isobutyl alcohol
50. Isopropyl acetate
51. Isopropyl alcohol
52. Methanearsonic acid, sodium salts
53. Methyl acrylate
54. Methyl alcohol
55. Methyl amyl alcohol
56. Methyl isobutyl carbinol
57. Methyl methacrylate
58. Monoethanolamine
59. Monoisopropanolamine
60. Morpholine
61. Paraformaldehyde
62. Polypropylene glycol methyl ether
63. n-Propyl acetate
64. n-Propyl alcohol
65. Propylene glycol
66. Propylene glycol methyl ether
67. Propylene oxide
68. Sodium alkyl
69. Sodium alkyl sulfates
70. Sorbitol
71. Sulfolane
72. Tetrahydrofuran
73. Triethanolamine
74. Triethylene glycol
75. Vinyl acetate

C. Chemicals Subject to Precipitation and/or Chelation

1. Antimony trifluoride
2. Cadmium chloride
3. Chromic anhydride
4. Copper sulfate
5. Ferrous sulfate
6. Latex, liquid synthetic
7. Nickel sulfate
8. Potassium dichromate
9. Potassium permanganate
10. Silver nitrate
11. Sodium ferrocyanide
12. Titanium tetrachloride
13. Zinc Chloride

Source: CG-D-16-77

RESPONSE TECHNIQUES

Overall Strategy

> The material in this section is based on a paper given at the Conference 1972 by G.W. Dawson, A.J. Shuckrow and B.W. Mercer of Battelle Pacific Northwest Laboratories. The title of the paper was "Strategy for Treatment of Waters Contaminated by Hazardous Materials".

Response procedures designed to mitigate the effects of hazardous material spills involve a complex blend of expertise and equipment. Personnel must conduct activities dealing with detection, identification and monitoring, water treatment or containment, environmental restoration, damage assessment, and regulatory review. Of these, identification, monitoring and damage assessment are the most intimately tied to the treatment phase of the response program.

Upon detection of a spill, prospects for an effective response rest in part on the ability to identify the materials released. Correct identification of the contaminants involved enables both an evaluation of the nature and extent of the hazard, and an assessment of what types of amelioration techniques may prove effective.

Although very sophisticated analytical techniques have been developed to perform this function, they often require large, sensitive equipment, trained personnel, and complicated procedures. Typically, these may be unavailable at or near a spill site. Consequently, time is lost in the identification phase and the prospects for minimizing damage by quick response are also lost. On the other hand, proceeding prematurely when little is known of the nature of the spilled material can result in wasted effort and injury to personnel.

Monitoring the spill to chart its diffusion and translocation through the aquatic environment may often be more easily accomplished than identification. Once the exact nature of the contaminant is ascertained, gross detection techniques can be employed to monitor the resulting plume. These may include TOC, pH, conductivity, or specific electrode measurement techniques. Although a degree of sensitivity is sacrificed, simplified detectors can provide rapid in situ methods to locate the areas to be treated and assess the effectiveness of the countermeasure employed.

Damage assessment analysis is required for considerations other than liability allocation. A properly designed assessment program can be used to evaluate the benefits of the countermeasure employed as well as any hazards inherent in the countermeasure itself. As information in this area is compiled, guidelines can be developed for better use of treatment techniques in the future.

Potential treatments which conceivably can be applied to ameliorate the effects of hazardous polluting substances in the aquatic environment are almost equal in number to the substances themselves. Obviously, it would be impractical to attempt to develop a specific countermeasure for each individual hazardous polluting substance. Moreover, since the resources which can be invested in developing countermeasures against hazardous polluting substances in the aquatic environment are limited, efforts must be directed toward those areas which promise to produce the maximum return on this investment.

The following criteria are suggested for evaluation of potential countermeasures:

- Countermeasures should be highly effective.
- Countermeasures should be applicable to a large number of substances.
- Countermeasures should be amenable to rapid, easy deployment. Highly specialized equipment and/or chemicals which require wide deployment and stockpiling prior to a pollution incident or which cannot be rapidly conveyed to the scene of an incident are undesirable.
- Countermeasures should be free from potentially harmful secondary effects in the aquatic environment, including noxious sludges.
- Countermeasures developed to combat spills of hazardous polluting substances should take advantage of existing technology, particularly that technology developed to combat oil spills, to the maximum possible extent.

Treatments which could be considered include chemical degradation, precipitation, neutralization, physical adsorption, complex formation, ion exchange, solvent extraction, foam separation, and biological degradation. Several of these can be eliminated because of the threat of secondary effects from the required additives and reaction products.

Chemical Degradation: This (transformation, oxidation-reduction, etc.) is generally unacceptable, since in most cases the agents that must be added to effect reaction pose an equal or greater potential hazard to the aquatic environment than the original pollutant. The situation is further complicated by the nature of the spill environment which does not allow close control of operating conditions. Concentration changes resulting from natural dilution processes could cause excess chemical addition, and so increase the damage to the aquatic environment.

Complex Formation and Neutralization: These suffer from similar difficulties related to not controlling the reaction environment and the inherent threat to water quality of the necessary additives.

Precipitation Techniques: These at first appear to be an attractive solution; however, the ultimate reaction poses a new type of threat. Lacking an effective means for removing the precipitate, the watercourse is subjected to a concentrated buildup of fine particulate matter which could damage feeding and spawning grounds with subsequent deleterious effects on aquatic life. The full impact of this type of damage has never been satisfactorily explained, but presumably it could rival the damages of the original pollutant to water quality. Further, the introduction of colloidal precipitates would still threaten gilled species and would decidedly reduce recreational and aesthetic benefits. While development of a subsurface collection device along the lines of a swimming pool vacuum would alleviate these effects, the enormous operational problems associated with such an approach are obvious.

Solvent Extraction and Foam Separation: These processes have several intriguing characteristics. Through proper choice of solvent, the product stream can be made to rise to the surface where ready collection with existing hardware should be possible. In addition, solvent extraction and foam separation processes are fairly broad-based, with a single component displaying applicability to many pollutant releases. Unfortunately, both processes require some control over the area to be treated, and treatment of any large body of water would necessitate use of several types of feed mechanisms. This implies high capital costs and storage facilities to stockpile such equipment regionally. Further, in the case of solvent extraction, a large quantity of solvent would be required to achieve only reasonable removal efficiency.

Vacuum or Dredging Systems: Vacuum or dredging systems for sunken materials are speculative. The swimming pool vacuum system may be a workable concept, but there is also a good chance that to effectively remove the contaminant, a large quantity of benthos would also have to be removed and handled. While heavy flocs and precipitates may prove innocuous enough to be left on the bottom, heavy liquids are a separate consideration. They pose a threat to water intakes which may be located near the bottom of a water source. For example, in the event of a spill, a substance such as ethylene dichloride would seek the bottom without dissolving.

Skimming and Booming: These techniques are available for oil spills. They could be applied to light insoluble materials as well as slightly soluble organics. Rapid response aimed at booming and skimming any undissolved portion of the contaminant could greatly reduce the severity of a spill.

Flow Augmentation: This can be applied only in certain circumstances. Mechanical mixers such as outboard motors can be used to aid diffusion. Flow augmentation may be utilized where stored water is available upstream. In any event, such a practice is reliant in part on good predictive models of the aquatic environment to determine the extent of contamination and expected duration of toxic concentrations. This information is available for a number of streams, as is time-of-travel data. Such data are not gathered in a single source nor properly referenced to allow for quick retrieval and application in practical emergency, however.

Burn-off Techniques: These are probably applicable only in very limited cases where the material remains confined in a small isolated area so as to minimize threats to safety and air quality. Many hazardous materials emit highly toxic vapors when heated to decomposition. Consequently, burning may intensify hazardous conditions and force evacuation of areas surrounding a spill site. The presence of heavy vapors which can travel along the ground and cause flash-backs further endangers safety when burn techniques are applied.

Oxygen Addition: This may be an attractive alternative available for substances posing a BOD problem. The oxygen is added to the water to maintain dissolved oxygen levels while a spilled material is being dispersed and degraded aerobically. There has been recent interest in injecting air or pure oxygen into water courses to achieve this end in areas of chronic low dissolved oxygen. The required apparatus is simple enough that mobilization for transport to a spill site should be possible. The problems would be those involved with supplying enough gas to accomplish the task and designing equipment that could efficiently transfer the gas into a variety of potential stream or reservoir configurations.

Biological Degradation: Biological degradation, while attractive in some respects, suffers from several difficulties. In order for degradation to proceed at a rapid rate, it would be necessary to have on hand large quantities of acclimated cultures. The problems associated with stockpiling many such cultures, each of which is specific to a particular substance, are obvious. Also, many hazardous materials are inherently resistant to biological degradation.

Physical Sorption and Ion Exchange: These appear to offer the greatest promise for treatment of acute spills in the aquatic environment. Sorption and ion exchange processes rapidly remove substances from solution and hold them in a relatively innocuous form for long periods of time. The use of exchange resins for acid, base, metal, or toxic salt spills and the use of activated carbon or sorption resins for organic matter could apply to a large number of hazardous polluting substances with varying degrees of efficiency. The two major problems associated with the application of mass transfer media to counteract hazardous polluting substances in a free-water environment are: (1) the mechanics of introducing the active media into the water; and (2) the subsequent removal of the media from the water.

Traditionally, such processes have been restricted to operation within columns. This method of application is limited by some very real concerns: the cost of stockpiling mobile columns for spill treatment; the problem of rapid deployment of the large equipment involved; and the practicality of pumping free waters into a contained environment for treatment. Some studies have been conducted with powdered activated carbon in which the sorptive media was dispersed in the water and allowed to settle. The basic premise of this work was that the carbon is innocuous to aquatic life and can be allowed to settle. Although this may be true in selected impoundments, in flowing streams the aesthetic damage would be extensive and gilled species would be subjected to the problems inherent in an environment highly saturated with particulate and colloidal matter.

Whatever mode of treatment is selected, consideration of spill dynamics is essential to maximize the benefits from a response. The dynamics of a release are dependent both on the mode of entry and the physical characteristics of the material.

- Gases may rise rapidly and disperse or hover over the water providing a continuous source for equilibrium levels of the material in surface waters.
- Liquids may float in slicks, slowly dissolving into the water; rapidly assimilate into the aquatic environment and go through normal dispersion patterns; or sink to the bottom and slowly dissolve into the water.
- Solids can follow the same pattern as liquids or can form fine suspensions in the water evidencing themselves as a turbid cloud.

The individual route or combination of routes experienced in a spill situation will affect the kinetics of any countermeasure employed. It is desirable, however, that the success of a response technique would not depend upon any single mechanism.

The nature of the body of water affected by a spill also has a great effect on the type of response that can be made. Small impoundments, where damage is believed to be complete (e.g., total fish kill), can be subjected to efficient but harsh chemical countermeasures since extensive restocking will be required anyway. Water from small impoundments can also be pumped through exterior facilities for treatment. Small streams and bodies of moving water can be dammed and routed through treatment facilities with the use of portable coffers and culverts. Large bodies of water pose the most challenging problems, since the entire volume affected by a spill cannot be practically controlled or managed. In these cases, response measures must be carefully scrutinized to avoid intensifying or augmenting damage to aquatic life through the use of improperly chosen techniques.

Ultimately, the most vital parameter in the determination of the success of a response is the time lag between the spill and the initiation of the treatment phase. Concentration-time lethality curves drawn for individual contaminants to aquatic life suggest that even high doses of hazardous materials can often be tolerated for short periods of time. This would be the case where transfer of contaminant across the gills of fish is the limiting step such that lethal amounts could not physically be absorbed in the given time period. Conversely, low concentrations can be lethal over prolonged exposure periods due to bioaccumulation or cumulative effects.

Most candidate response systems exhibit much higher efficiencies at higher concentrations. Hence, the benefits derived from a set level of treatment expense (e.g., a given quantity of chemical) are greater as the delay is shortened so that natural diffusion has not reduced concentrations to levels where treatment efficiency is poor. Rapid response also reduces the volume of water requiring treatment and offers greater assurance that few water supply systems will be jeopardized by intrusion of the spilled material.

Time is perhaps the most critical factor in spill response. Damage and time are linked through an exponential relation which puts a premium on response practices that can be enacted immediately. This requires a highly efficient reporting structure as well as countermeasures developed with quick deployment in mind. Programs must therefore be undertaken with speed of response as a major objective.

Decision on Spill-Handling

> This information in this section is based on *Manual for the Control of Hazardous Material Spills. Volume 1: Spill Assessment and Water Treatment Techniques,* EPA-600/2-77-227, prepared by K.R. Huibregtse, R.C. Scholz, R.E. Wullschleger, J.H. Moser, E.R. Bollinger and C.A. Hansen of Envirex for the Environmental Protection Agency.

The critical decision regarding handling of the spill is very difficult. Many variables affect the decision, and these variables must be considered by the user and altered to fit the specific situation. Only then can the final conclusion be reached. There are four ways the spill can be handled: (1) diluted and dispersed into the natural environment; (2) treated in situ with makeshift processes; (3) hauled to another site for disposal or treatment; and (4) treated in an on-site, but off-stream, treatment system. The following are presented to guide the thought processes of the OSC and allow him to decide on the best course of action for his situation.

The thought guide (Figure 1.1) is a graphical presentation of the thought processes required of the OSC when establishing the handling of a spill.

The boxed questions presented are merely summaries of many considerations which are to be answered by the OSC in establishing the final answer to the main boxed question. Once an answer of yes or no has been established, the arrows are followed to the box with the next applicable question. The result of following the thought guide will be determination of a feasible handling method for the specific spill situation. This guide is flexible and is intended to aid the OSC but not to make the decisions for him.

Only after all other possible alternatives have been investigated and found not to be feasible is the method of handling by dilution and dispersal to be considered. This method must be used only as a last resort to minimize local hazards. Care must be taken to determine if this method is feasible in that mixing the hazardous chemical with water does not cause undesirable side reactions or by-products.

Once it has been determined that dilution and dispersal is the only action available, then additional water sources must be brought to the spill site. Water should be added to the stream at a turbulent spot to allow complete mixing with the hazardous material.

Care should be taken not to exceed the capacity of the water body and extend the hazard past its natural boundary. Dispersion can also be induced by creating mixing zones in the waterway and reducing the pockets of concentrated contaminant which may exist.

Tables 1.12 through 1.16 summarize the advantages and disadvantages of a large number of spill-handling methods, most of which will be discussed in detail in subsequent chapters.

Table 1.12: Spills in Air

Technique	Method	Use	Advantages	Disadvantages
Mist knock down	Spray fine mist into air	Water-soluble or low-lying vapors	Removes hazard from air	Create water pollution problem and must be contained in solution
Fans or blowers	Disperse air by directing blower toward it	Very calm and sheltered areas	Can direct air away from populated areas	Not at all effective if any wind Need large capacity of blowers Hard to control

Source: EPA-600/2-77-227

Figure 1.1: Spill Handling Thought Guide

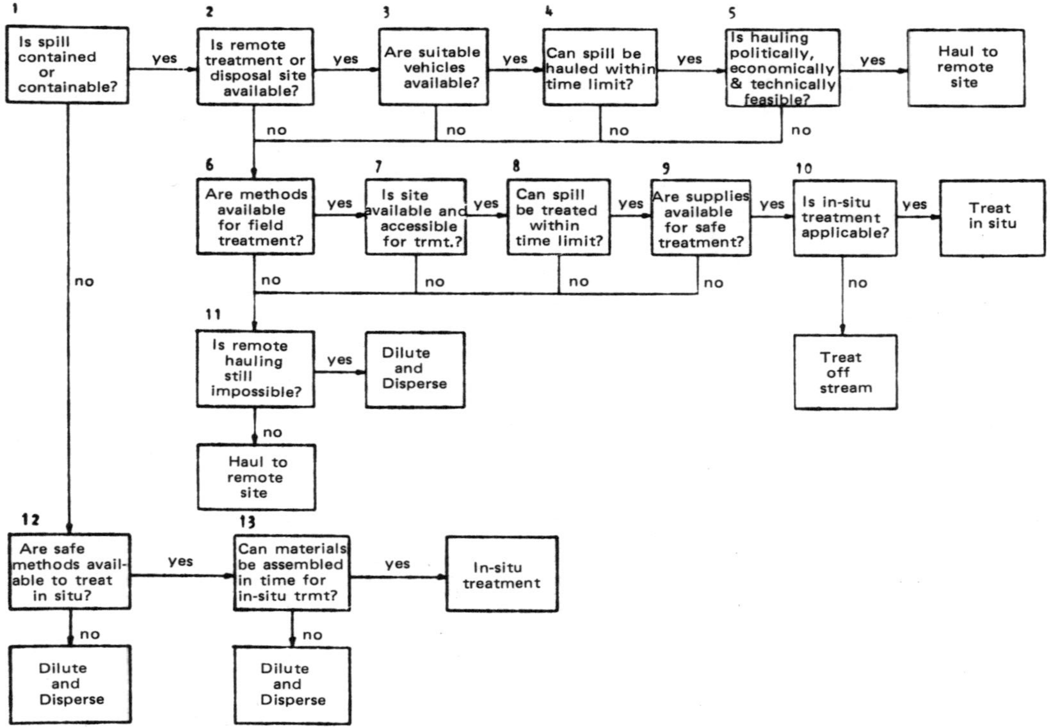

Source: EPA-600/2-77-227

Table 1.13: Spills on Land

Technique	Application or Construction Method	Use	Advantages	Disadvantages
Dikes Earthen	Create with bulldozer or earthmoving equipment to compact earth (height depends on earth type)	Flat or sloped surface	Material on site Construct with common equipment Construct quickly	Natural permeability of soil Seepage through ground Surface composition of soil not suitable in all cases
Foamed polyurethane	Use trained personnel to construct	Hard, dry surfaces	Hold up to several feet of water	Leaks on wet ground Hard to obtain dispersion device
Foamed concrete	Use trained personnel to construct	Flat ground Slow moving spill	Better adhesion to substrates (clay/shale/grass)	Hard to obtain foam and dispersion device Must set for a time period Will not hold high hydraulic heads
Excavation	Bulldozer or earthmoving equipment; line if possible	Soft ground Natural cavitation	Material on site Construct with common equipment	Move large amounts of material Natural permeability of soil Surface of soil not suitable in all cases
Excavation and dikes	Bulldozer or earthmoving equipment; line if possible	Soft ground	Need less space than separate Material on site Construct with common equipment	Move large amounts of material Natural permeability of soil Surface of soil not suitable in all cases

Table 1.14: Spills in Water—Heavier than Water Spills

Technique	Application or Construction Method	Use	Advantages	Disadvantages
Natural excavations and dikes	None	Where a natural barrier exists	No construction needed	Can't control the area which contains the spill
Construction of excavations and dikes	Dredges; hydraulic or vacuum pumps Divers with pumps then place concrete or sand bags to form dike if bottom material is not sufficient	If bottom can be moved	Material is on site	Hard to construct Stirred up bottom may cause dispersion and increased turbidity

Source: EPA-600/2-77-227

Table 1:15: Spills in Water—Soluble or Miscible Spills

Method	Application or Construction Materials	Use	Advantages	Disadvantages
Sealed booms	Boom Device to anchor	Contain depth Limited volumes Leaking containers	Contain entire depth of water	Deployment difficult Not used for large bodies Difficult to get good seal
Diversion of uncontaminated flow	Earthmoving equipment	Special area where topography is right	Can put cleaned water into diverted stream Used for flowing water	Difficult to move large amounts of earth Clear area needed Impermeability of ground
Diversion of contaminated flow	Block entrance with sandbags, sealed booms or dikes	Special area where topography is right	Can put clean water back into stream Used for flowing water	Difficult to move large amounts of earth Clear area needed Impermeability of ground Adverse environmental impact
Gelling agent	Gels, dispersion devices; use experienced personnel	If small volumes	Stop flowing contaminant Stop permeation	Hard to obtain Can't use in large area Must haul to dispose
Containment of entire waterbody	Diking materials Earthmoving equipment Sandbags, etc. Lining	For entirely contaminated area	Can allow containment of a large waterbody Materials on site Easily constructed	Not all waterbodies have containable overflow Permeability May be an unstable condition

Table 1.16: Spills in Water—Floating Spills

Method	Application or Construction Materials	Use	Advantages	Disadvantages
Booms	Varies; need deployment device	Not too much current	Used on large area Many varieties	Only in waves less than 2 to 4 ft Current speed less than 0.7 knots
Weirs	Weir and boat	Calm	Not easily clogged Collect and contain	Not used in rough water
Pneumatic barriers	Air compressor diffuser deployment method	Only shallow water	Do not create a physical barrier to vessels	Not in rough water Only shallow water Only thin layers or materials
Spill herding methods	Chemicals on water spray or property wash	To protect shore or other facilities	Useful in rough water	Not easily obtainable Not 100% effective

Source: EPA-600/2-77-227

Response Techniques for Representative CHRIS Chemicals

The information in this section is based on *Agents, Methods and Devices for Amelioration of Discharges of Hazardous Chemicals on Water,* CG-D-38-76, prepared by W.H. Bauer, D.N. Borton and J.J. Bulloff of Rensselaer Polytechnic Institute for the U.S. Coast Guard.

The objective of this research was to identify, evaluate and test the following: (1) off-the-shelf-type methods and devices in order to determine which may be most advantageously employed, either directly or in an easily modified form, to ameliorate a spill situation; and (2) high-potential, but undeveloped, sorption materials and application methods which may be most advantageously developed to ameliorate a spill situation for hazardous materials that are miscible or soluble. Amelioration was studied for only the chemicals that were selected as representative of the thirty amelioration response categories.

In Table 1.17 are listed the most advantageous agents for the representative hazardous chemical (RHC) in each of thirty amelioration categories and the hazardous chemicals in each category to which the agents may apply. The numbers given refer to the chemicals as listed within each category, as given earlier in Table 1.7.

Table 1.17: Ameliorating Agents

Agent for RHC	Hazardous Chemicals in Category to Which the Agent May Apply
Category I A	
Activated carbon	1-18, 20-59, 61-62
Polyurethane foam	1-18, 20, 23-28, 33-34 38-49, 53-59, 61-62
Polypropylene fibers	1-18, 20, 23-28, 33-34, 38-49, 53-59, 61-62
Oil herder	1-10, 14-16, 18, 20, 23-25, 53-55, 57-59
Category I B	
Activated carbon	1-30
Polyurethane foam	1-8, 11, 30
Polyolefin fibers	1-8, 11, 30
Calcium carbonate	17-25, 28
Calcium hydroxide	17-25, 28
Calcium oxide	17-25, 28
Calcium sulfate	24, 25
Category I C	
Activated carbon	1-2
Carbonized sawdust	2
Calcium carbonate	2
Calcium hydroxide	2
Calcium oxide	2
Water	1-2
Category I D	
None applicable	No amelioration: 1-2, 4-5
Category I E	
Activated carbon	1-47
Zeolite F (K form)	3, 12, 28, 31-33
Clinoptilolite	3, 12, 28, 31-33

(continued)

Table 1.17: (continued)

Agent for RHC	Hazardous Chemicals in Category to Which the Agent May Apply
Category II A	
Activated carbon	6, 8-13
Category II B	
Activated carbon	1-60
Polyurethane foam	1-36, 38-58, 60
Polyolefin fibers	1-36, 38-42, 49-60
Cellulose fibers	1-36, 37-60
Category II C	
Activated carbon	1-83
Polyurethane foam	1-83
Polypropylene fibers	1-83
Cellulose fibers	1-83
Oil herder	1, 7, 10, 11, 16, 17, 22, 29-31, 37, 40-42, 44-51, 53-57, 62-74, 76, 77, 78-83
Category II D	
Activated carbon	1-31
Polyurethane foam	1-31
Polyolefin fibers	1-31
Cellulose fibers	1-31
Zeolite 5A	2, 4, 13-15, 19, 20
Water	1, 2, 4-7, 9-11, 13-18, 21, 25, 28, 31
Category II E	
Activated carbon	1-107
Polyurethane foam	1-107
Polypropylene fibers	1-107
Cellulose fibers	1-107
Oil herder	1-107
Category II F	
Activated carbon	1-25
Polyurethane foam	1-25
Polypropylene fibers	1-25
Cellulose fibers	1-25
Amberlite XAD resins	1-25
Water	8, 13, 9, 14, 24, 25
Category II G	
Activated carbon	1-5
Polyurethane foam	1-5
Polyolefin fibers	1-5
Cellulose fibers	1-5
Amberlite XAD resins	1-5
Oil herder	1-5
Category II H	
Activated carbon	1-15
Polyurethane foam	1-15
Polyolefin fibers	1-15
Cellulose fibers	1-15

(continued)

Table 1.17: (continued)

Agent for RHC	Hazardous Chemicals in Category to Which the Agent May Apply
Category III A	
Activated carbon	1-6, 8-26, 28, 29, 32, 33, 35, 36
Polyolefin or polyisobutylene fibers	1, 20, 21, 23, 29, 33, 35, 36
Amberlite XAD-4 resin	1, 20, 21, 23, 29, 33, 35, 36
Category III B	
Activated carbon	1
Zeolite 5A	1
Category III C	
Activated carbon	1-16
Category III D	
Activated carbon	1-10
Category IV A	
Activated carbon	1-22
Dowex 1X10 or Amberlite IRA 402 Amberlite IRA 93 or De Sal process resin	1-22
Calcium carbonate	1-22
Calcium oxide	1-22
Calcium hydroxide	1-22
Sodium carbonate	1-22
Powdered ferrochromium slag	1-22
Water	1-22
Ultrafiltration	1-22
Category IV B	
Activated carbon	1-10
Amberlite IRA 93	1-10
Carbon dioxide	1-10
Sodium bicarbonate	1-10
Category IV C	
Activated carbon	1-30
Amberlite IRA 400	1-30
Water	1-30
Ozone or hydrogen peroxide or perhydrol	14, 19, 21, 27-29
Category IV D	
Activated carbon	1-13
Polystyrene sulfonate or Dowex 50WX8, Dowex 50WX4, Amberlite IR 252, or Amberlite IRA 900	1, 2, 4, 5, 6, 9, 12, 13

(continued)

Table 1.17: (continued)

Agent for RHC	Hazardous Chemicals in Category to Which the Agent May Apply
Category IV E	
Activated carbon	1-3
Zeolite 5A	1-3
Category IV F	
Activated carbon	1-7
Dowex 5WX4	1-7
Calcium chloride	6
Category IV G	
Activated carbon	1-42
Water	1-42
Category IV H	
Activated carbon	1-9
Polyurethane foam	1-9
Polypropylene fibers	1-9
Cellulose fibers	1-9
Amberlite XAD resins	1-9
Water	2-6, 8, 9
Category IV J	
Activated carbon	1-23
Amberlite XAD resins	1-23
Dowex 50WX8 or Amberlite IRA 900	1-9, 11-14, 16-21, 23
Category IV K	
Activated carbon	1-5
Polyurethane foam	2
Polypropylene fibers	2
Category IV L	
Activated carbon	1-4
Polyurethane foam	1-4
Polypropylene fibers	1-4
Category IV M	
Activated carbon	1
Styrene/divinylbenzene or methyl methacrylate/ethylene dimethacrylate copolymer	1
Amberlite IRA 93 or 400	1
Water	1
Category IV N	
Activated carbon	1-47
Polyurethane foam	5-7, 9-12, 14, 15, 18-20, 25-29, 31, 33, 35, 39, 42, 44-46
Amberlite XAD-1-8	1, 2, 4-12, 14, 15, 18-22, 25-29, 31-36, 38-42, 44-47
Amberlite IR 45 or Dowex 1 or polystyrene sulfonate	1, 2, 4, 14, 34, 36, 38, 40
Floridin XXF or Tonsil AC	6, 12-14, 38

(continued)

Table 1.17: (continued)

Not Clearly in Established Categories
Acetylene
Carbon bisulfide
Cyanogen bromide
1,4-Dioxane
Latex liquid synthetic
Malathion
Methanearsonic acid, sodium salts
Propylene oxide
Tetraethyl lead
Tetrahydrofuran
Tetramethyl lead

Source: CG-D-38-76

It was concluded that use of chemical agents for the amelioration of discharges on water of hazardous chemicals is feasible for 29 of the representatives of the 30 categories of hazardous chemicals and thus, by inference, for most of the CHRIS list of 400. No readily available chemical agents advantageous for amelioration of discharges of liquefied natural gas, LNG, were identified. The types of available agents evaluated to be most advantageous for amelioration were sorbents, ion exchangers, diluents, neutralizers, collectors, precipitants, reductants, oxidizers and ultrafiltration membranes. By far the most versatile type identified was that of sorbents, applicable to 29 of the representative hazardous chemicals. Ion exchangers and diluents were also versatile, applying to 7 or more representative chemicals. The remaining agents, such as neutralizers for sulfuric acid, were specific for 1 or 2 categories.

Sorbents of most general utility and effectiveness were activated carbons, polyurethanes, polyolefins and celluloses. The sorbents are available in many physical forms, from particles to foams, of high and low density and with oleophilic or hydrophilic characteristics. Because of wide-spread industrial use, much experience is available concerning the methods for application of sorbents to removal of dissolved chemicals from water and of sorption of floating chemicals.

In contrast, comparatively little information exists about the speed of action and relative effectiveness of individual ion exchangers, neutralizers, precipitants, reductants and oxidizers in amelioration. The practical available diluent is water, but forced dilution by water is appropriate for amelioration only when all other approaches are excluded by the circumstances of a discharge. It appears that use of chemical agents in amelioration of discharges of hazardous chemicals on water may be expected to be advantageous, especially so when sound practice has developed in the application of chemical agents to large discharges on water.

It was further concluded that available, off-the-shelf devices with demonstrated capacity for usefulness in containing, gathering and removing oil spills on water also will be advantageous in amelioration of discharges of floating liquid hazardous chemicals. The main requirement for use is that any polymeric construction materials be resistant to attack by the floating liquid hazardous chemicals. Components which contain plasticized polyvinyl chloride are unsuitable for long-term use but devices using this material may be readily modified by substitution of solvent-resistant materials. For the many representative chemicals of high flammability, forms of the devices which are ignition- and explosion-proof must be chosen.

Recommended Responses for Spills of Materials That Sink in Water

> The information in this section is based on *A Feasibility Study of Response Techniques for Discharges of Hazardous Chemicals that Sink,* CG-D-56-78 prepared by T.D. Hand, A.W. Ford, P.G. Malone, D.W. Thompson and R.B. Mercer of the U.S. Army Engineer Waterways Experiment Station for the U.S. Coast Guard.

Each of 70 sinking substances is classified and listed in Table 1.18 along with a recommended response. The chemicals have been grouped according to similar characteristics. The responses presented herein are given for each group of chemicals. Many of these techniques are untested and are presented only as potential responses. This table should be used along with Table 1.19 in choosing a response technique.

A number of factors were considered in developing the ratings of Table 1.19, including compatibility of the technique with the environmental setting, availability of necessary supplies and equipment, and overall efficiency of the technique in terms of response time and costs. Many of these factors were estimates since little information from actual field use of the techniques was available.

It must be emphasized that since many of the techniques are both chemical- and scenario-specific, the final choice of a technique must be based on actual conditions at the spill site. Tables 1.17 and 1.18 should be used only as an aid in determining the most appropriate response technique for a particular hazardous sinking chemical spill.

Recommended response procedures were chosen by considering both the nature of the chemical spilled and its interaction with sediments. The response options that are available include: (a) removing the spilled material and surrounding sediments using dredging and/or pumping techniques; (b) burying the material in place using a suitable covering material; (c) reacting the spilled material with a reagent that will render it safe or at least lessen the hazard; (d) reacting and removing the spilled material as the reacted product; (e) reacting and burying the spilled material as the reacted product; and (f) allowing the material to remain on the bottom of the waterway, taking no action.

Any decision as to response technique will be based on: (1) immediate and ultimate danger to local population and local biological communities from the spill; (2) difficulty and danger encountered with removal or burial; and (3) possibilities of producing a safe or safer product by reacting the spilled material with a suitable reagent.

In the majority of spill incidences involving these 70 materials, the high degree of danger associated with the material will require its recovery. Most of the materials considered are toxic liquids that would likely be dispersed by attempts at burial and whose burial may not effectively isolate the material from susceptible organisms in its vicinity. In most waterways, burial is an uncertain procedure that may be reversed by scour and erosion resulting from bottom currents and turbulence, despite counteractive efforts.

Table 1.18: Responses Other than Dredging Appropriate to Functional Classes of Chemicals

Characteristics	Chemicals Represented	Recommended Responses
Solid; inorganic; reactive with water	Aluminum chloride Calcium carbide Calcium oxide Lithium aluminum hydride	Containment; temporary cover; chemical treatment by neutralization
Solid; inorganic; reactive with air	Red phosphorus White phosphorus	Request special handling by manufacturer; oxidation in situ
Solid; inorganic; relatively insoluble and nonreactive in water	Aluminum fluoride Barium carbonate Calcium fluoride Calcium hydroxide Lead arsenate Sulfur (after cooling)	Covering; neutralization; precipitation of toxic ions
Solid; halogenated organic	Aldrin Benzene hexachloride DDT p-Dichlorobenzene 2,4-Dichlorophenol Pentachlorophenol Toxaphene Trichlorophenol	Containment; covering (mostly temporary); adsorption
Solid; nonhalogenated organic	Asphalts Benzoic acid Bisphenol A Carbaryl Dibenzoyl peroxide 2,4-Dinitroaniline 2,4-Dinitrophenol Diphenylmethane diisocyanate Naphthalene Phthalic anhydride	Containment; covering (mostly temporary); adsorption
Liquid; inorganic	Bromine Mercury	Containment; neutralization; adsorption
Liquid; halogenated organic	Benzoyl chloride Carbon tetrachloride Chlorobenzene Chloroform o-Dichlorobenzene Dichloromethane Dichloropropane Dichloropropene Epichlorohydrin Ethylene dibromide Ethylene dichloride Polychlorinated biphenyl Tetrachloroethylene Trichloroethane Trichlorofluoromethane Vinylidene chloride	Containment; adsorption; oxidation; biodegradation
Liquid; nonhalogenated organic	Acetophenone Aniline Benzaldehyde Camphor oil Carbolic oil Carbon bisulfide Cresols Dibutyl phthalate Dimethyl sulfate Dowtherm Furfural Glycidyl methacrylate Malathion Methyl parathion Motor fuel antiknock compounds containing lead alkyls Nitrobenzene Tetraethyllead Tetramethyllead Tricresyl phosphate	Containment; adsorption; oxidation; biodegradation

Source: CG-D-56-78

Table 1.19: Responses, Other than Dredging, Appropriate to Various Spills

	Land and Non-navigable Waters	. . Rivers . . Current		Ports and Harbors, Navigation .Channels .		. . . . . .Open Waters.	
		High	Low	In	Outside	High Current, Great Depth, or Both	Neither High Current nor Great Depth
Containment	H	M	H	M	H	L	M
Covering	H	M	M	L	M	L	M
Sorbents	M	L	M	M	M	L	M
Chemical	M	L	M	M	M	M	M
Biodegradation	L	L	L	L	L	L	L
Dispersion	L	M	L	M	L	M	M

Note: potential as a response technique—L = low or none; M = moderate; and H = high.

Source: CG-D-56-78

Burial is an acceptable first choice and final solution only in cases where the material is so innocuous that its reexcavation by tides or currents would not significantly endanger the surroundings. This review of response options suggests that in the majority of cases it will be necessary to depend on removal or recovery of material as the major technique for ameliorating the effects of a spill. In some circumstances combinations of these responses may be appropriate. For example, the major part of a spill may be recovered and minor amounts on the fringes of the spill area and/or any residual material left may be isolated by burial. There will doubtless be instances where recovery simply is not possible; in such cases burial, chemical reaction, containment, or even dispersion may be the only options possible. Obviously, such determinations must be made on a case-by-case basis.

Recommended Responses for Spills of Materials That Float on Water

> The information in this section is based on *Feasibility Study of Response Techniques for Discharges of Hazardous Chemicals that Float on Water,* CG-D-56-77 prepared by J.S. Grear of MSA Research Corporation for the U.S. Coast Guard.

Spills of floating hazardous chemicals usually involve smaller quantities than those of oils and petroleum products. Potential hazards are greater, however, in that these chemicals combine vapor toxicity with flammability hazards and are more difficult to detect and monitor. Most of the floating hazardous chemicals are colorless, and therefore present a difficult problem for visual detection, compared to opaque oils.

Consequently, response techniques involving confinement and immobilization must be modified to meet these different properties, when being transferred from oil spill to hazardous chemical response technology. Providing absolute safety protection against spills of all hazardous chemicals is probably impossible, but at best would involve a complex system which would be expensive and unwieldy for rapid responses.

The objective of this investigation has been to select those response techniques providing adequate protection against the greatest variety of chemicals.

This evaluation has been limited to those response techniques affecting the airborne vapor hazards associated with spills of these chemicals. Most spill response techniques have been developed for petroleum and petroleum products, where vapor hazards are less critical. The safety of response personnel and surrounding life becomes a primary objective, however, for spills of hazardous chemicals.

An evaluation of the feasibility of response techniques for the amelioration of spills of floating hazardous chemicals requires data concerning the safety, efficiency and compatibility of all components. Experience and expertise must be combined in evaluating the chemical and physical aspects of a spill to assess the candidate techniques.

Candidate response techniques may be collected into three basic categories: (1) those which reduce evaporation; (2) those which reduce vapor concentration; and (3) those which reduce vapor flammability.

Vaporization Rate Reduction, Task I: The objective of this task was to assess the capabilities of candidate spill response techniques for reducing evaporation from the surface of a spill of floating hazardous chemical. Techniques were ranked according to their ability to ameliorate toxicity and/or fire hazards associated with these vapors for the protection of response personnel and also the general environment.

The approach toward this objective involved identification of the factors influencing evaporation rates, investigation of response techniques having some effect on these factors, evaluation of their efficiencies and selection of the best response technique. Evaporation rates are controlled by five major factors: (1) heat transfer; (2) surface area; (3) surface turbulence; (4) surface cleanliness; and (5) vapor-saturated interfacial film.

Heat transfer and insulation are properties which have been thoroughly studied. A considerable body of literature was available for theoretical evaluation and practical application. It was sufficient for this evaluation to recognize that anything promoting heat transfer into the spilled chemical would aid vaporization while anything reducing heat transfer or removing thermal energy would inhibit it.

Surface area and surface turbulence are contributing factors in heat transfer. The total heat transferred is directly proportional to both, increasing their significance in determining evaporation.

Surface turbulence aids evaporation by minimizing the effects of the interfacial film. Surface convection, induced by either thermal gradients or gas turbulence, provides the natural means of increasing evaporation. The surface is cooled by vaporization and any disruption can provide a source of energy to offset evaporative cooling effects.

The influence of surface cleanliness is of considerable interest in this program since it reflects many of the parameters desired in the methods to be applied for vaporization rate reduction. The presence of surface contaminants disrupts the interfacial film, interferes with heat transfer and can also influence surface convection currents.

Vaporization rate reductions may be obtained by five candidate spill response techniques: (1) surfactant films; (2) foams; (3) sorbents; (4) gelling agents; and (5) cryogens.

Vapor Concentration Reduction, Task II: The objective of this task was to evaluate candidate spill response techniques concerning their capabilities for reducing vapor concentrations released from spills of floating hazardous chemicals. The criteria used to evaluate the various techniques were similar to those of Task I, ranking them according to their ability to reduce hazardous chemical vapor concentrations below the threshold limit or lower flammability limit.

The approach used to attain this objective was to identify the controlling factors, codify the techniques affecting these factors and select the ones having the best capabilities for reducing vapor concentrations.

Vapor concentrations are controlled by temperature, vapor pressure diffusion and dispersion. The influence of evaporation was investigated in Task I. In addition to the control of evaporation, vapor concentrations above spills of hazardous chemicals may be reduced by five general operations: (1) dilution; (2) sorption; (3) condensation; (4) encapsulation; and (5) reaction.

Vapor Flammability Reduction, Task III: The objective of this task was a critical evaluation of methods which could be used to ameliorate the fire hazards associated with floating hazardous chemicals. Potential techniques for reducing flammability were assessed on the basis of published, measured and estimated data for the pertinent controlling factors.

The approach used to evaluate vapor flammability reduction techniques was the preparation of a theoretical description of flammability hazards and the factors influencing their generation, the selection of response techniques having desirable effects on these factors and the consideration of their efficiency, effectiveness and ultimate fate (consequences) to prepare an estimate of potential applicability. Vapor flammability is controlled by four factors: fuel concentration; oxidizer concentration; activation or initiation energy; and reaction energy.

Fuel concentrations are determined by the vapor pressure of the spilled chemical and environmental conditions at the scene of the spill. The concentration is limited to the partial pressure at equilibrium of the individual chemicals, but for many hazardous chemicals this exceeds the lower flammability limit.

Oxygen concentrations of gas mixtures generated by chemical spills are set by the ambient atmosphere. Both fuel and oxidizer concentrations may be reduced by addition of inert gases, but this has limited applicability for some hazardous chemicals (e.g., tert-butyl hydroperoxide) which do not require additional oxidizer.

Activation energies demonstrate the broad range of properties exhibited by the floating hazardous chemicals. Energy required to initiate a combustion reaction may be increased by addition of energy-absorbing or reactive-intermediate-absorbing materials. Any response technique within this area should create an essentially inert gas mixture. The reaction energies from combustion of floating, hazardous chemicals are sufficient to make the reactions self-propagating. The strength of the reaction will depend upon the components and their relative concentrations in the reacting mixture. The energy released from combustion reactions may be controlled by dilution with inert materials or various reaction mechanism inhibitors. The method available to ameliorate combustion reactions all require the expenditure of a considerable amount of energy to prepare homogeneous gas mixtures. Flames propagating through heterogeneous mixtures, even containing suppressants, have greater strength than those of equivalent, homogeneous mixtures.

Amelioration of flammability hazards around spills of hazardous chemicals may be accomplished by five techniques: (1) flame inhibitors; (2) dilution; (3) cryogens; (4) foams; and (5) aerosols.

A summary of response techniques is given in Table 1.20.

Table 1.20: Spill Response Techniques

Containment	
Booms and barriers	Gelation
Liquid herders	Flocculation
Cleanup and Removal	
Weirs	Solvent extraction
Vacuum collectors	Centrifuge-separation
Skimming collectors	Bacteriological degradation
Adsorption	Catalytic oxidation
Burning	Chelation
Sinking or precipitation	Gelation
Neutralization	Ion exchange
Vaporization Reduction	
Thin films	Cryogenic cooling
Foam cover	Adsorption
Gelation	
Concentration Reduction	
Dispersion	Cryogenic condensation
Dilution	Encapsulation
Adsorption	
Flammability Reduction	
Inert foam cover	Cryogenic cooling
Dilution	Vapor suppressant addition
Liquid suppressant addition	

Source: CG-D-56-77

Recommended Responses for Dissolved Materials

The information in this section is based on *A Feasibility Study of Response Techniques for Discharges of Hazardous Chemicals that Disperse Through the Water Column,* CG-D-16-77 prepared by E. Drake, D. Shooter, W. Lyman and L. Davidson of Arthur D. Little, Inc., for the U.S. Coast Guard.

The report examines the feasibility of treating spills of chemicals which disperse throughout the water column and which are inherently more difficult to treat than those spills, such as oil, which form a separate phase in the water body. There are, at present, no effective methods to treat this type of spill and any proposed technique will encounter a number of theoretical and practical difficulties. The authors have examined in detail the feasibility of using four different methods as response techniques for amelioration of chemical spills which disperse in the water column. These are neutralization, solvent extraction, precipitation, and chelation. Each technique is applicable to a different class of compounds. For example, neutralization can only be applied to acids and bases, solvent extraction is potentially useful for organic compounds, and precipitation and chelation are potentially useful for certain inorganic compounds such as toxic heavy metals.

Neutralization: Strong bases are the most economical reagents for the neutralization of strong acids and vice versa. However, if excess reagent is applied or if the reagent is applied in an area away from the spill, then the result is a large deviation in pH in the opposite direction, thus compounding the effects of the original spill. Therefore, the most promising reagents are weaker acids and basis which will not cause the pH to fall outside the range of 6 to 9 if excess reagent is used or if the reagents are applied in the wrong position. Sodium bicarbonate appears to be the most promising agent for treatment of acid spills, and sodium dihydrogen phosphate the most promising agent for the treatment of alkaline spills. Both of these reagents are available in solid form, are inexpensive, are readily soluble in water and pose no handling or storage problems.

A summary of the treatment recommendations for the list of spill chemicals subject to neutralization is given in Table 1.21. Some of the spill chemicals listed are very weak acids or bases which would not cause deviation of pH outside the range of 6 to 9 if they were spilled into a water body. For these chemicals, no treatment is recommended.

Table 1.21: Spill Chemicals Subject to Neutralization

Acids—Recommend Treatment with Sodium Bicarbonate	
Acetic acid	Hydrogen fluoride
Acrylic acid	Nitric acid
Formic acid	Oxalic acid
Hydrochloric acid	Phosphoric acid
Hydrofluoric acid	Propionic acid
Hydrogen chloride	Sulfuric acid
Sulfuric acid (spent)	
Compounds That React in Water to Give Acids—Recommend Treatment with Sodium Bicarbonate	
Acetic anhydride	Oleum
Aluminum chloride	Phosphorus oxychloride
Benzoyl chloride	Phosphorus pentasulfide
Bromine	Phosphorus trichloride
Chlorosulfonic acid	Polyphosphoric acid
Maleic anhydride	Sulfur monochloride
Nitrogen tetroxide	Sulfuryl chloride
Nitrosyl chloride	Titanium tetrachloride

(continued)

Table 1.21: (continued)

Bases–Recommend Treatment with Sodium Dihydrogen Phosphate

Aminoethylethanolamine	Hydrazine
Ammonium hydroxide	Monoethanolamine
Caustic potash solution	Monoisopropanolamine
Caustic soda solution	Morpholine
Cyclohexylamine	Potassium hydroxide
Diethanolamine	Sodium hydroxide
Diethylamine	Triethanolamine
Diethylenetriamine	Triethylamine
Diisopropanolamine	Triethylenetetramine
1,1-Dimethylhydrazine	Trimethylamine
Ethylenediamine	Hexamethylenediamine

Compounds That React in Water to Give Bases–Recommend Treatment with Sodium Dihydrogen Phosphate

Anhydrous ammonia	Sodium
Ethyleneimine	Sodium amide
Lithium aluminum hydride	Sodium hydride

Compounds for Which no Neutralization Treatment is Recommended

Acids: Hydrogen cyanide–high toxicity, vapor and liquid.
Bases: Aniline, dimethyl formamide, hexamethylene tetramine, methyl ethyl pyridine, pyridine, and urea–very weak bases, will not violate pH criteria, therefore treatment by neutralization is not necessary.

Source: CG-D-16-77

Most of the chemicals listed in Table 1.21 can be neutralized in situ; treatment appears to be worthwhile for spills of 50 to 100 gallons or more in confined areas, lakes, rivers and even estuaries if treatment can be done before a tide reversal. A treatment skid, weighing about 1 ton, consisting of a pump, a system for metering the powdered neutralizers into solution, and a spray discharge nozzle would be suitable for treatment of spills as large as 10,000 gallons and even for partial treatment of still larger spills.

The skid and the chemicals can be air transported to the spill site. Deployment using the skid equipment on a boat such as a buoy tender appears to be simple, although a fair amount of manual labor is needed in moving 50 to 100 lb bags of treatment chemical.

As a response technique, neutralization appears to be feasible, suitable for amelioration of a large group of soluble chemicals, and environmentally desirable. Misapplication of the selected agents has minimal adverse impact.

For successful implementation, improved data on in situ reactions and further development of analytical techniques for predicting dispersion through water columns are needed as well as development of techniques for following or locating contaminated zones prior to treatment initiation. Monitoring of pH during treatment and active interaction with on-line dispersion models is also desirable for most effective application of treating agent. Nevertheless, these are problems that are within the state of the art and it appears that the U.S. Coast Guard could develop effective response strategies for neutralizing acidic or basic spills of chemicals which dissolve in the water column.

Solvent Extraction: If an organic compound is soluble in water, there then exist structural groups which are compatible with the highly polar nature of water. Typical of such groups are the –OH of alcohols and acids and the $-NH_2$ of amines. To transfer soluble organic compounds to a phase immiscible with water, one must consider the nonpolar or hydrophobic groups in the material. This also indicates that the solvent itself should be nonpolar.

Other criteria which are important in the choice of a solvent relate to its solubility in water and to its toxicity.

Many potential solvents were considered and vegetable oil (e.g., soybean oil) was finally selected. This material is inexpensive, readily available, nontoxic, and only slightly soluble in water. The study examined the potential of soybean oil to extract soluble organic chemicals and equipment with five equilibrium stages assuming equal volume flows of fresh solvent and contaminated water in each stage. For some chemicals, such treatment would be sufficient to reduce the concentration of a chemical in water to a level below the toxic limits; in other cases, the extraction was less effective. Of the 75 chemicals considered, only 18 appear to be promising for removal by extraction. Another 27 soluble chemicals have such low toxicity limits that no treatment is necessary. Twenty-two compounds were found to be so polar that no extractive solvent appears reasonable. Finally, for 8 materials, insufficient data were located to allow one to determine the applicability of solvent extraction. A listing of the compounds in the various groups is presented in Table 1.22.

Table 1.22: Spill Chemicals Subject to Solvent Extraction

Treatment with Vegetable Oil Is Suggested	
n-Amyl alcohol	n-Butyl acetate
n-Butyl alcohol	Carbon bisulfide
Chloroform	Dichloromethane
Ethyl acetate	Formaldehyde solution
Isoamyl alcohol	Methyl acrylate
Methyl amyl alcohol	Methyl isobutyl carbinol
Methyl methacrylate	Paraformaldehyde
n-Propyl acetate	Sodium alkylbenzene sulfonates
Sodium alkyl sulfates	Vinyl acetate
No Effective Solvent	
Acetone cyanohydrin	Allyl alcohol
1,4-Butanediol	1,4-Butenediol
1,4-Butynediol	Chlorohydrins (crude)
Diethylene glycol dimethyl ether	Diethylene glycol monoethyl ether
Diethylene glycol monomethyl ether	Diisopropanolamine
Ethylene cyanohydrin	1,4-Dioxane
Ethylene glycol monoethyl ether	Ethylene glycol monobutyl ether
Ethyleneimine	Ethylene glycol monomethyl ether acetate

(continued)

Table 1.22: (continued)

No Effective Solvent	
Hexylene glycol	Ethylene glycol monomethyl ether
n-Propyl alcohol	Isobutyl alcohol
Polypropylene glycol monomethyl ether	Triethanolamine
No Treatment Necessary	
Aminoethanolamine	sec-Butyl alcohol
tert-Butyl alcohol	Corn syrup
Dextrose solution	Diacetone alcohol
Dichloromethane	Diethanolamine
Diethylene glycol	Dimethyl sulfoxide
Ethoxytriglycol	Ethoxylated dodecanol
Ethoxylated pentadecanol	Ethoxylated tetradecanol
Ethoxylated tridecanol	Ethyl alcohol
Ethylene glycol	Glycerin
Isobutyl alcohol	Isopropyl acetate
Methyl alcohol	Monoethanolamine
Monoisopropanolamine	Propylene glycol
Propylene oxide	Sorbitol
Triethylene glycol	
No Data Available to Make Recommendations	
Dimethyl sulfate	Dipropylene glycol
Epichlorohydrin	Ethyl acrylate
Methanearsonic acid, sodium salts	Morpholine
Tetrahydrofuran	Sulfolane

Source: CG-D-16-77

Although there are 18 chemicals suitable for solvent extraction treatment with vegetable oil, the equipment for treating large spills of these chemicals is cumbersome and complicated both to operate and maintain. For 10,000 gallon spills, an apparatus weighing 10 to 20 tons appears necessary. The equipment could be airlifted in a C130, but its size and weight would require deployment from a barge (for spills in a lake or river) or a heavy truck (confined spills). The contaminated portion of the water would be pumped on-board, treated by the apparatus, and returned to the water at a point far enough away from the contaminated zone to prevent dilution of the water remaining to be treated.

This equipment is sufficiently specialized so that the practical implementation of a solvent extraction treatment technique presents many problems, except perhaps for small, confined spills where a much smaller apparatus could be used for gradual amelioration treatment.

The development of solvent extraction treating-equipment might be made more attractive, however, if it is built in modules that can be used for other types of responses. For example, the first module, designed for filtering solids, might also serve as a treatment skid for chemical spills subject to amelioration by precipitation. The entire system might also be suitable for use in conjunction with treatment response using a sequestering chelating agent which can then be recovered and recycled by solvent extraction. However, basic data are lacking for meaningful design of such treatment modules at present. The concept appears to be technically feasible but the potential for developing it into a practical field response method is not clear. Further analysis of the extent of environmental damage associated with this class of spills and comparison with treatment methods other than those included in this study should be conducted to determine whether a solvent extraction response system justifies further development.

Precipitation: No single agent is capable of treating all of the chemical compounds in the list of chemicals suggested for treatment by precipitation. It is not possible for the same type of treatment to be equally effective for both cations and anions. Although a variety of reagents can be employed to precipitate each of these compounds, many of the reagents do not reduce the concentration of the toxic metals to a sufficiently low level. In some cases, the use of a precipitating agent could introduce a new toxic element into the system.

A summary of the recommended treatments for the chemicals subject to precipitation is given in Table 1.23. Dibasic sodium phosphate (Na_2HPO_4) is the preferred agent for six of the thirteen chemicals. The others require individual treatment. No treatment is recommended for silver nitrate in marine waters because the high concentration of chloride ion which exists (0.5 M) is sufficient to precipitate the silver. Titanium tetrachloride reacts with water to liberate hydrochloric acid and, therefore, should be treated by neutralization. At the same time a relatively nontoxic precipitate of white titanium is formed. Treatment of synthetic latex is not recommended because no simple treatment can be guaranteed effective. Also, the effects of treatment may compound the adverse effects of the spill and may depend on the particular latex formulation.

Table 1.23: Spill Chemicals Subject to Precipitation

Spill Chemical	Reagent
Cadmium chloride	Dibasic sodium phosphate
Copper sulfate	Dibasic sodium phosphate
Ferrous sulfate	Dibasic sodium phosphate
Nickel sulfate	Dibasic sodium phosphate
Zinc chloride	Dibasic sodium phosphate
Antimony trifluoride	Dibasic sodium phosphate
Silver nitrate	Sodium chloride (no treatment in marine waters)
Titanium tetrachloride	Treat as hydrochloric acid by neutralization
Potassium dichromate	Sodium sulfite
Potassium permanganate	Sodium sulfite
Chromic anhydride	Ferrous sulfate
Latex	No treatment
Sodium ferrocyanide	Ferric chloride

Source: CG-D-16-77

It is difficult to assess whether or not an in situ precipitation is possible for the various chemical/treatment combinations since the settling characteristics of the precipitate depend on its particle size distribution and the turbulence of the water column. If settling does occur, then environmental impacts of precipitate removal by dredging or of leaving the material on the bottom may not be acceptable in some cases. If the precipitate does not settle out, it may continue to move with the water as a colloidal dispersion.

These difficulties can be overcome by pumping the contaminated water into a treatment system where precipitation and precipitate collection occur before treated water is returned to an uncontaminated region of the water body. The equipment required for this purpose is fairly simple, consisting of a pump, a precipitation tank, a solids separator, and a precipitate storage vessel. As discussed previously, this equipment might also serve as a module for treating the inlet water with a solvent extraction system.

Chelation: Chelate compounds are potentially capable of ameliorating spills of some chemicals because the complexes formed with toxic metals have a lower toxicity than the original spilled chemical. There are two types of chelating agents: sequestrants which bind the metal ion but remain soluble within the water column, and precipitants which, as the name implies, cause precipitation of the chelate-metal complex and, in this manner, remove it from the water column.

A summary of the recommended treatment by chelating agents is given in Table 1.24. The most promising sequestrant is ethylenediaminetetraacetic acid (EDTA). This compound forms stable chelate complexes with copper, nickel, zinc, cadmium and iron and can reduce the concentration of these metals to below the water quality criteria at pH of 6 or higher. EDTA is a widely used reagent and has a relatively low toxicity.

Table 1.24: Spill Chemicals Subject to Chelation

	Recommended Chelating Agent	
Chemical	**Sequestrant**	**Precipitant**
Antimony trifluoride	. . . no suitable reagent	
Cadmium chloride	EDTA	Oxine
Chromic anhydride	. . . no suitable reagent	
Copper sulfate	EDTA	Oxine
Ferrous sulfate	EDTA	Oxine
Latex, liquid synthetic	. . . no suitable reagent	
Nickel sulfate	EDTA	Oxine
Potassium dichromate	. . . no suitable reagent	
Potassium permanganate	. . . no suitable reagent	
Silver nitrate	. . . no suitable reagent	
Sodium ferrocyanide	. . . no suitable reagent	
Titanium tetrachloride	. . treat by neutralization . . .	
Zinc chloride	EDTA	Oxine

Source: CG-D-16-77

EDTA is probably not effective with the metals silver, antimony, and titanium and is not suitable for the other compounds such as potassium dichromate, potassium permanganate, chromic anhydride and sodium ferrocyanide.

The most promising precipitant is 8-hydroxyquinoline (oxine) which precipitates copper, nickel, zinc, cadmium, and possibly silver, antimony and titanium. Suitable chelating precipitants for the anions have not been demonstrated. Additional information is needed on reaction rates and effects of pH on reactions before a final conclusion on feasibility can be drawn.

Treatment by sequestering with EDTA might be implemented in the field using the same equipment as that developed for the solvent extraction treatment method. The EDTA could probably be recovered and recycled using solvent extraction with vegetable oil. For the precipitating chelation scheme, the precipitation equipment described previously might be appropriate.

CLEANUP IN INDUSTRIAL AREAS

The material in this section is based on a paper presented at Conference 1976 by Thomas F. Dalton of Metropolitan Petroleum Petrochemicals Co., Inc. The title of the paper was "Hazardous Material Cleanup in Industrial Areas."

Introduction

Recent experience in cleaning up oil and hazardous materials in industrial areas has shown that there are problems unique to these metropolitan areas.

Traffic: In heavily industrialized areas there is generally a problem of traffic jams, accidents, road repair and construction which can pose a problem to companies responding to hazardous material spills. It is therefore important to realize that minimum response time during peak traffic periods is virtually impossible. If it can be arranged, the companies should communicate by radio with their equipment and/or have alternate routes determined in advance when the major arteries are congested.

Mobilization of Equipment: Another basic problem confronting cleanup companies is the mobilization of manpower and equipment with particular emphasis on special types of equipment. In order to effectively respond and clean up a hazardous material spill it may be required periodically to use a front end loader, bulldozer, large waste disposal dumpsters, dump trucks, cherry pickers, cranes, winches, and a host of other miscellaneous equipment. Companies and/or regional planning groups should compile a list of this specialized equipment and its availability on short notice.

Disposal: One of the biggest immediate problems confronting a hazardous spill cleanup company is how to dispose of the product picked up, especially if it is not compatible with the normal disposal methods readily available to that company. For example, 10,000 gallons of low flash naphthenic solvent cannot be disposed of by dumping it into a high temperature separator, nor can it be stored in a slop tank where the wasted oil is subsequently used for supplemental boiler fuel. Statewide and regional planning should be done to provide alternate means of disposal for major hazardous materials. These alternate means should include sanitary land fill and high temperature incineration.

Debris (wood and leaves) is normally picked up with the oil or hazardous material in rivers and inland waterways. This debris must also be disposed of in a land fill or an incinerator.

Safety Assessment: One of the immediate areas of concern upon arriving on a hazardous material spill scene is an assessment of exact hazard and/or subsequent safety precautions to be taken. In most industrial firms, especially those where there is a full-time safety or environmental committee, the physical and chemical properties of the spilled product are readily available. In large industrial areas, where there are professional fire departments who can quickly consult their engineering department for an evaluation, not only the hazard, but also the suggested control and safety measures will be recommended.

A third input can be obtained from the EPA or U.S. Coast Guard, who can, through one of the nationwide networks and by means of their good communications, rapidly identify and obtain feedback on necessary corrective measures.

Lastly, the cleanup company itself, either through its own technical competency and/or experience, can provide some input as to the exact nature of the hazard and recommended safe cleanup procedure program.

There is a wide variety of substances spilled in an industrial area ranging from fuel oils, solvents, plasticizers, fatty acids, mineral acids, chlorinated solvents and organic intermediates. In many instances, the company manufacturing or transporting the products can handle a spill, especially if it is small in quantity and readily accessible. But when the volume of the hazardous material or oil is large (over 500 gallons), then a third party cleanup contractor is normally called in.

Comprehensive Planning for Hazardous Material Spills

In order to plan effectively for mitigating and cleaning up hazardous material spills, there should be some specific plans set forth at both the federal and municipal level. The EPA regulations covering SPCC (Spill Prevention Control and Countermeasure Plan) are an excellent example of a positive approach to establishing responsibilities and procedures for spills. (EPA Regulation, Title 40, Chapter 1, Subchapter D, Part 112 "Oil Pollution Prevention"). Similar legislation of a specific nature should be put out for hazardous materials.

Municipal Spill Planning: The SPCC Plan of the EPA was a good example of planning at the federal level. Consider now, planning at the municipal level.

One of the most comprehensive plans for handling spills is the New York Fire Department Regulations (9-72) covering "material and equipment requirements for oil spill control at bulk storage plants and petroleum product pipe lines." In this plan, the NYFD has established a list of minimum amounts of equipment and material to be maintained on-site, where any petroleum product is stored. The amount of material and equipment that should be available is predicted on the total capacity or gallonage stored at the site.

The minimum requirements for 1,000,000 gallons of storage or less, are as follows:

(1) 2,000 lb of absorbent material–This absorbent material must be of a type approved by the Fire Commissioner and have the ability to absorb 0.5 gallon of petroleum product for each 1 lb of absorbent material.

(2) 300 ft oil spill containment boom–Must be of a type acceptable to the Fire Commissioner.

(3) 55 gallons dispersant chemicals–This dispersant must be of low toxicity and may only be used when directed by the On-Scene Commander.

(4) one skimmer–This equipment must be of a type acceptable to the Fire Commissioner.

In addition, each facility is required to have on hand adequate manpower, boats, transportation and storage facilities to properly utilize the materials and equipment specified on a 24 hour per day basis. These minimum recommended quantities were determined with the knowledge that the bulk terminal or pipeline operation has an approved back-up oil spill cleanup service. The Fire Commissioner reserves the right to vary the quantities and types of material and equipment consistent with the requirements of public safety.

The municipal plan does not delve into explanations or theory, but merely confines itself on a pragmatic basis to what is needed as a minimum. This, of course, is the best place to start. More cities should adopt plans such as New York's Public Law #70. Granted, all municipalities may not all have the same situation, but at least an effort should be made to make everyone aware of their responsibilities to protect the public.

REFERENCES

(1) U.S. Congress, *Federal Water Pollution Control Act As Amended,* Public Law 92-500, p. 89, 92nd Congress, S.2770, Washington D.C., (Oct. 18, 1972).

(2) Cargo Information Cards, (Chemcards), Manufacturing Chemists Association, 1825 Connecticut Ave., N.W., Washington D.C. 20009.

(3) *U.S. EPA Field Detection and Damage Assessment Manual for Oil and Hazardous Materials Spills,* EPA Contract No. 68-01-0113, (June 1972).

(4) U.S. Coast Guard, *Chemical Hazards Response Information System (CHRIS),* 4 volumes, CG-446, Washington, D.C., (January 1974).

(5) Lyman, W., Nelson, L., Partridge, L., Kalelkar, A., Everett, J., Allan, K., Goodier, J.L., and Pollack, G., *Survey Study to Select a Limited Number of Hazardous Materials to Define Amelioration Requirements,* Report No. CG-D-46-75, Arthur D. Little, Inc., Cambridge, Mass., (March 1974).

(6) Bennett, G.F., ed., *Proceedings of the 1974 Conference on Control of Hazardous Material Spills,* San Francisco, Calif., Inst. Chem. Engrs. and EPA, (August 25-28, 1974).

(7) Pilie, R.J., Baier, R.E., Ziegler, R.C., Leonard, R.P., Michalovic, J.G., Peck, S.L., and Bock, D.H., *Methods to Treat, Control, and Monitor Spilled Hazardous Materials,* EPA-670/2-75-042, U.S. Environmental Protection Agency, National Environmental Research Center, Cincinnati, Ohio, (June 1975).

(8) Bauer, W., Borton, D.N., and Bulloff, J.J., *Agents, Methods, and Devices for Amelioration of Discharges of Hazardous Chemicals on Water,* CG-D-38-76, U.S. Coast Guard, Washington, D.C., (August 1975).

(9) Drake, E. et al., *A Feasibility Study of Response Techniques for Discharge of Hazardous Chemicals that Disperse Through the Water Column,* CG-D-16-77, U.S. Coast Guard, Washington, D.C., (July 1976).

(10) *Feasibility Study of Response Techniques for Spills of Hazardous Chemicals that Float on Water,* CG-D-56-77, MSA Research Corporation for the U.S. Coast Guard, Washington, D.C., (October 1976).

MECHANICAL CLEANUP METHODS

DISPERSION AND DILUTION

Dispersion

The information in this section is based on *A Feasibility Study of Response Techniques for Discharges of Hazardous Chemicals that Sink*, CG-D-56-78 prepared by T.D. Hand, A.W. Ford, P.G. Malone, D.W. Thompson and R.B. Mercer of the U.S. Army Engineer Waterways Experiment Station for the U.S. Coast Guard.

Dispersion, whether by chemical or mechanical means, should not be considered as a first choice response to any hazardous material spill. Indeed, containment, rather than dispersion, should first be considered when a spill occurs. The U.S. Coast Guard CHRIS manual, while recognizing the use of dispersion techniques, suggests using such only sparingly, and even then only when all other corrective methods have been eliminated from consideration. Table 2.1 presents a list of those hazardous chemicals in CHRIS which have been suggested as possible candidates for dispersion in selected instances (1).

The principle behind dispersion is to spread out the spilled chemical so that the concentration of the material is reduced to below recommended limits within a relatively short transport distance from the spill site. Failure to do this would result in spreading the effects of the spill over a wider area. In some instances, for example with organic liquids, this might permit increased biodegradation of the material. Typical instances where dispersion could be used are as follows:

- In open water, where rapid dilution of the spilled chemical might be expected upon dispersion.
- In small streams flowing into larger rivers. In this instance, a spill dispersed in the stream would be carried to a nearby river where, presumably, the greater flow would result in quick dilution below recommended limits.
- At the mouth of a harbor with fast tidal currents. These currents

could make techniques such as covering difficult. Dispersants could be applied to the spill on an outgoing tide, whereupon the chemical would be carried to open water and diluted below recommended limits.

The use of dispersants must be subject to decisions made on a case-by-case basis. This decision would depend on environmental conditions at the site, the amount of chemical spilled, the chemical's toxicity, and similar considerations. Both chemical and mechanical means of dispersion are available. Technology for application of these to spills of sinking chemicals is not well developed, the greatest amount of experience coming from the control of surface spills. Chemical dispersants, for example, are available for spills of hydrocarbons and have been used on a number of spills. A discussion of the different types of dispersion mechanisms follows.

Table 2.1: CHRIS Chemicals Subject to Dispersion

Aniline	Dichloromethane
Benzoyl chloride	2,4-Dichlorophenol
Bromine	Dimethyl sulfate
Calcium carbide	Epichlorohydrin
Calcium oxide	Furfural
Carbolic oil	Lithium aluminum hydride
Cresols	Toxaphene

Source: CG-D-56-78

Mechanical Dispersion: Depending on the size and location of a spill, a number of mechanical dispersion devices are readily available. These can be deployed with only small capital hardware outlays. Water streams, from a fire hose connection on either ship or shore, could be effective in relatively shallow waters and for small spills. Propwash from a boat could also be used. Depending on the depth of water and size of spill, anything from an outboard motor to the wash of a tugboat could be used. The use of compressed air, perhaps in conjunction with easily deployable and commercially available lagoon aeration systems such as the Schramm "Aqua Puss" Lagoon aeration system (2), could be used to disperse selected materials from the bottom.

The Aqua Puss system consists of a central manifold surrounded by radiating diffuser arms, consisting of three-tube polymerized PVC hose. Air is supplied by a surface compressor. As air rises in the water column, bottom water is entrained and pushed toward the water surface. This allows currents in the water column to carry away the spill. It is possible that this could be used in conjunction with propwash or water-jet dispersion to effect more complete dispersion. It is possible that a temporary system similar to the above could be made from readily available PVC hose and a gasoline-driven compressor. This setup was suggested in CHRIS for use in air barriers (1).

Procedures for use of propwash, water jets, and compressed air in oil spills are usually associated with movement of a slick from obstructed areas and towards skimmers (this is not dispersion, but uses many of the same techniques). Booms could be used in conjunction with such methods to assist in deflection of oil towards a desired location. It is equally likely that the hazardous material spill barrier could be used in a similar mode for the deflection of sunken spills. This might enable the spill to be cleared from obstructed areas in a manner similar to oil. Both solids and liquids could be moved in this manner, though the procedure would work best with liquids due to their flow characteristics. An additional means of mechanical dispersion, probably useful in small streams flowing into larger rivers, consists of flow augmentation using water stored upstream of a spill. This could be released to wash the spill into the river. This method, suggested by Dawson et al (3), would be partially dependent on the availability of a means of predicting duration of toxic concentrations, flow rates, etc., for the water body. These may not always be available.

Chemical Dispersion: Chemical dispersion methods have not been developed for use on spills of sinking hazardous materials. However, a number of dispersants are available for use on oil spills. These dispersants are molecules with an oil-soluble (lipophilic) end and a water-soluble (hydrophilic) end. The dispersant will locate and orient itself at an oil-water interface, reducing surface tension and permitting dispersion into the upper portion of the water column of fine droplets of oil.

Application of such dispersants has been accomplished using hand-operated pumps and pressure units, portable pump eductor systems, specially prepared vessels with attached spray booms, and helicopters or other aerial spray methods. Mixing, possibly using the propwash of a surface workboat, may be necessary to aid in dispersion, though self-mix dispersants are available. It should be noted that these dispersants are less dense than water and, in their present form, would not be useful for bottom spills.

Any application of dispersants to spills of sinking liquid organics would likely be limited to open-water areas. Dispersant use is considered by the EPA on a case-by-case basis under the authority of Annex X of the National Oil and Hazardous Substances Contingency Plan (40 CFR 1510). This will generally be conducted by the EPA in consultation with appropriate State and Federal agencies consistent with reduction of hazards to vulnerable species of waterfowl or which will result in the least overall environmental damage. In any case, dispersants will have to be accepted by the EPA prior to any use.

Use of dispersants for sinking organic spills will depend on formulation of a heavier-than-water surfactant compound proven to be applicable to the chemicals under consideration. A pump eductor system using a water pickup hose and dispersant metering device attached to drums of the material might serve the purpose. The pump outlet could be connected to a suitable dispersing head similar to a spray bar, this being positioned slightly above the spill at the ocean bottom. Water volume should be large in relation to the spill and there should be high currents available to aid in dispersion.

It should be noted that supply of the dispersants is not great in the United States, due primarily to difficulty in obtaining permission for use. Even if a heavier-than-water dispersant could be formulated, manufacture and storage of it in quantities sufficient for use in emergency situations would be delayed until product acceptance was obtained from the EPA.

Summary: Dispersion should generally be avoided in most spill situations and used only as a last resort. The only large-scale field testing of dispersion techniques has been in the area of oil spills.

Dilution with Water

> The material in this section is based on *Agents, Methods and Devices for Amelioration of Discharges of Hazardous Chemicals on Water*, CG-D-38-76 prepared by W.H. Bauer, D.N. Borton and J.J. Bulloff of Rensselaer Polytechnic Institute for the U.S. Coast Guard.

The only practical liquid spill diluent is water. Forced dilution will dissolve miscible or soluble chemicals quickly and it may disperse some immiscible or insoluble ones. Unlike sorption, which concentrates toxic or flammable material and increases fire hazards, dilution tends to attenuate such dangers, especially if the water is applied in a mode that lowers vaporization rate and vapor concentration. Forced dilution has been generally advocated as a means of amelioration for hazardous chemicals that mix with water, especially those that dissolve acidically or basically (4). Dilution is the usual consequence of a spill into natural waters, and it has undoubtedly been involved in the greatest proportion of past spills, reported and unreported.

Water for forced dilution of a soluble or miscible spill might come from the intruded water body or from a nearby water supply. Whenever natural dilution is ecologically tolerable for the treatment of a spill, forced dilution offers the benefit of a more immediate, more rapid and more certain attenuation of hazards posed by initial high concentrations of spilled chemicals. Whether dilution is acceptable is determined by the use of the water involved: drinking, fishing, bathing, feedwater, recreational, navigational, etc.; the nature of the water: flowing stream, lake with or without a natural outlet, metropolitan shore, remote shore, open sea; and the size of the water body and the nature and size of the spill in any particular case. Thus, forced dilution, like every other kind of amelioration, can only be discussed on an ad hoc basis until spill classification and amelioration level standards are established. It is feasible where alternative ameliorations cannot be adequate or timely.

Water dilution is considered to be a possible response in the case of the following representative CHRIS chemicals which were classified in Table 1.6.

Hydrogen Fluoride: Dilution is generally advocated for soluble acids or bases.

Sulfuric Acid: Forced dilution has been considered as an amelioration technique for spills of sulfuric acid (4). It is most advantageous for large spills in water bodies capable of adjustment to acid intrusion.

Methyl Alcohol: Forced dilution would be most advantageous for large water bodies, especially for large spills therein, particularly where site conditions favored continued natural dilution, oxidation or photo-oxidation and biodegradation.

Potassium Cyanide: Forced dilution has been considered for sodium cyanide spills (4). For small spills in waters remote from ingestive, processing or other cyanide-susceptible use, massive dilution is most advantageous, especially if no other amelioration resources can be given adequate or timely deployment.

Sodium Alkylbenzenesulfonates: Forced dilution might be advantageous for use against sodium alkylbenzenesulfonates in water bodies not destined for drinking or feedwater use, or where natural dilution cannot be awaited. In situations where biodegradation might be possible (5), there would be added reason for considering it.

Use of Air as a Diluent for Hazardous Vapors

> The information in this section is based on *Feasibility Study of Response Techniques for Discharges of Hazardous Chemicals that Float on Water*, CG-D-56-77 prepared by J.S. Grear of MSA Research Corporation for the U.S. Coast Guard.

Simple dilution provides a direct approach toward vapor concentration reduction. The dilution technique involves the transport and mixing of uncontaminated air with the vapors released from a chemical spill. The volume of uncontaminated air must be large enough to maintain the concentration of hazardous chemical vapors below their threshold limit value or lower flammability limit.

Performance specifications can be calculated from data generated with a natural dispersion model. This model predicts the evaporation rates for spills of floating, hazardous chemicals to be in the range of 1 to 3.5 m^3 of vapor released per hour per m^2 of spill surface (3.3 to 11.5 $ft^3/hr/ft^2$).

Typical spills are anticipated to cover between 378 and 3,780 m^2 (4,069 to 40,690 ft^2) and release between 378 and 13,230 m^3 of vapor per hour (13,350 and 467,200 ft^3/hr). If an average threshold limit value of 10 ppm is assumed, then between 3.78×10^7 and 1.3×10^9 m^3 of uncontaminated air must be added each hour (1.3×10^9 to 4.7×10^{10} ft^3/hr) to keep the concentration of the hazardous chemical vapor at this limit. If an average lower flammability limit of 1% by volume is assumed, then between 3.78×10^4 and 1.3×10^6 m^3 of uncontaminated air must be delivered each hour (1.3×10^6 to 4.7×10^7 ft^3/hr) to keep the vapor concentration at this limit.

These maximum estimates, approximately 10^9 m^3 of clean air per hour (about 5×10^{10} ft^3/hr) to achieve the threshold limit and approximately 10^6 m^3 of clean air per hour (5×10^7 ft^3/hr) to achieve the lower flammability limit concentrations, indicate the necessity of very large equipment. The dilution technique for responding to spills of hazardous chemicals may be considered to be a man-made wind. Such large gas volumes can be handled by blower equipment incorporating surplus jet engines. Such blowers are used by railroads to remove snow, and by airports to remove fog.

Some typical examples of this equipment are manufactured by Railway Maintenance Corporation, Pittsburgh, Pennsylvania. Their equipment is designed to remove snow from the road beds and switches to permit smooth operation in winter. An example of particular interest to this application is their Hurricane model blower, which is based upon an Allison J-35 jet engine. This blower can generate a 650 mph air blast, which would deliver approximately 2.9×10^5 m^3/hr (about 10^7 ft^3/hr) of diluting air over the spill. The Hurricane model jet blower is fabricated as a self-propelled railway car assembly complete with the accessories necessary for 5 to 6 hours of independent operation. It occupies a space of approximately 10 m^2 (about 108 ft^2), weighs 12,250 kg (27,000 lb) and costs approximately $75,000. Modifications, such as flat-bed mounting or a skid assembly, for shipboard operation are conceived as resulting in reductions of over 50% in weight and cost.

Sporadic use against hazardous chemical spills, coupled with proper maintenance and storage precautions, can be expected to yield a long (10 to 20 year), useful system life, with occasional (2 to 5 year) major overhaul programs. Operation would involve 1 to 2 men for start-up, followed by occasional adjustments and periodic refueling during long-term spill response applications. Assuming the spill conditions are calm to stagnant, as many as 5,000 jet blowers would be required to ameliorate the toxic properties of chemical vapors released from a large spill. If it were required only that the lower flammability limit not be exceeded, then up to 5 jet blowers would be required. In either case, any natural wind would reduce the number of blowers required.

CONTAINMENT–AN OVERVIEW

> The material in this section is based on a paper given at Conference 1972 by I. Wilder and J.E. Brugger of the Environmental Protection Agency. The title of the paper was "Present and Future Technology Requirements for the Containment of Hazardous Material Spills."

From the standpoint of environmental protection, the most effective way to control a spill of a hazardous material is to contain it in its most concentrated form. The advantages of prompt containment of spills in general and acute releases of hazardous materials in particular are: (1) minimization of damage to the total environment, (2) facilitation of on-site cleanup operations, and (3) prevention of dispersion into watercourses. Almost every spill is a special case by reason of (1) the diversity of hazardous materials, their containers, and the spill situation (during storage or in transit) and (2) the unpredictability of geographic location, time of day, and weather conditions. The problem of controlling hazardous material spills is made more difficult by the lack of extensive experience in spill containment and by the absence of readily available specialized countermeasure equipment near the spill site.

Within the context of total environmental protection, the prevention of spills of hazardous materials is the most important line of ecological defense. The danger to the public health and welfare and the insult to the ecology are only infrequently limited to the immediate spill area, even when prompt containment is achieved. Realistically, the damage from a spill cannot be confined exclusively to the water, land or atmospheric ecosystems. Further, in some fortunately rare cases, occurrence of a spill means that, in order to preserve the human lives endangered by the spill, damage to the environment must be allowed.

For those directly involved in a hazardous material spill, personal safety is the first consideration, followed by warning and assistance to others. Attention must then be given to halting or containing the spill along with notification of appropriate Federal, State, and local authorities, affected water and sewage treatment plant supervisors, industrial plant managers and safety crews, and cognizant police and fire departments. The very diversity of those who must be alerted points out the totality of the environmental impact of a spill.

Basically, containment is needed to insure environmental protection: the more efficient the containment, the greater the protection of the environment. Spills, as well as chronic discharges, of hazardous materials at significant levels are unnecessary releases that pollute the air, water, and land environment. Such discharges result from accidents, carelessness, bad procedures, poor attitudes, and other practices that are not responsive to the public interest. Technically, this paper is addressed to the problem of the containment (or recontainment) of hazardous materials following a spill or acute release; the case of chronic, long-term, slow release is considered only incidentally. While the need to prevent a spilled material from entering a watercourse is emphasized, the requirements for protecting the atmosphere and earth are given explicit attention.

Importance of Containment

The advantages of prompt containment of spills in general and of crisis discharges of hazardous materials in particular are: (1) minimization of damage to the total environment, (2) facilitation of on-site cleanup operations, and (3) prevention of dispersion into watercourses through flow into sewers, run-off into streams, and percolation into the earth with subsequent seepage into subsurface water supplies.

The fundamental importance of containing a spill is basically the need to maintain lateral and vertical control of the material. Loss of lateral control implies that spreading will occur. The more spreading, the more widespread the surface damage to the environment and the greater the difficulty of detoxification and cleanup. Vertical spreading, of which percolation and seepage of liquids serve as examples, leads to contamination of well and subsurface water supplies and makes the process of natural environmental restoration much less rapid because of long-term retention and slow release. The spill of a gas or of a volatile material is another example of an accident in which vertical control is lost; as in the case of percolation and seepage of a liquid, lateral control is also eventually lost.

Control, then, generally depends on passive or natural countermeasures for environmental protection or restoration. The rapid dispersion of a spilled gas makes containment impractical; damage is fortunately less intense but also more widespread. Containment of a seeping or percolating liquid is difficult not so much because diffusion or dispersion is rapid, but rather since the spilled material becomes more or less inaccessible.

Failure of lateral or vertical control leads to damage to the environment. Some of the consequences of this loss are worthy of discussion. For gases and volatile materials, air pollution is perhaps the most obvious consequence. Likewise, for spills of liquids on the earth, damage to the area affected will be quite apparent. There is, unfortunately, more subtle and less immediate damage to the environment, particularly to the water ecosystem. When a spilled hazardous material is washed into a sanitary sewer, the toxic effects of chemicals can upset the biological process of the municipal waste treatment plant, resulting in both the toxic spilled material and untreated sewage being discharged into a receiving stream or body of water. In the event that a spill is flushed into a storm sewer, the material usually flows unimpeded to the nearest stream. Thus, one sees that accidents in municipalities cause direct and indirect insults to the water ecosystem. Further, spills that occur some distance from watercourses can be washed or flow into streams directly. On-land spills can seep into ground waters and endanger the public health of individuals and communities utilizing wells and springs. Toxic effects of land spills can be realized months after the actual occurrence of the spill when the contaminated ground water recharges the surface streams.

To the extent that the spilled or converted material in a watercourse is hazardous, fish kills and other less noticeable damage to aquatic biota will occur. Use of polluted river water for irrigation may cause damage to crops and soil and may also introduce the pollutant into a food chain. Contamination of water supplies results in numerous sublethal effects on all living things, including man. The effects of long-term exposure to sublethal dosage is generally unknown, but carcinogenesis and mutagenesis could occur. Perhaps the most common effect of pollution in watercourses is the oxygen depletion effect (enhanced BOD).

Aesthetic damage and general nuisance effects of pollution are aspects that are less readily quantified but are ones that cannot be disregarded or neglected. In fact, damage to water use as a place for sports such as boating, fishing, and swimming is a factor that has contributed strongly to public concern about environmental protection. Disagreeable conditions include visible floating material, suspended or settled solids, discoloration or excessive turbidity, evolution of gases, especially hydrogen sulfide, blooms or high concentrations of plankton, and surfactants that foam when water is agitated or aerated.

General Aspects of Containment

In discussing the containment of hazardous material spills, one must consider what is included in the class of hazardous materials, whether the material is liquid, a solid or a gas or liquefied gas, what the weather conditions are, where the spill occurred geographically, and whether fire or explosion hazards exist. To add perspective on the containment of spills of hazardous materials, one might profitably consider various aspects of containers and of containment before a spill has occurred. In a rather general and not mutually exclusive categorization, containers can be classified by:

(1) Site: indoors or out-of-doors;
(2) Facility: in plant, in the field, in storage, in an open refuse pile or pit, behind a dike;
(3) Mobility: fixed, in a moveable container, in transit;
(4) Mode of transportation: truck, railroad, barge, ship, pipeline, airplane;
(5) Relative to ground level: on, above, or below;
(6) Nature of surrounding surface: prepared (concrete, blacktop, compacted) or natural (earth, water);
(7) Physical or processing conditions: (a) pressurized, atmospheric, or under vacuum; (b) heated, cooled, stirred, aerated;
(8) State: in process, in storage, moving in pipes, in transit;
(9) Container type: tank, barrel, bag, box, reactor vessel, package, dike, Dewar flask;
(10) Type of container material: metal, glass, plastic, paper, burlap;

(11) Structure of container: (a) open-top, sealed-impervious, mesh; (b) encased, single wall, composite or multiple wall; and
(12) Operation: in process, during loading or unloading.

This listing is intended to show the great diversity of situations under which spills can and do occur. Obviously, there is a great disparity in ease of containment following a spill depending on whether it occurs: (1) at a plant during processing, loading or unloading where countermeasure equipment should be readily available or (2) in the field or during transit where containment or control devices or equipment are mostly unavailable immediately. Additionally, during the actual course of spillage, one should consider in what quantity and for how long the material will: (1) remain in the original concentration in the original container, (2) be confined in an equally concentrated form in an auxiliary container or behind a dike, (3) flow over a surface and collect in natural depressions, (4) reach and be diluted by a watercourse or be dispersed by wind, rain, or explosion.

Complete collection of the concentrated hazardous material from a spill and confinement of the material without hazard to human life or to the ecosystem is an ideal in any containment scheme. In practical cases, except under unusual conditions, the priority of active countermeasures would reflect the ideal: halt the flow, contain as much of the material in the most concentrated form as soon as possible, contain or inactivate spreading material or material that has entered a watercourse.

For coping with the wide spectrum of spills, based on container, material, location, and quantity, a catalog of spill containment techniques is advantageous. The list should include tried and proven methods, as well as novel approaches. An endeavor that can be undertaken to uncover new techniques is to list operations that can be performed on the two elements of containment, namely, the container and the contents. The results of such an attempt follow. Operations that can be performed on the container include:

(1) Rebuilding or reintegrating the container by reassembling, shoring up, reenforcing, pulling separated parts together;
(2) Building a substitute container by:
  (a) Forming dikes from earth, sandbags, water-inflatable bags;
  (b) Assembling a knock-down container such as a plastic swimming pool or an inflatable boat-like device;
  (c) Using collection or confinement devices (booms, curtains, skimmers, portable dams) when a spill reaches a watercourse;
  (d) Digging, and preferably lining, a pit; and
  (e) Blocking a sewer downstream from point of entry of spilled material;
(3) Repairing the container by:
  (a) Patching with foam or boiler patches;
  (b) Solidifying or freezing the contents at the leak; and
  (c) Adding something to the material in the container that collects at the leak (to make an in-

ternal patch or barrier from rags or wadding added to the liquid in the container) or that congeals at the leak (a commercial anti-freeze reportedly has this property);

(4) Making geometric changes in position: up-end or rotate a leaking liquid or solids container so that the leak is on top or adjust the position of a container filled with a fluid so that the rupture is in contact with plastic earth (clay, wet concrete);

(5) Bagging or otherwise entirely enclosing or encasing the container; and

(6) Attaching a collection bag at the place of the leak.

Operations on the contents of the container can be more diverse. In the particular case of liquids escaped from the container, some operational procedures on the fluid are:

(1) Changing fluid properties (physical changes) by:
  (a) Converting to a solid through gelling or freezing;
  (b) Vaporizing or gasifying (possibly followed by burning); and
  (c) Increasing viscosity to retard spreading;

(2) Reticulating by:
  (a) Soaking up or immobilizing with sponges, adsorbents or absorbents; and
  (b) Adding magnetizable particles and applying a field;

(3) Converting chemically by sequestering, neutralization, precipitation (see chapter on chemical methods);

(4) Diverting to a sump, or pump to a suitable container;

(5) Blocking flow at the entrance to a sewer or watercourse through diking or dam building;

(6) Consuming on site by burning (subject to stringent requirements of no significant air pollution); and

(7) Layering or covering over to retard evaporation.

Obviously, this cataloging of operations is quite incomplete but serves to demonstrate a technique for developing containment methods.

As emphasized before, except under unusual conditions of public health and welfare, concentrated effort should be expended to contain a spilled material. In so far as is practical, containment methods should be designed so that they facilitate cleanup and restoration over a minimum area.

Containment of Spills of Volatile Substances

Generally, spills of gases or of readily volatile substances cannot be easily contained, though some attempts can be made to plug the leak in the container, transfer the material to another container, or render the substance less hazardous through some chemical reaction as the material is released. For spills of flammable materials, precautions must be taken to prevent unintentional ignition by flames, hot objects, or sparks from arcing or discharge of static electricity. Foam covering may be of value in retarding the evaporation of liquids. In some cases and with due consideration for the maintenance of air quality, the material can be consumed by on-site burning, with the probability and consequences of explosion properly evaluated. When burning is chosen, it is usually the lesser of evils but not a really positive, desirable procedure. The case of confinement or inactivation of spills of volatile hazardous materials requires almost individual discussion, as the following remarks indicate. A chlorine spill could possibly be mitigated by reaction with photographer's hypo.

Acidic oxides like those of sulfur and nitrogen could be converted to acids by treatment with water followed by subsequent inactivation with bases. Adsorption on charcoal may be possible in some cases. Spills of gaseous or liquefied hydrogen cyanide, cyanogen and cyanogen chloride might be treated by reaction with basic oxides or with alkaline solutions, preferably chilled. Hydrogen fluoride could be reacted with silica gel; ammonia, with an acidic material; carbon monoxide, with some copper compounds. Carbon monoxide, as well as the more volatile hydrocarbons like methane, ethylene, and even ethylene oxide could be burned and, with proper precautions, would not pose a significant air pollution problem. Chemical methods such as these will be discussed in the following chapter.

Many volatile substances, such as carbon monoxide, chlorine and hydrogen cyanide, are toxic to humans and to other living species. The most prudent procedure for spills of these substances is often the evacuation of people and sealing off of the area. This is a type of inverse containment, where the material is unconfined but people are contained instead. One can regard the donning of gas masks and of protective equipment in a similar vein. There certainly is loss of control of the spilled material when people have to be contained while the hazardous material is free.

Containment of Spills of Solids

In the absence of bad weather, fire, explosion, and contact with other spilled reactive materials, solids will usually be self-containing. The lack of inherent mobility, which is characteristic of solids as contrasted with fluids, improves the prospects for maintaining lateral and vertical control. Further, solids are more frequently packaged and shipped in smaller containers, bags and drums, rather than in tank cars or truck tankers or by pipeline. Indeed, there might be advantages from the standpoint of environmental protection, as well as from the view of preserving human life, in requiring that all highly toxic substances be shipped in small, sturdy containers, even though the materials are used in large quantity.

Containment of a spilled solid in the most concentrated form usually requires only collection by shovelling or vacuuming. Suitable protective equipment for personnel may be required. There are obvious complications in spills of solids such as phosphorus pentoxide, phosphorus pentasulfide, dyes, ammonium nitrate, mercurials, arsenicals, and certain insecticides, rodenticides and pesticides, which are chemically reactive, explosive or toxic.

When a solid is fluidized, containment becomes more difficult. In situations where solids are dispersed by wind or explosion or solubilized by contact with rain or with the water used in flushing debris or in fighting fire, practical containment efforts may be limited to preventing run-off into sewers or streams. When the spill of a solid hazardous material is accompanied by fire, containment can be quite difficult and the consequences grave. Volatilization and combustion are frequently incomplete. Particulates and noxious reaction products are formed and dispersed. The heating or burning of solids such as ammonium nitrate, inorganic chlorates and perchlorates, and highly nitrated aromatics (picric acid, dinitrophenols) may result in explosion.

While it appears at first sight that containment of spills of solids should be relatively simple, the preceding discussion shows that environmental conditions, carelessness, and bad practices can at times make the spill of a solid as harmful to the environment as that of a gas or a liquid, and containment as difficult.

Containment of Spills of Liquids

It is with spills of hazardous liquids, and the class includes slurries and fluidized mixtures in process, that the more difficult but nonetheless often tractable problems arise for containment and localization in concentrated form. As previously mentioned, the general advantages of containment are essentially three-fold: (a) reduction in local environmental damage, (b) reduction in extended damage, such as to watercourse, and (c) greater ease of collection, conversion or detoxification. Approaches to achievement of these purposes will be discussed for general and for specific cases.

Containment on Land: Retention in the most concentrated form of liquid hazardous materials during actual spilling can be approached in several ways. There are two fundamental options: (a) stop flow from the leaking original container, and (b) collect the material in an auxiliary tank or sump, or confine the material in an auxiliary tank or sump, or confine the material with a barrier either on land or in a watercourse, as required. In an almost trivial case, the mere closing of a valve or shut-down of a pump may stop the flow, as may application of a suitable Band-Aid to a leaking hose or application of some sort of putty or packing to a joint or valve.

Another way to stop the leaking of liquid from a container is to plug the leak, possibly with a crude boiler patch. The Rocketdyne Division of North American Rockwell Corporation, under Environmental Protection Agency Contract No. 68-01-0106, is addressing itself to this problem of stopping leaks in containers and is developing field-use, foamed or resinous, quick-setting materials that can be applied over leaks and ruptures in a wide variety of configurations of containers and under a broad range of circumstances, even underwater.

When the leak cannot be stopped at the source, the spilling liquid can be allowed to flow by gravity to a natural sump (creating a flow channel may require hand shovelling or machine trenching). On the other hand, some sort of artificial containing barrier can be created to confine the liquid, such as a dike on land. Prevention of entrance of a hazardous material, especially a soluble one, into a watercourse is extremely important.

When the spilled liquid is held in a natural sump or confined by a dike, the hazardous material can subsequently be pumped into a suitable auxiliary container. Residuals can be physically or chemically treated in place, as is appropriate. Treatment of the material that seeps into the earth poses special problems and emphasizes a common disadvantage of diking, whether achieved by forming a wall with portable or mechanized earth-moving equipment or with water-inflated sand bags. The disadvantage is the absence of an impervious bottom barrier layer to prevent percolation into the earth (vertical spreading). In some fortuitous circumstances, the spill can be confined on a prepared surface (concrete, blacktop). Earthen surfaces and dikes, however, are prone to absorb spilled liquids. In special cases, it may be possible to install plastic sheeting or shielding to minimize this seeping tendency. In this regard, Cornell Aeronautical Laboratory, Inc., under Environmental Protection Agency Contract No. 68-01-0110, is developing a sprayable, impervious sealant for preventing percolation into the earth.

The actual construction of earthen dikes is in itself a weakness in their use, for suitable dike-forming equipment may not be available and close approach to many spills by unprotected personnel can be hazardous. Removal of the material confined by an earthen dike generally requires that a sizeable quantity of earth must be removed as well. Disposal and/or reclamation presents a serious problem.

Under Environmental Protection Agency Contract No. 68-01-0100, MSA Research Corporation is developing portable equipment for construction of highly expanded, foamed plastic dikes. The rationale for using foamed dikes is chiefly ease and rapidity of setup. A truck tanker operator equipped with a back-pack type of unit could build such a containment wall, which has the advantage of confining the spilling liquid in its most concentrated form. Preliminary investigations by MSA indicate that concrete-based material can also be foamed in this approach to emergency containment. Adhesion of the foamed dikes to wet or dry earthen or prepared surfaces, reactivity of the foam with the spilled material, and operator protection are all important factors under consideration. MSA is also investigating the use of foamed plastics as a cover to blanket and inactivate pooled spills. Many of these containment methods are discussed in greater detail in subsequent sections of the book.

Containment in a Watercourse: When a spill of a hazardous material occurs in such a way that the material (liquid or fluidized solid) enters a watercourse, the problem of containment in the most concentrated form is exceedingly difficult.

For insoluble floating hazardous materials, the techniques applicable to oil spills may be suitable. These methods include the use of booms, flow guides, curtains, absorbers, and skimmers. However, when the spilled floating material is flammable, like benzene or some cyclic hydrocarbons, the ever-present dangers of explosion and fire must be recognized. These hazards also exist when a flammable material is flushed into a sewer, which can be looked upon as an enclosed watercourse. Even when a flammable liquid can be kept out of a watercourse and confined on land, laterally to halt spreading and vertically to prevent evaporation, dangers of accidental ignition and explosion remain. Knowledge of flash points and explosion limits are of value, but evacuation is the most practical countermeasure.

For materials that sink in water, certain types of underwater dikes may be useful. Long-term holdup of creosote and of chlorinated hydrocarbons on the bottom of streams may produce serious ecological damage as may vacuuming or dredging. When a water-soluble, hazardous material enters a watercourse, containment can be difficult. Cornell Aeronautical Laboratory, Inc., Rex Chainbelt, Inc., and Battelle Memorial Institute are studying alternate approaches such as in-place reaction and adsorption under EPA contracts.

Ocean Systems, Inc., under Environmental Protection Agency Contract No. 68-01-0103, is developing a containment device that can be installed in a watercourse. The unit, which may be regarded as a sealed boom, consists of a plastic curtain with a floatable collar and has provision for attachment to the bottom of a stream or body of water. Floatable material is confined above the watercourse surface, sinking material is immobilized at the bottom, and the solution resulting from dissolution of a soluble hazardous material is confined within the sealed boom. Appropriate inactivation or collection techniques can then be applied. Time and ease of deployment of the curtain are critical. Behavior in swiftly moving or turbulent streams or under impact with floating debris may limit applicability.

The problems of collection, reconcentration, and detoxification of a spilled hazardous material that has entered a watercourse especially point out the need for initial containment of the spill in the most concentrated form. The ounce of prevention is costly, but the pound of cure is much more so, particularly when damage to the water ecosystem is considered.

Summary

In this discussion, emphasis has been placed on the advantages of containment of hazardous material spills in their most concentrated form. The idea of containment was correlated with the concept of control, and that of a spill, with the loss of control, both lateral and vertical.

The advantages of prompt containment of spills are: (1) minimization of damage to the total environment, (2) facilitation of on-site cleanup operations, and (3) prevention of dispersion into watercourses through flow into sewers, run-off into streams, and percolation into the earth with subsequent seepage into subsurface water supplies. Generally speaking, the greater the spreading of a spilled material, the more extensive the environmental damage. That cleanup and restoration are facilitated by containment of a spill in concentrated form becomes quite clear when one reviews the significant problems of collecting or inactivating a material that has been dispersed in a watercourse.

EXISTING DEVICES FOR FLOATING HAZARDOUS CHEMICALS

> The information in this section is based on *Agents, Methods and Devices for Amelioration of Discharges of Hazardous Chemicals on Water*, CG-D-38-76 prepared by W.H. Bauer, D.N. Borton and J.J. Bulloff of Rensselaer Polytechnic Institute for the U.S. Coast Guard.

The objective of this study was to determine what existing devices may be most advantageously employed in the amelioration of spills of eight floating representative chemicals. The following rationale was used to identify the most advantageous devices:

> The devices must be available either in an off-the-shelf or easily modified form.
>
> The devices must have demonstrated capacity for usefulness in containing, gathering and removal of oil spills on water.

The devices must be resistant to attack by the hazardous floating chemicals or must be easily modified to be resistant.

Existing devices which may be most advantageously employed to ameliorate spill situations involving eight representative floating chemicals, as classified in Table 1.7, p 40, are listed in Table 2.2. Devices for containment use were divided into categories according to whether the devices were off-the-shelf and suitable, or were off-the-shelf and easily modified for use. This distinction was not applied to the devices identified for removal use. Devices were classified as easily modifiable if components, which would be deteriorated by contact with the representative hazardous chemicals, could readily be replaced by off-the-shelf solvent-resistant components.

Table 2.2: Existing Cleanup Devices

For Containment Use, Off-the-Shelf
- Vikoma Seapack, The British Petroleum Co., Ltd.
- Offshore Boom, Clean Water, Inc.
- Boa-Boom I, and Boa-Boom II, Environetics, Inc.
- Sea Sentry, The Goodyear Tire and Rubber Co.
- High Seas Barrier, Offshore Devices, Inc.
- Pollution Boom, A.G. Peterson & Sons, Inc.
- Minisealdboom, Uniroyal, Inc.
- Standard Sealdboom, Uniroyal, Inc.
- Boom, Types E, T-8 and T-16, William Warne & Co., Inc.
- Standard Oil Boom, Gamlen Chemical Co.
- Heavy Duty Boom, Gamlen Chemical Co.

For Containment Use, Easily Modified
- Acme O.K. Corral, Acme Products Co.
- Optimax, American Marine, Inc.
- Simplex, American Marine, Inc.
- Inshore Booms, Bennett Pollution Controls Ltd.
- Offshore Booms, Bennett Pollution Controls Ltd.
- CWI Harbour Boom, Clean Water, Inc.
- Coastal Oil Boom, Coastal Services, Inc.
- Anti-Pollution Boom, Gamlen Chemical Co.
- Standard Boom, Hurum Enterprises, Inc.
- Type I and Type II Oil Barrier, The Marsan Corporation
- Calm Waters Sea Curtain, Kepner Plastics Fabricators, Inc.
- Fast & Choppy Water Sea Curtain, Kepner Plastics Fabricators, Inc.
- MP Boom, Metropolitan Petroleum Petrochemicals Co., Inc.
- Harbour Containment Barrier, Ocean Systems, Inc.
- Megator Mini-Boom, Megator Corporation
- Aluminum Boom, Kepkan, Inc.
- Aqua Fence, Pacific Pollution Control, Inc.
- Mark V and Mark VI Harbor & River Booms, Slickbar, Inc.
- Sea Boom 3PF, Submarine Engineering Assoc., Inc.

For Removal Use, Off-the-Shelf
- Inflexible Plane Skimmers
  - Vikoma Seaskimmer, B.P. North America, Inc.
  - Clean Sweep, Lockheed Missiles and Space Co., Inc.
- Moving Belt Skimmers
  - Belt Oil Skimmer, Aerodyne Development Co.
  - Inshore-Offshore Skimmer, Bennett Pollution Controls Ltd.
  - Rex Belt Skimmer, Envirex Inc.

(continued)

Table 2.2: (continued)

Dip Skimmers, JBF Scientific Corp.
Marco Skimmers, Class I, II, III, Marco Pollution Control Corp.
Oil Mop, Oil Mop, Inc.
Reclam-Ator Skimmer, Peabody Welles, Inc.
Slicklicker, R.B.H. Cybernetics (1970) Ltd.

Suction Head Skimmers
Acme Models FSV5 and FSV40, Acme Products Co.
Megator Delta Mark II, Megator Corp.
Beta Mark II, Megator Corp.
Puddle Mop, Megator Corp.
Oilhawg, Parker Systems, Inc.
Slickskim, Slickbar, Inc.

Weir Skimmers
Acme Models SK-39T and SK-51T, Acme Products Co.
Slurp Skimmer, Coastal Services, Inc.
Crisafulli Aqua-Sweeper and Mini-Sweeper, Crisafulli Pump Co.
Oela III, Industrial Municipal Engineering
Sea Broom, Sunshine Chemical Corp.
ORS, Models 125, 1000 and 2000, Ocean Systems, Inc.

Floating Sorbent Moving Belt Skimmer
Sop System, Ocean Design Engineering Corp.

Source: CG-D-38-76

Representative Floating Hazardous Chemicals

The eight floating representative chemicals with which this work was concerned were: (II A) pentadecanol, (II B) ethyl ether, (II C) kerosene, (II D) n-amyl alcohol, (II E) naphtha:solvent, (II F) ethyl acetate, (II G) hexane and (II H) n-butyraldehyde. Of these chemicals, three have the general properties of the saturated hydrocarbon oils. These are kerosene, naphtha:solvent and hexane. They may be considered to be similar in properties to gasoline, differing primarily in vapor pressure at a reference temperature. The n-amyl alcohol has high solvency power. This is also true of ethyl acetate, ethyl ether and n-butyraldehyde, although the materials which dissolve readily in these liquids are not all the same.

As a group, however, the last four listed chemicals are among those liquids which exhibit high solvency powers as compared to the three saturated hydrocarbon chemicals, kerosene, naphtha:solvent and hexane. The saturated hydrocarbons are practically insoluble in water. The ether is very slightly soluble, and the n-amyl alcohol, ethyl acetate and butyraldehyde all have small but significant solubilities in water. All are combustible and although ease of flammability varies widely, all must be considered serious fire hazards, except the floating solid, pentadecanol. Although pentadecanol is combustible, the solid alcohol, having very low vapor pressure, is not readily ignited.

The chemical and physical properties of the eight representative floating hazardous chemicals are such that all devices such as booms, barriers, skimmers, hoses and auxiliary equipment suitable for use in controlling oil spills involving inflammable and combustible oils are immediately applicable to the amelioration of spills of hazardous chemicals with the following major exception. Some of the representative hazardous chemicals are representative of classes of chemicals which have the characteristic of exhibiting high solvency for some of the materials in certain plastic, polymer-based materials which may be used in construction of the devices for oil control. The materials in these devices have been chosen for resistance to solution in oils, but have not been chosen for resistance to the special solvency exhibited by some of the representative hazardous chemicals. These hazardous chemicals, in general, do not affect the metals such as galvanized iron, stainless steel, brass and aluminum, in the amelioration devices. However, all portions of the devices and systems which contain certain polymeric materials may be severely affected by the solvent action of the hazardous chemicals.

Some polymeric materials, such as Buna-N, Neoprene and other elastomers, are highly resistant to solvent action by certain of the representative hazardous chemicals, as well as by hydrocarbon solvents of the type of gasoline, kerosene and naphtha:solvent. Other polymeric materials, such as those fabricated from polyvinyl chloride, may contain, in addition to the polymer, processing aids such as plasticizers, dibutyl phthalate and di-octyl phthalate, for instance. The plasticizers are generally chosen because of very low vapor pressure and consequent minimum evaporation rate. They are not dissolved by the action of water, saltwater and petroleum hydrocarbons to a significant extent.

In contrast, the high solvency representative hazardous floating chemicals generally will remove plasticizing agents from polymeric materials on continued contact. As a result of this action, the tensile and elastic properties of the material deteriorate, leading to eventual loss of utility. On short contact damage resulting from these high solvency chemicals would not be readily noticeable, but on continued contact the materials would be expected to fail. In the itemized lists in Table 2.2, those devices and systems which did not use construction materials expected to contain plasticizer were listed as off-the-shelf and immediately useable, while those devices and systems generally known to contain plasticizing materials were listed as easily modified. These latter devices and systems were considered to be easily modified because no change in design would be required to make them solvent resistant. All that would be required would be to use readily available solvent-resistant materials which did not contain the easily leachable plasticizers.

In considering feasibility, in detail, of using the devices itemized in Table 2.2 for containment and removal of floating hazardous chemicals, characteristics of the particular representative floating chemicals must be taken into account, and individual descriptions follow.

Pentadecanol: Pentadecanol usually is in the form of white, waxy, easily deformed crystals, although it may be handled in solution in solvents such as ethyl alcohol. The solid form will be assumed to be the form in which spills of the hazardous chemical are likely to occur. While in many respects pentadecanol is similar to such long chain hydrocarbons as paraffin wax, it is different in its capacity to spread rapidly on contact with water across large surfaces in a layer one molecule deep.

The formation of the surface layer occurs because the hydroxyl group is very strongly attracted to water while the long paraffinic hydrocarbon chain is very strongly repelled by water, resulting in the spreading of an oriented film. The film formed on the surface of quiescent water is continuous and rigid but very weak. As long as there is crystalline pentadecanol floating on the surface, molecules will spread until stopped by a solid, nonaqueous boundary. The more rapidly a barrier is placed in the way of the expanding film of pentadecanol on a water surface resulting from spilled crystals, the greater is the quantity of the pentadecanol that will be present as floating crystals which may be recovered by numerous devices. The surface film of pentadecanol is more difficult to recover. It is only one molecule thick and thus transparent if viewed from above.

Because of its high molecular weight, the vapor pressure of pentadecanol is very low at the normal temperatures expected in water bodies. A very slight odor is noticeable. Hazard of fire is low although the material is combustible. The monomolecular surface film may be broken up and dispersed on a flowing body of water, but the chemical does not evaporate into the air at an appreciable rate. In a confined area, the material may, in time, leave the water surface and adhere to adjoining soil or containing structures. The chemical is not generally reactive and is very stable with respect to air oxidation.

n-Amyl Alcohol: There is an extremely small solubility of n-amyl alcohol in water. The molecular weight of n-amyl alcohol is sufficiently low that the liquid is flammable because of the high vapor pressure at ambient temperature. In very low concentrations in air, the odor resembles that of whiskey, but when the alcohol is present in higher concentrations the odor is irritating. n-Amyl alcohol is a very good solvent for many polymeric substances.

Ethyl Ether: Ethyl ether has a molecular structure resulting from the linkage of two ethyl groups by an oxygen atom. Spreading on water is due primarily to gravitational forces, but may be rapid because of its low viscosity. Ether has a very small solubility in water. The vapor pressure is very high at ambient temperatures and ether may be expected to vaporize rapidly. Because of its high vapor pressure and combustibility, it is very flammable. The chief hazard in handling this chemical, aside from fire, is that if inhaled it will cause loss of consciousness. Ethyl ether is a very good solvent for many polymeric substances.

Ethyl Acetate: Ethyl acetate has a viscosity that is higher than that of ethyl ether and a lower vapor pressure. Evaporation, however, is rapid from a surface floating on water, and flammability is high. Ethyl acetate has an appreciable solubility in water. The dissolved material decomposes in water very slowly, forming acetic acid and ethyl alcohol. At the concentrations formed, these materials are not hazardous. Like all esters, when the vapor is diluted, it has a strong perfume odor, this one characteristic of over-ripe bananas. Ethyl acetate is a very good solvent.

n-Butyraldehyde: Like ethyl ether and ethyl acetate, butyraldehyde is a modified hydrocarbon containing oxygen. It differs in that it has a very irritating and offensive odor and is more chemically reactive. Both liquid and vapor are very irritating to eyes, nose and throat, and harmful if inhaled or swallowed in appreciable amounts. The vapor pressure is high and thus flammability is high. While not strongly reactive with water, butyraldehyde has a limited solubility and is a reactive chemical compared with ethyl acetate and ethyl ether. It has a low viscosity, and under gravitational head spreads rapidly on water. It is an excellent solvent for a great many materials.

Naphtha:Solvent, Kerosene and Hexane: The members of the group of chemicals, naphtha:solvent, kerosene and hexane, are very similar chemically. All are composed primarily of straight or branched chain hydrocarbons containing only hydrogen and carbon. Naphtha:solvent is a mixture of hydrocarbons, mostly aliphatic, and may be considered to be intermediate to gasoline and kerosene in most of its properties. Naphtha:solvent may be contained by booms and recovered by devices used for oils such as crude oil. It is combustible, but because of its comparatively high boiling point it is not flammable.

Hexane is a single hydrocarbon similar to the hydrocarbons present in the two mixtures, gasoline and kerosene. It may be regarded as similar to gasoline in which it is present. The vapor pressure is so high that the flammability and fire hazard is great when this chemical is spilled on water. The viscosity of the chemical is low and spreading occurs rapidly.

The hydrocarbons in kerosene have a sufficiently high molecular weight that the vapor pressure is comparatively low over the liquid when spilled on water at ambient temperatures. While not flammable, the liquid is combustible and a fire hazard exists in handling it, though the hazard is much less than is the case for hexane and naphtha:solvent. The chemical is irritating to skin and eyes and harmful to swallow as are hexane and naphtha:solvent. It has a somewhat higher viscosity than hexane, but the viscosity is sufficiently low that gravitational spreading will rapidly occur when large amounts of the chemical are spilled on water. All three of the hydrocarbon chemicals are similar in solvency to gasoline, and all are only fair solvents. All devices, suitable for oils, are expected to be somewhat more resistant to contact with naphtha:solvent, hexane and kerosene than to contact with n-amyl alcohol, ethyl ether and butyraldehyde.

The particular characteristics of the eight representative hazardous floating chemicals, which have been discussed above, are especially pertinent to evaluation of the feasibility of amelioration by existing oil control devices of spills of the representative hazardous floating chemicals.

Evaluation Criteria

The overall feasibility for use in ameliorating a spill situation by means of a device and its auxiliary equipment was based upon favorable judgement as to the proven effectiveness shown by the device or system in dealing with spills of hydrocarbon oils, the recovery rate with which a spill of specified size may be contained or removed, the stability of the construction materials used in the device when in prolonged contact with the representative hazardous floating chemicals and the mobility of the device and related system and speed of deployment. It was assumed, in general, that fire fighting equipment, fire control apparatus, protective clothing and gas masks would be supplied and used as necessary, according to good safety practice. That form of a device or system in which it is provided with spark-proof motor, pump and auxiliary equipment was the only form considered for use in control of the hazardous liquid chemicals, because of the fire hazards they generally present.

In the following, feasibility for use in amelioration of spills of representative hazardous floating chemicals on water is discussed for each of the containment devices and removal devices itemized in Table 2.2.

Evaluation of Off-the-Shelf Containment Devices

Vikoma Seapack, The British Petroleum Co., Ltd.: The Vikoma Seapack contains 1,600 ft of flexible, inflatable double chamber boom with automatic inflation after laying of the boom. When deployed, the boom consists of a tube immersed in water, filled with water, of 17-inch draft, and a tube fastened above filled with air, of 27-inch freeboard. Fastened above the air tube is a 3-inch flotation tube. The boom fabric is a laminated Neoprene-nylon-Neoprene sandwich which is vulcanized. The boom material was tested in all the representative hazardous floating liquids, and resistance to solvency was found to be good, with no visible deterioration in any solvent. Equipment used for transportation to the scene of a spill and the mechanical systems used to provide inflation must be explosion-proof to minimize the hazard of fire in deployment. For spills of appropriate large size, the Vikoma Seapack is applicable for the containment of all the representative hazardous floating chemicals.

Offshore Boom, Clean Water, Inc.: The Offshore Boom consists of a 30-inch skirt supported by a series of inflatable floats, 10 inches long and 12½ inches in diameter. The boom is manufactured in 55-ft sections, hinge-joined for interconnection. A galvanized chain is fastened in the base of the skirt to provide ballast. The galvanized iron chain might furnish a spark fire hazard on deployment, but should be safe when it is in the water. The boom is made from a Chemigum rubber-reinforced nylon fabric. The stability, when the boom material was tested in the representative hazardous floating liquids, was good except for ethyl acetate in which it exhibited a weight loss of 15% in the test period, discolored the solvent and showed curling, swelling and softening.

In ethyl ether, the solvent was discolored and the fabric stiffened when dry. In butyraldehyde, there was solvent discoloration, marked curling, swelling and stiffening. In naphtha, there was solvent discoloration and the fabric stiffened when dry. Hexane and kerosene showed negligible effect. Use of the Offshore Boom is feasible for the amelioration of spills of the representative hazardous floating chemicals, but stability of the boom fabric is borderline in all but hexane and kerosene. It is indicated that, while the boom is suitable for immediate use with the representative hazardous liquids, it would be expected to have a limited life.

Boa Boom I and Boa Boom II, Environetics, Inc.: Boa Boom I consists of a flexible-type skirt, floated by means of air chambers. Boa Boom II is similar except that flotation is accomplished by means of a polyfoam-filled bladder. Both booms are comparatively light and may be rapidly deployed. Boom skirts are made from a polyurethane fabric which is highly resistant to solvent action of all of the floating hazardous liquids. Booms of prior construction from Environetics which contain polyvinyl chloride components are not satisfactory for use with the hazardous liquids, but may be easily modified by substitution of polyurethane or other solvent-resistant skirt fabric.

Sea Sentry TM, The Goodyear Tire and Rubber Company: The Sea Sentry TM boom is a heavy-duty boom suitable in offshore applications for containment of the eight chemicals under consideration without modification. The curtain material and inflatable flotation elements are made of solvent-resistant rubber-impregnated nylon fabric. Joints are made by vulcanization. The boom is heavy, weighing 550 lb/55-ft section, and it requires 7 men for deployment and inflation at a rate of 1,000 ft/hr. Feasibility of containment is high for the representative hazardous floating chemicals.

High Seas Barrier, Offshore Devices, Inc.: The High Seas Barrier has a 4-ft high flexible curtain with rigid vertical struts on 6 ft centers. When deployed, it has a 21-inch freeboard and a 27-inch draft. The barriers are furnished in 306-ft lengths normally, but any desired length is available. At intervals, inflated floats provide stability for the deployed barrier. An external tension line is provided. The curtain is made of two-ply nylon fabric with an elastomer coating. The boom float uses a vulcanized elastomer. Tension lines are made of propylene and nylon or polyester. The boom skirt, the tension line and the retainer line materials showed no deterioration in solvent stability tests. However, the boom float material showed stiffening when dry after testing in ethyl ether, ethyl acetate, naphtha:solvent and hexane. In spite of this, it is believed that the boom as a whole is applicable without modification for the amelioration of spills of the eight representative hazardous floating chemicals.

Pollution Boom, A.G. Peterson & Sons, Inc.: The Pollution Boom consists of a 36-inch skirt; draft, 24 inches; and freeboard, 12 inches. Flotation units 18 inches long and 3 inches wide are fastened at each side of the barrier. The floating units are made of plastic foam. The boom is furnished in 50-ft standard lengths with lead ballast and stainless steel cable. The stainless steel cable would furnish some fire hazard in deployment, but should be safe when wet. The model of this boom with Neoprene boom skirt is suitable for use with all the representative hazardous floating chemicals. The polyvinyl chloride coated nylon boom skirt material, alternatively furnished by A.G. Peterson & Sons, Inc., should not be employed.

Mini Sealdboom and Standard Sealdboom, Uniroyal, Inc.: Sealdbooms consist of a flexible ballasted skirt with sealed foam flotation; hardened spring steel strips are used for stabilization. All components are fully enclosed with vulcanized Paracril-OZO coated fabric. Ballast consists of lead weights. Sections are bolted together or fastened together with quick-connect fittings. These connections might present a spark possibility during deployment, creating a fire hazard, but should give no difficulty once the boom is in the water. The Sealdbooms are furnished in a useful variety of heights, 18 inches, 36 inches and 72 inches, with drafts of 12, 24 and 48 inches, respectively. The Sealdboom is supplied in a basic section length of 40 ft. Stability tests were made on the boom skirt Paracril-OZO coated fabric in all the floating hazardous liquids, and solvent resistance was found to be good in each.

Boom, Types E, T-8 and T-16, William Warne & Co., Inc.: The Type E boom consists of a flexible ballasted skirt and an air-inflated float chamber with 15-inch draft and 6-inch freeboard. The skirt material is made of Terylene fabric impregnated with Chloroprene rubber. The T-8 and T-16 types have similar flexible ballasted skirts, but flotation is accomplished by plastic foam-filled buoyancy chambers. The boom skirt material is like that of Type E. Chain ballast is used for all types of the Warne booms, and metal connectors are used for the standard 50-ft section lengths. The chain and the link hinges may furnish a spark hazard on deployment, but should be safe when in water. The boom skirt material is expected to be highly solvent-resistant. The various types of Warne booms are applicable without modification to amelioration of spills of all the representative hazardous floating chemicals.

Standard Oil Boom and Heavy Duty Boom, Gamlen Chemical Co.: The Standard Oil Boom is a fire-resistant boom with a flotation element made of rigid aluminum-magnesium alloy units. The skirt is made of 3-ft lengths of rigid aluminum-magnesium panels connected by reinforced asbestos fabric for flexibility. Tension lines are stainless steel or galvanized steel, covered with a silicon resin. The boom is supplied in unit sectional lengths of 16.4 ft with a draft of 12 inches and freeboard of 8 inches. The Heavy Duty Boom is of similar construction, but with 14-inch draft and 10-inch freeboard. The Standard Oil Boom and the Heavy Duty Boom, since they contain no material that can be damaged by the hazardous chemicals, are suitable without modification for use in the amelioration of all the representative hazardous floating chemicals.

Evaluation of Easily Modified Containment Devices

The devices described below are all considered to be applicable to the amelioration of spills of the representative hazardous floating chemicals if solvent-resistant materials are substituted as indicated. No change in design would be necessary.

Acme O.K. Corral, Acme Products Co.: The O.K. Corral Boom consists of a skirt, weighted with ballast chain sealed into the trailing edge and fitted with floats 9 ft long and from 4 to 12 inches in diameter, made from Dow Ethafoam. Booms are furnished in units of a minimum 50-ft standard length, with 6-to 24-inch draft and 4-to 12-inch freeboard. The floats are resistant to organic solvents, but the skirt material has poor stability in all the representative hazardous floating chemicals. In each case, extreme curling and marked stiffening was exhibited in the test solvents and when dried.

Optimax and Simplex, American Marine, Inc.: The Optimax boom consists of a skirt, weighted with chain ballast and floated by cylinders of Dow Ethafoam in 6-inch widths and 7-ft lengths. The draft is 12 to 18 inches and freeboard, 7 inches. The boom is furnished in 100-ft lengths with 7-ft fold segments. Tension lines are steel cables, sheathed at the top. Because of the externally mounted 5/16-inch galvanized iron tension chains, a spark hazard would exist when this boom is deployed, but once the boom is wet it should be safe for hazardous liquids with respect to fire hazard.

The Simplex Boom is similar in size and construction, but uses ¼-inch galvanized ballast chain. The boom float material is resistant to organic solvents, but the fin and hem fabric of the skirt was found not suitable for use with any of the representative floating hazardous liquids. From 22 to 27% weight loss results from leaching by the representative hazardous chemicals. In n-amyl alcohol, ethyl ether, naphtha:solvent, hexane and kerosene, the polyvinyl chloride coated nylon boom skirt material samples exhibited marked curling and stiffening. In ethyl acetate and butyraldehyde, the skirt fabric samples delaminated, exhibited extreme weakening, curling, swelling and softening, and showed extreme stiffening and curling when dried.

Inshore and Offshore Booms, Bennett Pollution Controls Ltd.: The Bennett Pollution Controls booms have flexible skirts, Dow Ethafoam round floats of 4'6" lengths and 8" diameter or rectangular 2" x 6", made with a draft of 24" and freeboard of 12". They are supplied in 50-ft standard sections, which are interconnecting. Ballast consists of lead weights on a steel cable. The tension line is steel cable. Spark hazard would exist in deployment of these booms, but when the booms are in the water the spark hazard should be eliminated. The curtain material supplied is polyvinyl chloride coated nylon fabric, which is not satisfactory for use with the hazardous floating liquids because the material is unstable in organic solvents.

CWI Harbour Boom, Clean Water, Inc.: The CWI Harbour Boom consists of a flotation section and a skirt, weighted with ¼-inch galvanized chain ballast, which serves as a tension member along the bottom of the skirt. Float cylinders of polyethylene are housed in a continuous tube, with a ½-inch polypropylene line threaded through the float cylinders serving as a tension member. Fittings are marine brass or hot-dipped galvanized iron. Seams are stitched with solvent-resistant polyester thread. The boom is furnished in 50-ft sections with 8-inch freeboard and 24-inch draft.

The design of the Clean Water Harbour Boom is excellent for reduction of fire hazard because a continuous covering encases all elements of the boom. The polypropylene tension lines are highly resistant to organic solvents. Experimental tests were made on the resistance to organic solvents of the polyethylene float material and of the polyvinyl chloride coated nylon fabric boom skirt. No deterioration of the boom float was detected in the representative hazardous floating liquids. Weight loss of from 17 to 21% and general stiffening and curling of the skirt, when dry, were exhibited after stability tests in the representative hazardous floating liquids.

Coastal Oil Boom, Coastal Services, Inc.: The Coastal Oil Boom is a flexible skirt barrier with external flotation. It may be furnished with 12- or 24-inch draft and corresponding 6- or 12-inch freeboard in 100-ft standard lengths. Flotation units are cylindrical, 9 inches long and 4 inches in diameter, made of polyethylene. Ballast is made of galvanized iron chain. The exposed chain would involve a fire hazard from possible sparks in deployment, but the boom should be safe when wet with water. Experimental stability tests showed that the polyethylene boom float material had good stability in all the representative hazardous chemicals. The polyvinyl chloride coated nylon fabric of the boom skirt was extremely unstable in the chemicals, showing from 5 to 31% weight loss in leaching. In all cases, extreme curling was shown in the solvent, and in the case of the tests in ethyl acetate and butyraldehyde, delamination occurred.

Anti-Pollution Boom, Gamlen Chemical Co.: The Anti-Pollution Boom is a semirigid boom made with a float consisting of polystyrene foam beads enclosed in a polyvinyl chloride fabric chamber and a flexible ballasted skirt, made of polyvinyl chloride coated nylon fabric. The draft is 12 inches and the freeboard is 6 inches. The boom is furnished in 100-ft standard lengths. The ballast is made of 2" x 2" rigid polyvinyl chloride or stainless steel cable. The stainless steel cable would furnish some spark hazard on deployment, but would be safe when wet. The cover of the flotation element and the skirt material are both made of polyvinyl chloride coated nylon fabric, a type which has been shown to be very unstable on contact with the high solvency representative hazardous liquids.

Standard Boom, Hurum Enterprises, Inc.: The Hurum Standard Boom has a flexible skirt, floated by integral polyvinyl chloride tubing filled with Dow Ethafoam. The 36-inch boom has a draft of 24 inches and a freeboard of 12 inches. It is supplied in 50-ft lengths. Booms may be furnished in 14- to 72-inch sizes.

The skirt material is polyvinyl chloride impregnated nylon fabric. The ballast is made of lead, crimped to the bottom of galvanized steel cable. This cable and the tension line steel cables might cause a spark hazard in deployment, dangerous to flammable liquids, but when deployed the boom should have little fire hazard. The Dow Ethafoam is resistant to organic solvents, but the tubing and skirt fabric would not be stable with respect to the solvent action of the representative floating liquid hazardous chemicals.

Type I and Type II Oil Barrier, The Marsan Corporation: The Marsan booms, Types I and II, are applicable to inshore and offshore use, respectively, but are otherwise similar. The Type I Oil Barrier has a draft of 12 inches and a freeboard of 8 inches, and the Type II Oil Barrier has a draft of 24 inches and a freeboard of 16 inches. Both types are furnished in 50-ft standard lengths. Each of the booms has a flexible skirt with enclosed foam flotation and ballast. Enclosed galvanized ballast chain should offer little spark hazard. A tension line, made of Dacron and nylon-reinforced webbing, is compatible with the representative hazardous floating liquids, as is the urethane foam flotation element. The polyvinyl chloride coated nylon fabric, however, would be unstable when wet with the representative hazardous floating liquids. This material has been shown to be drastically damaged in contact with high solvency chemicals.

Calm Waters Sea Curtain and Fast & Choppy Water Sea Curtain, Kepner Plastics Fabricators, Inc.: The Kepner Plastics Sea Curtains are made with skirts fitted with closed cell plastic foam flotation elements. The booms are normally furnished in 100-ft lengths and vary in strength according to whether a calm or disturbed water environment is involved. The various booms are furnished in total heights of from 13 to 34 inches, and vary in strength according to the type of water environment which is expected. Closed cell plastic foam is used to provide flotation. Draft of the 13-inch boom is 9 inches and the freeboard is 4 inches. The largest boom has a draft of 21 inches and the freeboard is 13 inches. Stability tests were performed on the polyvinyl chloride coated nylon fabric skirt material normally furnished; stability was found to be poor in all the representative hazardous floating chemicals to such an extent that the booms would have a very short life when used with these chemicals.

The Kepner booms are also furnished with polyethylene construction materials. The polyethylene material would be resistant to organic solvents. Without change in basic design, but with specification of polyethylene or other organic-solvent-resistant fabric, the Kepner Sea Curtain booms are suitable for use in amelioration of spills of the representative hazardous floating chemicals.

MP Boom, Metropolitan Petroleum Petrochemicals Co., Inc.: The MP Boom flotation element is a continuous fabric tube filled with foam beads. This supports a 12-inch flexible skirt, and a ballast rod of high-density polyvinyl chloride with stainless steel cable also serving as tension line. The boom is furnished in 100-ft standard lengths. All the materials of this boom are resistant to attack by organic solvents, except the polyvinyl chloride coated nylon fabric, which is used throughout. This fabric is highly unstable toward deterioration in the high solvency representative hazardous floating liquid chemicals.

Harbour Containment Barrier, Ocean Systems, Inc.: The Harbour Containment Barrier has a flexible skirt supported with outrigger floats (draft, 24"; freeboard, 24"). The barrier is supplied in 50-ft section lengths. Flotation lenghts of 2½ ft on each side of the boom skirt are 8 inches in diameter and of cylindrical shape. The floats have Dow Ethafoam elements. No ballast is used. The tension line is made of ½-inch steel rope or of 1⅛-inch Dacron line. The Dacron line would be preferable because of a possible spark hazard when deploying the barrier with the steel rope. The polyvinyl chloride coated nylon mesh fabric used for the boom skirt was tested for stability in the hazardous floating liquid chemicals; while this sample of polyvinyl chloride coated fabric was distinctly more resistant to solvent action in most of the representative hazardous floating liquids than most polyvinyl chloride fabrics tested, its stability in ethyl ether, naphtha: solvent, ethyl acetate, butyraldehyde and benzene was poor.

Megator Mini-Boom, Megator Corporation: The Megator Mini-Boom is a small boom, supplied in 16.5-ft lengths with 7-inch draft and 2-inch freeboard. The boom consists of an enclosed polyethylene foam float, a skirt and enclosed steel weights as ballast. The tension line is made of nylon rope. The enclosed metal ballast should not provide a spark hazard. All parts of this boom are resistant to the solvent action of the representative hazardous floating liquid chemicals, except the fabric used in the skirt and in enclosing the float. The polyvinyl chloride coated nylon reinforced material has been shown to deteriorate drastically in the representative hazardous floating liquids.

Aluminum Boom, Kepkan, Inc.: The Kepkan Aluminum Boom is constructed of corrugated aluminum sheet curtain material, and flotation is accomplished by acrylonitrile-rubber base, closed-cell foam floats attached to the boom skirt. Two models are supplied with 10- and 21-inch draft and with 5- and 7-inch freeboard, respectively. The boom is supplied in 100-ft section lengths. Ballast is round aluminum bar stock. Stability tests were performed on the float material; in most of the representative hazardous materials, a small weight loss occurred after storage in the chemical, indicating some solubility. More serious were the weight gains shown in some solvents. In butyraldehyde, for instance, a weight gain of 141% was found in the test period and, in n-amyl alcohol, a 21% gain was noted. It is believed that such weight gains indicate the solvent is penetrating through the foam cells with consequent large changes in density likely to affect the flotation ability. Solvent-resistant float materials are available and, with substitution of such for the flotation material, the Kepkan Aluminum Boom is useful for amelioration of spills of the representative hazardous floating chemicals.

Aqua Fence, Pacific Pollution Control, Inc.: The Aqua Fence booms are made in a number of models, and vary in draft from 8 to 24 inches and in freeboard from 8 to 24 inches. Booms are supplied in seamless lengths as desired from 25 to 1,000 ft. Flotation is accomplished by means of outriggers, using closed-cell polyurethane foam-filled floats. Ballast and tension lines are not used.

The Aqua Fence foams are of rugged construction. The skirt is 3/16-inch thick, made of a synthetic resin impregnated woven fabric base. The outrigger flotation elements are closed-cell polyurethane foam, covered with high-impact synthetic resins. Skirt and float materials were not available for testing during the period of this work, but it is believed because of the heavy construction that solvent penetration and reduction of stability should be a long-term process. The booms could be applied to containment of the representative hazardous floating liquids, off-the-shelf, with the reservation that the long-term stability toward solvent action is not known. Should the long-term stability prove poor, the boom could be easily modified by substitution of organic-solvent-resistant resins.

Mark V and Mark VI Harbor & River Booms, Slickbar, Inc.: The Mark V and Mark VI Harbor and River Booms are essentially of the same design, but with different sizes. Drafts of 6, 8, 10, 12 and 24 inches, and freeboards of 4, 6½ and 12 inches are furnished in continuous lengths from 50 to 500 ft. The booms are constructed with a flexible ballasted skirt with polyethylene foam flotation elements covered with continuous polyethylene skin. The curtain material is polyester woven fabric, impregnated with polyvinyl chloride. Ballast is hardened lead. Tension lines are from ¼- to ½-inch steel cables. The cables might give rise to sparks in deployment, but should be free from fire hazard when in the water. The polyethylene boom float material was found to be stable in all the representative hazardous floating liquid chemicals, but the polyvinyl chloride coated nylon curtain material showed poor stability in all the representative floating liquids, generally losing on the order of 20% of the material through solution in the test period, and marked stiffening and curling generally occurred when dry after the test. While the boom material would be suitable for short periods in contact with the test liquids, it would not be suitable for continued use.

Sea Boom 3PF, Submarine Engineering Assoc., Inc.: The Sea Boom 3PF is a heavy-duty barrier furnished in 24-ft sections with a draft of 24 inches and a freeboard of 12 inches. The boom is constructed with a polyvinyl chloride base, ¼-inch thick curtain and a flotation element of polyethylene foam-jacketed float. Boom ribs are made of Adiprene, a polyurethane polymer. The rib and float materials are resistant to the solvency effects of the representative hazardous floating liquids. The curtain material, as shown in the results of tests of stability, showed very poor stability in prolonged contact with n-butyraldehyde, ethyl acetate, ethyl ether and naphtha:solvent. When used for these chemicals, the boom would be expected to have a limited life, the extent of which is difficult to judge for the ¼-inch thick curtain material. Use of the Sea Boom 3PF for amelioration of spills of the eight hazardous floating liquids is feasible for an undetermined life.

Evaluation of Removal Devices

Removal devices, unlike containment devices, are extremely varied in principle of operation. All devices discussed in this section have been successfully applied to the amelioration of oil spills and are in manufacture. Because of the diversity of the construction, the existing devices were classified according to the recovery method employed by the device. In one class, moving plane skimmers, the pickup element consists of a sheet of metal or plastic which in motion descends through the oil slick into water and rises, containing adhered oil which is wiped off and drained into a reservoir while the blade returns to the oil slick and water system.

In a second class, the belt skimmers, the initial pickup element is an absorbing material in the form of a moving belt. The belt may enter the water and rise carrying oil, or it may enter the water and descend carrying surface oil. In some cases, the belt may be fastened to an element such as a drum which allows the belt material to enter the water and rise containing oil. In each of these cases, oil is removed and led to a containment reservoir.

A third class is formed by the weir skimmers. While these have some features in common with the suction head skimmers to be described, the weir skimmer's initial pickup is made by flow of oil from the slick on the water surface over a lip from which it can descend into a collecting vessel, from which the oil is removed by suction from a remote pump or removed by an in-head motor driven impeller. Suction head skimmers form a fourth class. As in the case of weir skimmers, oil enters the suction head pickup area by flow from the slick and, instead of being collected in a reservoir, suction applied to the pickup area of the suction head removes the oil as collected to a separate container.

Inflexible Plane Skimmers: Vikoma Seaskimmer, B.P. North America, Inc. – The Vikoma Seaskimmer is an inflexible moving plane type. The device is transportable and put into position by means of a crane attached to an operating base which may be deck-mounted or dock-mounted. No modification should be required for a Vikoma Seaskimmer to be used in the amelioration of spills of the floating hazardous chemicals as far as damage to its components is concerned. However, it is not expected that the Vikoma Seaskimmer may be profitably applied to the recovery of the representative hazardous chemicals. The reason for this is that the rate of pickup is dependent upon the viscosity of the liquid being recovered, and the lower the viscosity of the spilled material, the more rapid the drainage as the moving disc lifts the adhering liquid. The apparatus has recovered oils with a viscosity of from 8 to 100,000 cs. The viscosities of the eight representative chemicals range from 0.2 to 5 cp. Because of its poor expected rate of pickup for liquids of such low viscosity, the Vikoma Seaskimmer is not evaluated as highly practical. Its capacity to pick up pentadecanol solid is expected to be too low for practical use.

Clean Sweep, Lockheed Missiles & Space Co., Inc. – The Clean Sweep moving plane skimmer is part of a high seas oil recovery system which was built for the United States Coast Guard. The recovery device picks up oil by the action of a disc drum unit containing thin aluminum discs supported by floating hulls. In operation, as a disc element arises from contact with oil in a surface slick, adhering oil is carried up by the wiper blades which are scraped at the 12:00 o'clock position, allowing the recovered oil to flow into a sump from which it may be pumped to storage.

When furnished with optional pumps, controls and monitoring instruments which are explosion proof, the device is expected to be useable for the representative floating hazardous chemicals, since all parts are relatively resistant to solvent action. The Clean Sweep is furnished mounted on pontoons or catamarans and is portable. This skimmer may be used in rough water according to manufacturer claims. It is not expected that the Clean Sweep recovery unit will rapidly pick up the representative hazardous chemicals because of their low viscosities.

However, until actual tests are made to determine the effective rate of pickup of low viscosity liquids, and until a minimum tolerable rate of pickup is established, the Clean Sweep must be considered as a possible device for use in amelioration of spills of the representative floating hazardous chemicals.

In a letter of August 7, 1974, the Lockheed Missiles & Space Co., Inc. reported that Clean Sweep has been demonstrated to recover floating gasoline, kerosene, benzene and toluene. Two of these liquids are among the representative hazardous chemicals which are the concern of this discussion. They are benzene and kerosene with viscosities of 0.65 and 1.2 respectively, in the neighborhood of 20°C. No data as to the recovery rates when these materials are being picked up was given by Lockheed. In the absence of other information, the pickup rate for such materials is estimated to be a maximum of 5 gpm.

Moving Belt Skimmers: Belt Oil Skimmer, Aerodyne Development Co. — This unit is not expected to be effective for cleanup of the floating hazardous chemicals. Poor adhesion and rapid gravity drainage in the reverse direction of the movement of the rising belt would be expected to make the recovery rate too small for practical use. A serious disadvantage is that the skimmer is intended for fixed mounting and not for a mobile unit.

Inshore-Offshore Skimmer, Bennett Pollution Controls Ltd. — The Bennett Pollution Controls moving belt skimmer is mounted on a vessel which is self-powered. Oil from the surface of the water flows into a weir from which it is picked up by motion of a polyurethane sorbent belt. The unit may be used at a stationary location in a current, or may move under self-power through a slick. The unit, as manufactured, is not suitable for use with the representative floating hazardous chemicals because of the fire hazard involved in the power system.

The company reports that the equipment could be adapted to be driven by air motors, totally enclosed electric motors or a remote power source for the hydraulic system. If the fire hazard is eliminated, the device is suitable for use with the floating hazardous chemicals. The collection system involves polyurethane which is highly resistant to organic solvents. No information is given on which recovery rate estimates can be based. It is expected that, when applied to the floating hazardous chemicals, the Inshore-Offshore Skimmer would be most useful in a harbor location where the spilled material could be brought to the device by closing a boom containing the slick.

Rex Belt Skimmer, Envirex Inc. — The Rex Belt Skimmer uses an endless high-polymer belt. The lower end of the belt loop moves freely and is immersed below the surface of the water body. As the belt revolves, an adherent oil film is picked up and elevated to a point near the drive drum where the oil is scraped off and flows into a reservoir.

Furnished with explosion-proof motor and control devices, the Rex Belt Skimmer should be applicable to recovery of the representative hazardous floating chemical liquids. The equipment is not mobile and is designed for fixed installation at a captive body of water. The polymer rubber belt material is resistant to solvents, but slow deterioration of the belt would be expected. This skimmer is not considered highly practical for the recovery of the representative hazardous floating chemicals.

Dip Skimmers, JBF Scientific Corp. — The Dip Skimmers are descending belt-type machines. A 1000 Series device is mounted on a vessel equipped with twin propulsion, a moving collecting belt, a collection well and a pump to transfer oil from the recovery unit to storage. Remote controls may be provided and the units are air operated. The 2000 Series units are designed to be used in either a sweeping or a stationary mode. They may be moored where wind and current bring the oil to the unit, or they may be fastened to the side of a vessel and used in a sweeping mode. The 3000 Series units are fully self-contained oil recovery systems, diesel powered with twin screws, and all pumping propulsion and belt functions are hydraulic.

In JBF skimmers, the operating principle is motion of a continuous belt which moves from the slick surface downward into the water from a drive drum, carrying with it oil forced to follow the surface of the moving belt inclined plane to a collection well under the unit. The belt returns to the surface after passing over an immersed hydropulley. Buoyant forces cause the oil to surface in the well, displacing water out the bottom. As the oil collects, it is pumped off to storage. The heavy-duty conveyor belt may carry oil or sorbents or both down to the collection well. Recovery in excess of 90% of oil-sorbent mixtures presented to the system in one pass of the unit has been reported. When used with a set of rigid articulated sweeps, the effective pickup width may be increased and the oil collection rate does increase.

The dynamic descending plane system of oil collection has many advantages, among which an important one is that the material removed from the surface of the water is collected under water, reducing the fire hazard. Once a hazardous chemical is transferred from the belt to the collection well, hazards due to flammability would be present. The storage well and transfer pumping systems should be made proof against explosion and fire.

The device may be moved into an oil slick at the speed of motion of the belt which moves in a retrograde direction. This enables pickup to occur with very little flow across the surface of the belt, and consequent loss of oil. The skimmers are useful in rough water, though not as efficient as in calm water. All parts are compatible with the high solvency floating liquid hazardous chemicals with the exception of one of the two belts furnished.

The belt made from polyvinyl chloride showed extremely poor resistance to solvent action of the representative hazardous floating chemicals. The sample taken from a polyurethane belt was very stable, showing no detectable deterioration. The JBF Scientific Corp. also may furnish a stainless steel belt which would be compatible with the representative hazardous floating chemicals. A unit, which is remote-controlled, may be obtained. Such a unit would be very useful, considering the explosion and fire hazard attendant with the collection of the hazardous floating liquids. All units may be supplied with explosion-proof machinery and controls. Hoses and other auxiliary parts may be supplied from solvent-resistant materials.

The various JBF skimmer units are expected to be applicable to the recovery of the representative hazardous solid and liquid floating chemicals without modification and with the specification of the choice of a solvent-resistant pickup belt and of the choice of all the optional explosion proofing equipment. The skimmers should be highly advantageous for the amelioration of any substantial spill of the representative hazardous floating chemicals.

Marco Skimmers, Class I, II, III, Marco Pollution Control Corp. — The Marco oil skimmer uses the principle of a rising continuous inclined conveyor belt which lifts oil from the surface of water. Entrained oil is wrung from the belt by a squeeze roller into a sump from which it may be pumped to storage. An impeller system is furnished to draw oil and water through for mobile use. The bow fairings serve in the mobile case to channel floating oil to the belt surface. The device may be trailer mounted for delivery to the point of spill. The system is available in three classes of vessels for various size spills. Oil collection booms are furnished to sweep up to a 40-inch swath in order to increase the rate of oil feed to the belt. The belt is made of reticulated polyurethane foam, whose pore size has been adjusted to allow flow-through drainage of water and retention of oil. Stability tests on the belt material show that, although there is some weight loss on contact with the eight representative hazardous floating liquids, the condition of the belt material remains good and no visible deterioration is noted.

The oil recovery modules of the Class I, II and III skimmers are explosion-proof and the Class III model has an explosive gas detection system on board with a CO_2 system in the engine rooms. If the engine rooms are sufficiently explosion-proof, the Marco skimmer is suitable for the recovery of the representative hazardous chemicals without modification. Good recovery would be expected for all the representative hazardous floating chemicals except possibly solid floating pentadecanol.

Pore sizes have been adjusted for liquids of viscosities of 8 cp or higher. It is possible that a modification of the belt pore size could be found which would be more effective for the low viscosity representative hazardous floating chemicals. It is believed that it would be feasible to use the Marco skimmer unit as furnished for recovery of all of the hazardous floating liquid chemicals without modification. Fire-proof transport systems would have to be furnished for the mobile units.

Oil Mop, Oil Mop, Inc. – The Oil Mop belt-type skimmer recovers oil from water surfaces by means of motion of a continuous belt rope mop which floats in the oil slick on contaminated water. The mop travels around a floating or moored pulley and is drawn into a wringer device mounted on shore or boat-mounted. At the wringer, collected oil is squeezed out of the rope and the clean portion of the belt is continuously returned to the water surface. The effectiveness of the mop is ascribed by the developers to a configuration of feather-like fibers woven into a plastic rope with a thick nap. A high pickup of oil per unit volume is ascribed to the oleophilic and hydrophobic properties of the polypropylene plastic fibers.

The ropes are manufactured in diameters from 4 to 35 inches and can be produced in any length desired. Mop engines can be furnished in a size appropriate to the rope size. In operation, the rope mop is weighted to keep it in the water. It shows high capacity for containing oil. If the slick size is small enough, the company advises containing the oil inside the rope system. Oil also may be picked up from the edge of a moving slick moving toward the mop. The devices for wringing and moving the belt are available in explosion-proof electric motor construction, which must be used if the representative floating hazardous chemicals are to be recovered. The polypropylene plastic used in the belt would be stable for long periods against solvent action by the hazardous chemicals.

Except for consideration of viscosity effects on pickup rate, Oil Mop in its various forms should be applicable to the containment of the representative hazardous floating liquid chemicals, but not to pentadecanol. It is believed that the wringing system would not permit the collection of pentadecanol solid particles. The low viscosity exhibited by the hazardous chemicals would be expected to greatly reduce the pickup capacity per unit weight of belt because of flow-out of the system as the belt is removed from the water surface.

The general applicability of the Oil Mop units to the representative hazardous floating liquid chemicals warrants considering them as feasible without modification for amelioration of the hazardous chemical spills, even though the recovery rate for the chemicals is diminished from that for higher viscosity oils. The polypropylene belt would possibly have a reduced life in contact with the representative hazardous liquid chemicals, but the life should be substantial. Small sizes in which Oil Mop units may be furnished would be especially applicable to small spills in confined areas, a point which is favorable in considering the feasibility of Oil Mop systems for amelioration of the spills of the representative floating hazardous chemicals.

Reclam-Ator Skimmer, Peabody Welles, Inc. – The Reclam-Ator Skimmer recovers oil by means of a continuous belt mounted around a rotating drum partially immersed in water. The foam-covered drum absorbs oil as it rises through the slick of oil from the water, and the oil is then continuously squeezed from the drum into a recovery sump from which it can be removed to storage. The recovery system includes a float mount. The unit may be furnished with explosion-proof fittings which would be necessary for use with the representative hazardous floating chemicals.

Peabody Welles, Inc. reports considerable information on the relation of recovery rate to viscosity for various foam pore sizes which are available ranging from No. 10 to No. 80. As the viscosity of the material goes down, the effective pore size number increases. In the lower viscosity ranges below 1,000 SUS, the recovery rate rises very sharply as the viscosity is lowered. According to the graph of gallons per hour recovered versus oil viscosity, given in 1973 Peabody Welles brochure, a recovery rate of 2,500 gph is predicted for an oil of 100 SUS viscosity when the No. 80 pore size foam is used. A recovery rate of 4,500 gph is predicted for the same viscosity oil when the No. 60 pore size foam is used. These recovery rates are based on a continuous oil slick of minimum ⅛-inch thickness flowing toward the recovery drum at a velocity of 3 fps, and are given for the 50-inch wide unit.

From tests with polyurethane foam, it is expected that the stability of the foam with respect to the representative hazardous floating liquid chemicals would be sufficient to permit the use of the Reclam-Ator for recovery purposes without modification, insofar as can be determined from the company's descriptions of the parts. Use of the Reclam-Ator would not be expected to be feasible for the solid floating particles of pentadecanol, since these would not be absorbed into the foam belt. According to the manufacturer, the device is limited to use in calm water at currents of up to approximately 1 knot.

Slicklicker, R.B.H. Cybernetics (1970) Ltd. – The Slicklicker is a vessel-mounted, belt-type skimmer. In operation, an endless belt moving over rollers is extended from the vessel so that the belt is partially immersed. The upper surface of the belt continuously moves up through the slick, picking up oil which is removed in a squeeze roll at the upper end of the rotating belt. The oil is collected in a tank on the vessel from which it may be pumped to shore storage. A cellulose cloth belt is used, originally impregnated with oil, which makes the belt resistant to water and attractive to oil. The material of the belt was tested in the hazardous chemicals and no deterioration was found. It is possible, however, that continued pickup of the hazardous liquids and removal by squeezing would reduce the capacity of the belt for liquid pickup because the hydrophobic oil with which the belt is treated would be removed during the repeated squeezing and wetting operation with the hazardous liquids.

Suction Head Skimmers: Acme Models FSV5 and FSV40, Acme Products Co. – Acme Products Co. supplies two suction head skimmers, differing in size. Both operate on the principle of providing a shallow weir collection head from which the oil may be recovered through suction applied to a vacuum tube connected to a vacuum source. Level of collection for the weir is set by adjustment of the floats. The Acme vacuum head skimmers are furnished with Neoprene hose and with polyurethane floats. Both of these are highly resistant to solvent action by the representative hazardous floating chemicals. For spills on water bodies of appropriate size, the Acme Products Co. vacuum skimmers are suitable without modification for recovery of spills of the eight representative floating chemicals, including pentadecanol. Vacuum trucks, or vacuum storage tanks must be supplied with explosion-proof motors and fittings which are available from the company. These skimmers appear best suited for small spills and relatively calm water.

The liquid chemicals would be expected to flow into the weir sump under their own gravitational head, with proper adjustment of the floats. Recovery of solid floating pentadecanol would be more difficult because the low density solid particles would not readily move into the weir unless carried by considerable water flow. If the suction heads were used to pick up pentadecanol, it is expected that much water would have to be handled with the pentadecanol, with subsequent recovery of the pentadecanol through gravity separation in the collection tank.

Megator Delta Mark II, Beta Mark II and Puddle Mop, Megator Corp. – Each of the three Megator weir-type skimmers consists of a suction collecting head and a mobile pump to provide suction through a hose linked to the head. The three skimmers depend upon gravity flow of oil from a slick for filling the head. The Beta Mark II skimmer has a shallow weir fitted with stainless steel screen through which oil flows into a stainless steel suction bowl supported in the water by a flotation element. The Puddle Mop has a shallow, circular intake weir fitted with a screen but without a flotation element. Both skimmers are intended for hand held use, and they are deployed from a dock edge or other firm support.

The Delta Mark II skimmer has a similar intake weir to that of the Puddle Mop, but it is equipped with three float arms and floats which enable it to ride free after adjustment of the weir intake level by means of the floats. Collected oil is recovered, as with the other two skimmer heads, by suction created by the Megator mobile pump which may be furnished with a flame-proof motor and a starter. From the pump a delivery hose transfers collected oil to appropriate storage.

The Megator skimmers are designed for use in spill areas free from debris, such as leaves, weeds, paper and so forth, and are applicable mainly to calm water surfaces. The hose is of spiral reinforced polyvinyl chloride construction. Both Vacuflex and the Heliflex suction hoses supplied are of spiral reinforced polyvinyl chloride construction and are not expected to be suitable for the representative hazardous floating liquids because of the solvent action of the liquids. The hoses would be satisfactory for short term use, but they would be expected to deteriorate severely on long-term contact. Both the floating suction hoses and a Surflex hose, also supplied, contain polyvinyl chloride and are not recommended. When modified by supply of organic-solvent-resistant hoses, which are readily available, the Megator skimmers and pump systems are applicable to the amelioration of spills of the representative hazardous floating liquid chemicals. Collection of floating solid pentadecanol particles would be severely impeded by the presence of the screens. With screens removed, the three heads would be suitable for recovery of floating solid pentadecanol.

Oilhawg, Parker Systems, Inc. – The Oilhawg skimmer is a part of an oil recovery system which includes a trailer, recovery tanks, air-operated diaphragm pumps and appropriate hose. The system is highly mobile and may be driven to the site of a spill. The skimmer is deployed at the end of a long pipe held in appropriate position by an operator from a pier, barge or boat. The skimmer head is self-leveling and its position is controllable and adjustable from shore. It is of aluminum construction and is fully resistant to the solvent action of the representative hazardous floating chemicals. Three heads are used in tandem to increase the rate of pickup. Heads are equipped with strainers to hold back floating debris. When supplied with solvent-resistant hose and explosion-proof pumping systems, use of the Parker Systems' Oilhawg skimmers and pickup system is feasible for the representative hazardous floating chemicals without modification.

Slickskim, Slickbar, Inc. – The Slickskim skimmer head is a general purpose weir-type suction head, flexible enough to follow small wave patterns, and oil enters from a spill through ports along the perimeter and passes through a suction hose to pumps and storage. Heads are furnished with ½- and 1-inch inlets. The skimmer may be deployed and used by one operator. The flexible head is made of an elastomeric substance in which marked swelling occurs on long contact with butyraldehyde, benzene and ethyl acetate. Such swelling might change the opening size and impede the pickup of hazardous liquid spills. Heavy absorption of the three liquids which swell the material of the skimmer head would be expected to change the density of the head.

An advantage of the Slickskim suction head is that it can operate in as little as 3 inches of water, making it especially applicable to spills in shallow water bodies. A rigid head or an aluminum head also furnished would be resistant to the solvent action of all the representative hazardous floating liquids and would be generally applicable.

If supplied with explosion-proof motor, solvent-resistant hoses and solvent-resistant suction heads, the Slickskim and pump system is applicable without modification to the amelioration of spills of the representative hazardous floating chemicals. The recovery system, of which the Slickskim skimmers are a part, is highly mobile and readily transported to a spill site, requiring only storage capacity to be supplied. It is especially applicable to small spills in calm water.

Weir Skimmers: Acme Models SK-39T and SK-51T, Acme Products Co. – The Acme SK-39T and SK-51T units are weir-type skimmers with pumps mounted directly on the skimmer. In operation, the skimmer is placed into position, skimming depth is adjusted and oil that collects in the weir by gravity flow is pumped out to storage through a discharge hose. The hose normally furnished is made of polyvinyl chloride coated nylon fabric, and this would have to be replaced with readily obtainable solvent-resistant hose. The remainder of the unit is metal and urethane-filled fiber glass. With proper hose and optional explosion-proof motor, these skimmers should be directly applicable to the amelioration of spills of the representative hazardous floating liquid chemicals.

Slurp Skimmer, Coastal Services, Inc. – The Slurp Skimmer is a weir-type device with a rectangular opening and a horizontal lip over which oil from a spill can flow into a suction well. The shape of the collection well is such that high pumping rates with correspondingly small amounts of oil in the weir lower the horizontal lip to apply to thick oil slicks. When the slick is thin, the pumping rate is slowed down, leaving a higher level of oil in the collection portion, and the consequent upward tilting reduces the immersion of the lip of the weir conforming to the oil thickness. According to the manufacturer, in choppy waves, the weir automatically rises until only the wave crests are collected, as controlled by the pumping rate governed angle of inclination.

All parts of the Slurp Skimmer are metal and the device is applicable without modification to recovery of the representative hazardous floating chemicals. Hoses furnished may be made of polyvinyl chloride, Neoprene and stainless steel. Because of expected solvent deterioration of the polyvinyl chloride hose, only the Neoprene and stainless steel hoses should be used with the floating hazardous liquid chemicals. Any pumping system which is used with this skimmer must be explosion-proof because of the fire hazard exhibited by the hazardous chemicals. Since the flow rate of pentadecanol into the Slurp area may be slow because it is a particulate solid, not a liquid, the rate of pickup for pentadecanol may be very slow unless means are provided, such as barriers, to herd solid pentadecanol particles into the pickup throat of the skimmer.

Crisafulli Aqua-Sweeper and Mini-Sweeper, Crisafulli Pump Co. – The Crisafulli weir-type skimmers admit oil from a slick over a horizontal lip into a weir which serves as a reservoir for collected oil. From this the recovered oil is transferred to storage. In the Aqua-Sweeper unit, the weir is mounted in a self-propelled catamaran which includes a sump and a hydraulic pump unit. The unit is mobile and may be deployed by use of a crane, or launched from a tilt-bed trailer. The skimmer should be able to pick up the representative hazardous floating liquids, as far as the mechanical arrangements are concerned. However, no information is available on the explosion rating and the system is powered by a gasoline engine unsuitable for the fire hazard posed by the hazardous liquids. Modifications would be required for this equipment to be used with flammable and combustible liquids. With the additional modification of substituting solvent-resistant hoses for those furnished, the Aqua-Sweeper would be suitable for amelioration of spills of the hazardous floating chemicals, including pentadecanol solid.

In addition to the Aqua-Sweeper, a Mini-Sweeper is supplied which is not self-propelled. It may be transported to a site where needed and where it would be supported by its catamaran mount. The horizontal lip of the collecting weir is 25 inches in width. A pump is provided on the catamaran to pump oil from the collection sump to storage. No information is given as to whether the motor supplied is explosion-proof, but such a motor should readily be obtainable. With the modifications of an explosion-proof power system and hoses which are resistant to organic solvents, the Mini-Sweeper is applicable to amelioration of spills of the representative hazardous floating liquid chemicals.

Oela III, Industrial Municipal Engineering or Coastal Services, Inc. — The Oela III has a circular weir floated in water at a height adjusted by means of four floats to the desired level of immersion for entry of oil. In use, oil flows over the circular edge into the weir, from which it is removed by pumps externally located. A fitting for 2-inch hoses is attached to the base of the weir. The Oela III skimmer is easily deployed by two men and is located by the length of the hose which connects it with the pumping equipment used to transfer the recovered oil to storage. Except for a plastic ring around the lip of the weir, the Oela III skimmer is made entirely of metal. It is, therefore, applicable to recovery of all the representative hazardous floating liquids without modification. For use with hazardous liquids, specifications should be made that hoses are organic-solvent-resistant or are metal.

Sea Broom, Sunshine Chemical Corp. — The Sea Broom is a weir-type skimmer. In operation, oil from the slick enters a very shallow weir from which it is pumped out as collected by a pump mounted centrally in the floating skimmer. With the modifications of explosion-proof motors and solvent-resistant hose, the Sea Broom skimmer is applicable to the amelioration of the representative hazardous floating chemicals.

ORS, Models 125, 1000 and 2000, Ocean Systems, Inc. — The ORS 125 skimmer is dependent upon motion into the slick and is normally expected to be towed at a speed of from 1 to 3 knots. The 2,000 gpm recovery system weighs 25,000 lb, including auxiliary equipment and packaging. A ship having a 5-ton boom is used to transport the device. The scale of this equipment appears very large considering the comparatively smaller spills expected for the representative hazardous floating chemicals, as compared to the large spill possibilities for oil. A requirement arising from the high volatility of most of the representative hazardous floating liquids is that recovery be rapid. It is estimated that the deployment time of the ORS 1000 and ORS 2000 integrated systems would be so great as to make them of little use for ameliorating spills of the representative hazardous liquids.

Floating Sorbent Moving Belt Skimmer: "SOP" System, Ocean Design Engineering Corp. — The "SOP" pickup system captures floating liquid in particulate sorbent material which is distributed over the spill by air jet into a wide dispersal pattern between two articulated harvesting booms. Floating saturated sorbent material is picked up by a moving belt conveyor deployed from a catamaran hull and is transferred to a series wringing device which squeezes the sorbed liquid from the saturated sorbent. The liquid is transferred by pumps to storage tanks. The recovered dried sorbent is transferred forward by conveyors and again is broadcast onto the oil spill.

Use of a device of this type, involving floating sorbent, is considered to be advantageous for use with the representative hazardous floating liquids for several reasons. Absorption of floating hazardous liquid into the floating sorbent material may tend to reduce the evaporation rate. Capture and immobilization of the floating liquid in the sorbent would facilitate containment by booms. Hazardous liquid contained in the sorbent lumps would be expected to be less readily moved under the barrier because of water currents than the freely floating liquid. These points favor the use of floating sorbent for initial capture of hazardous floating liquids. Such use of sorbents, however, demands that a recovery system be present in which sorbed liquid and sorbent may be separated as rapidly as possible. The feature of the "SOP" System that the sorbent is recovered and reuseable reduces the amount of sorbent required. However, in some cases, the maximum benefit of floating sorbent use would be derived if the whole body of liquid were sorbed into sufficient quantity of the floating sorbent to capture the spill. Regeneration of the sorbent by the wringing process would eventually recover the sorbent that had been dispersed. From the literature search and consultations with firms engaged in pollution control, no identification was found for the use of such a system for the representative floating hazardous liquids, though its effectiveness for oil pickup has been demonstrated. In order to evaluate the "SOP" pickup system for possible use with the representative hazardous floating liquids, it was necessary to determine whether sorbents existed which would be suitable for use with materials having the properties of the hazardous liquids. Sorption methods are discussed in detail in a subsequent chapter.

Experimental work established that sorbents exist off-the-shelf which are highly suitable for use in the circumstances required by the "SOP" System. In all other respects, the "SOP" pickup device is applicable to recovery of the hazardous floating liquids with the proviso that fire hazards be eliminated in the motive equipment, pumps and transfer systems. The system is not suitable for the recovery of the representative floating hazardous solid, pentadecanol.

DREDGING

> The information in this and the following section is based on *A Feasibility Study of Response Techniques for Discharges of Hazardous Chemicals that Sink*, CG-D-56-78 prepared by T.D. Hand, A.W. Ford, P.G. Malone, D.W. Thompson and R.B. Mercer of the U.S. Army Engineer Waterways Experiment Station for the U.S. Coast Guard.

Though largely untested in the role, dredging can generally be considered to be a feasible means of removal of hazardous materials from the bottom of most bodies of water; in the case of land spills, conventional excavation and/or pumping techniques would be analogous. There are a number of different types of dredging systems in operation throughout the United States, each adapted to specific kinds of dredging in specific circumstances. No generalization can be made as to which type of dredge is the best for hazardous material recovery; such a determination must be based on the amount of bottom material which must be recovered, on the environmental setting in which the spill occurred, and in some instances on the character of the material that was spilled.

The information in this section attempts to sort out this complex task. Though all dredges have certain advantages in certain situations and no dredge should be ruled out, especially if it is available, two specific types of dredges can be singled out as having high potential in a variety of situations: the Mudcat (and any competitors that may arrive on the American market, i.e., Amphidredge) because of its availability and portability; and pneumatic dredges because of their versatility, portability, past use in related roles, and purportedly good operating characteristics (high solids content, low turbidity).

In long-range planning for the use of dredges to recover sinkers, the Coast Guard should be especially aware of the long lead times necessary in conventional dredging due to surveys, contract arrangements, environmental impact assessments, etc. If immediate response of either private or Corps of Engineers dredges is contemplated, particularly for high risk areas, consideration should be given to establishing standby agreements or contracts.

Selection of a dredge system for hazardous material recovery may be strongly influenced by considerations involving the temporary storage, transportation, treatment, and disposal of the contaminated dredged material. Backwards planning, starting with the contemplated disposal method and site, may be the best approach.

Applicability to Hazardous Material Recovery

Conventional dredges are not specifically designed or intended for use in recovering spilled hazardous materials resting on the bottom, but are the logical and perhaps only feasible means to this end. The factors normally considered in planning a normal dredging operation are no less important in planning a chemical spill response; however, the following additional factors must be considered because hazardous materials are involved:

- Need for precise determination and marking of boundaries of area to be dredged.
- Need for very precise lateral and vertical control of dredging head.
- Requirement for special precautions tailored to specific chemicals.
- Requirement for temporary storage, transport, and treatment prior to disposal of dredged material.
- Need to predict the likely damage to aquatic and benthic organisms to be caused by the dredging operation and its effect on resuspension of any contaminant.

With few exceptions, the particular chemical involved in the spill will not be an important factor in choosing the best dredging equipment for the job, although on-site precautions and treatment/disposal arrangements must fit the specific nature of the chemical. In many instances, the contaminating material will be incorporated into the local sediments and will not have an identity of its own; it will be recovered along with the sediment-water slurry as contaminated dredged material. In those cases where the chemical is still present on the bottom as an intact mass, then its properties (principally whether it is a solid or liquid) may influence dredge selection.

Dredging alone can seldom be expected to accomplish total recovery of spilled chemicals. Accordingly, it should be considered in combination with other measures, such as burial or in-place chemical treatment. These methods would be particularly applicable around the ill-defined periphery of a spill where contaminant concentrations may be too low to justify continued dredging due to the amount of material that would be involved. Dredging should also be considered as a possible follow-up measure after the contaminant is buried with an inert or active material or immobilized by covering with a man-made sorbent material such as activated carbon. A two-stage response such as that described would have the advantages of quickly mitigating the acute hazards by burial, but, nevertheless, ultimately removing the chemical from the environment. Moreover, dredging of a chemically or physically tied-up material would result in less hazard from resuspension.

In addition to recovery of the contaminant, dredging could play additional roles in an overall response effort. In attempts to contain the dispersion of a chemical moving on the bottom under the influence of gravity or currents, dredges could be used to excavate trenches or build up underwater dikes, both to act as barriers. The same functions could also be useful in controlling or containing the spread of a cover material. These applications are discussed later in this chapter.

Previous experience with the use of dredges to clean up hazardous chemical spills is extremely limited. The only major incident from which significant lessons have been learned was the accidental discharge and nearly complete cleanup of 265 gal of polychlorinated biphenyls (PCBs) in a barge spill in the Duwamish Waterway in Seattle harbor. The cleanup effort is well documented in the after-action report and summary (6) prepared by the Seattle District, Corps of Engineers, the agency tasked with cleaning up the spill, and in a case history (7) of the incident written by James C. Willmann of the EPA's Region X. The lessons learned in this episode, both technical and practical, have been heavily drawn upon for many of the assessments and judgments found it this report.

In related applications (8), the Japanese have extensively dredged harbor sediments that were dangerously contaminated with heavy metals and organic chemicals. In these cases, there were no spills as such and thus there was no sense of responding to an accident. The primary value to be derived from the Japanese experiences is in the precise, tightly controlled dredging operations and in the effective in-line dredged material treatment, chemical fixation, and disposal scheme. The proprietary name for one of these systems is TST (Takenaka Sludge Treatment) (9).

Another Japanese firm has designed, fabricated, and demonstrated a special dredging head for cleaning up contaminated sediments; the device minimizes resuspension and turbidity and follows the contours of the bottom, taking a constant-thickness cut (10). On a much smaller scale, the EPA has conducted tests using the Mudcat dredge to recover spills of simulated solid hazardous materials (11). The EPA has also sponsored the dredging of a reach of the Little Menomonee River (Wisconsin) bed that had been contaminated with creosote wastes (12). This demonstration study featured a trailer-mounted physical chemical treatment train to process the contaminated sediments.

As a general statement, it appears at this point that existing dredging technology could be effectively used to recover spilled hazardous chemicals with adverse environmental impacts greatly exceeded by the beneficial environmental impact of removing the offending material. However, lack of experience in this role and the need for careful planning of not only dredging, but treatment and disposal operations as well, will mean very long lead times. Even the Japanese, who have the most experience in this area, do not undertake such operations as a response.

Description of Dredging Equipment

For purposes of this report, any equipment or device that could be used to remove material from the bottom of a waterway or from a surface spill is considered dredging equipment. This includes handheld suction equipment and land-based construction equipment (e.g., power shovels, draglines, backhoes, front loaders, dozers, etc.).

Dredging equipment and its nomenclature resist neat categorization. As a result of specialization and tradition in the industry, numerous descriptive, often overlapping, terms categorizing dredges have developed. For example, dredges can be classified according to: the basic means of moving material (mechanical or hydraulic); the method of storage or disposition of dredged material (pipeline, sidecaster, hopper); the device used for excavating sediments (cutterhead, dustpan, plain suction); the type of pumping device used (centrifugal, pneumatic, airlift); and others. Dredges do not come in standard models, but in fact are more likely to be designed for a specific type of job and repeatedly modified to suit the requirements of new owners or new tasks. Dredges range in size from the Mudcat, only 30 ft long, to 350-ton, barge-mounted cutterheads and 350 ft, ocean-going hopper dredges.

For purposes of this report, dredges will be classified as mechanical, hydraulic, pneumatic, and special purpose. It is recognized that these categories are neither mutually exclusive nor parallel; however, with proper explanation they will prove to be convenient and cause a minimum of confusion.

Mechanical dredges remove bottom sediment through the direct application of mechanical force to dislodge and excavate the material at almost in situ densities. Most mechanical dredges deposit the dredged material into scows or barges for transportation to a disposal site. Specific types of mechanical dredges in common use are the dipper, the bucket ladder, and the grab or clamshell.

Hydraulic dredges remove and transport sediment in liquid slurry form. They are usually barge-mounted and carry diesel or electric-powered centrifugal pumps with discharge pipes ranging from 6 to 48 inches in diameter. The slurries containing 10 to 20% solids by weight are often transported several thousand yards through pontoon-supported pipelines to a water or land-based disposal site. Other methods of handling dredged material include sidecasting, loading into barges or scows, and direct loading of on-board hoppers. Dredge types included in this category include plain suction, dustpan, cutterhead, and hopper dredges.

Pneumatic dredges are treated as a distinct category only because of their comparative novelty in this country. Originally developed in Italy under the trade name Pneuma, these systems feature a pump that operates on compressed air and hydrostatic pressure and that is reported (13) to be capable of handling slurries up to 70% of the in situ sediment density. Reliable data on the percent solids by weight are unavailable. Pneumatic dredges are otherwise hydraulic pipeline systems.

The special purpose category of dredges includes the Mudcat, a readily available, scaled-down hydraulic dredge, any of the various hand-held suction devices or portable pumping systems that could be used for hazardous material recovery, and the whole range of land-based construction equipment that would be invaluable in the case of a land spill (front loaders, backhoes, power shovels, scrapers, draglines, etc.).

The National Car Rental Agency has designed and developed the Mudcat, a small hydraulic dredge for excavating sediments up to 18 inches thick per pass from depths up to 15 ft. The machine is pontoon-mounted and features an auger-like cutting device that feeds the sediment to the suction intake of a diesel-driven centrifugal pump. The auger is mounted along the base of a bulldozer-type blade. The whole arrangement with suction pipe attached is controlled by a hydraulic boom. The dredge is not self-propelled, but is moved along on an anchored cable during each traverse of the excavation, and the dredged material is discharged ashore through a float-supported pipeline. The width of the cut is approximately 8 ft, and applications to date have included cleaning out small reservoirs and streams (14).

Evaluation of Equipment

In the following sections, a summary is given of the principal advantages and disadvantages of each dredge.

Mechanical Dredges: Mechanical dredges (clamshell, dipper, and bucket) are designed for hard or soft materials and normally are not self-propelled crafts. No provision is made for material containment; thus, these units must work alongside the disposal area or be accompanied by disposal barges during the dredging operation. Principal advantages of mechanical dredges:

- The clamshell is capable of deep-water excavation.
- They can be controlled and maneuvered in small and confined areas and would be useful in areas with obstructions and debris.
- They excavate materials at nearly in situ densities; thus, a smaller volume of dredged material must be handled and disposed of.

Principal disadvantages of mechanical dredges:

- They are capable of only modest production rate ($\leqslant$500 cubic yards per hour).
- They require separate disposal vessels and equipment.
- They cause a great deal of turbidity and sediment (contaminant) resuspension.
- They would be ineffective against a free or unadsorbed liquid contaminant.

Hydraulic Dredges: Barge-Mounted Dredges – Barge-mounted types, such as the plain suction, dustpan, and cutterhead dredges, will have difficulty in rough water (greater than 3-ft waves). Excessive vertical movement of the ladder can cause the head to be forced into or bounced off the bottom leading to excessive impact loads on the ladder, digging equipment, and transmission. Additionally, large differential movement between the barge and sections of the float-supported pipeline could result in undesirable stresses and failures in rigid pipe connections. Principal advantages of conventional barge-mounted hydraulic dredges:

- Depending on size, they are capable of the highest production rates of any dredge (up to 15,000 cubic yards per hour).
- Pipeline directly to treatment/disposal area could minimize handling of and exposure to contaminated dredged material.

Principal disadvantages of conventional, barge-mounted hydraulic dredges:

- They cannot be employed in rough waters.
- The large volume of dredged material is 80 to 90% water, requiring major dewatering and consolidation operations for efficient disposal.
- Anchoring cables and pipelines present temporary obstructions in navigable water channels.
- Cutterheads and suction lines are hindered and possibly damaged by underwater debris, large rocks, and other obstacles.

Hopper Dredges – The ocean-going hopper dredges are self-propelled vessels with self-contained storage of up to 8,000 yd^3. Principal advantages of hopper dredges:

- Self-contained storage of dredged material eliminates need for separate storage barge, scow, or pipeline. Some possess pump-out capability.
- Hopper dredges can operate in rough, open waters and relatively strong current.
- They operate without anchors and other restraints and can be used in shipping channels without causing excessive interference with normal traffic.

Principal disadvantages of hopper dredges:

- Deep draft precludes use in shallow waters, including barge channels.
- They cannot work continuously, but must alternately load up, move to disposal site, dump or pump out, and return.
- For hazardous materials, the full hopper capacity cannot be used to preclude overflow.
- Hopper dredges excavate with less precision than other dredge types.
- Open-water dumping, though quick and efficient, cannot be used for contaminated material.

Pneumatic Dredges: Pneumatic dredges are basically hydraulic pipeline systems which use a compressed air-operated pumping device. Developed in Italy and Japan, they have been used successfully in a number of cleanup operations including the Duwamish PCB spill. Principal advantages of pneumatic dredges:

- They are crane-supported and thus can be operated in close and restricted areas and can be mounted on barges, seagoing vessels, as well as dockside.
- They can be operated in shallow or deep water with no theoretical maximum depth.
- They can be relatively easily dismantled and transported by truck or air.
- They may be able to yield denser slurries than conventional hydraulic dredges.
- With passive excavating heads, they cause little turbidity or resuspension of solids.

Principal disadvantages of pneumatic dredges:

- They are capable of only modest production rates (up to 390 cubic yards per hour).
- Cables and pipelines present temporary obstructions in navigable water channels.
- Pneumatic systems are not in widespread use in the United States and, therefore, may not be as readily available as other types.

Special Purpose Dredges: Special purpose dredges include the Mudcat, handheld suction devices, as well as conventional earth-loading and moving equipment. In most cases, their greatest value lies in small or surgical cleanup operations, or, in the case of earth-loading equipment, land spills.

Mudcat – An EPA study (11) concluded that the Mudcat dredge was effective in removing undesirable particulate matter from pond bottoms. In the course of tests made during this study, two significant observations were made. First, the Mudcat was shown to have a greater efficiency in the removal of sediment during a backward cut than in a forward cut. The explanation given was that the mud shield was fully extended over the cutting auger in the backward direction and thus was more effective in decreasing the resuspension of any bottom sediments into the surrounding water column.

Secondly, the specific gravity of the spilled material had a definite effect on recovery efficiency. It was observed that as the specific gravity of the target material decreased, the material recovery rate decreased and resuspension of the spilled matter into the water body markedly increased. No reason was given for this phenomenon, but it is surmised that lighter material is more easily disturbed and, therefore, more readily resuspended. Principal advantages of the Mudcat:

- It is compact and readily transportable by truck or air.
- It can be operated in confined and isolated areas and in very shallow waters.
- It is compatible in production rate with an existing trailer-mounted treatment unit developed by the EPA.
- It would be readily available for lease from the National Car Rental Corporation and might be able to be activated on shorter notice than other dredges.

Principal disadvantage of the Mudcat:

- Its size and production capacity would limit it to small jobs.

Handheld Devices – No attempt will be made to describe the many types of portable, handheld suction devices that could be used in many instances as minidredges. A suction hose manipulated by a diver with pump and storage tank on board a barge, boat, or land-based truck could be invaluable in the precision dredging of intact masses of a solid or liquid contaminant. Practical use would be limited to very small spills or well-defined concentrations in difficult locations. Principal advantages of handheld suction devices:

- They are extremely mobile and universally available.
- With manual positioning, they are capable of surgical cleanup work.

- They would be particularly effective in vacuuming identifiable masses of pure contaminant, particularly liquids and free-flowing solids.

Principal disadvantages of handheld suction devices:

- They are limited to very small quantities of material.
- They are ineffective against consolidated sediments.

Earth-Loading Equipment – Land-based, earth-loading equipment ranging from farm tractor-size loaders and backhoes to giant shovels and draglines used in strip mining could have valuable application to spill response, particularly in the land and nonnavigable water scenarios. No attempt will be made to describe the various types and sizes of equipment that are available in most locations. An important source of many of these type items of equipment on short notice, depending on the location, may be engineer units of the U.S. Army, active and/or reserve.

Location and Availability of Dredges in the U.S.

The status of the U.S. dredging fleet as determined in the Corps of Engineers National Dredging Study is comprehensively summarized in a paper of the same title by Murden and Goodier (15). The U.S. Army Corps of Engineers operational fleet is comprised of 15 hopper, 3 sidecaster, 11 cutterhead, 8 dustpan, and 5 mechanical dredges. With the exception of two hopper units, all were constructed prior to 1949. In comparison with many of the modern, automated dredges operated in Europe and Japan, the Corps' dredges are labor-intensive operations requiring 2½ to 3 times the manpower to operate. Though the Corps' fleet is well maintained (at great expense in downtime and labor), it is, nevertheless, obsolescent and inefficient relative to the state of the art. Even the highly touted Corps hopper dredge fleet is badly in need of modernization.

The industrial fleet is composed of approximately 450 plants plus approximately 250 portable Mudcat dredges owned and available for lease from the National Car Rental system. The industrial fleet has only four barge-mounted hopper dredges and no sidecaster capability (three privately owned ocean-going hoppers were expected to be operational in 1978). Primarily, the industrial fleet is composed of 264 cutterhead, 161 clamshell, 13 dipper, and 19 hydraulic plain suction dredges.

In general, the industrial fleet suffers from the same maladies as the Corps' fleet. The condition of the total dredging fleet can be traced to the great uncertainty in the dredging industry. Business is spasmodic and contractors are unwilling to undertake the financial burden of updating the dredging fleet with new equipment. With few exceptions, the National Dredging Study (15) revealed that industrial dredging organizations have not been actively involved in research and development to improve dredging capabilities and have not, until recently, built any new dredges incorporating advanced designs.

In addition, the study implies that the industry is highly fragmented and has been undergoing a process of contraction and consolidation for several years. This decline in size of the industry has been caused by the high capital cost of maintaining and replacing equipment, the impact of inflation on operating and repair costs, and the restraint on dredging activities resulting from environmental restrictions. Nevertheless, an urgent need exists to modernize and replace much of the equipment in Corps and industrial inventories.

Table 2.3 summarizes the distribution by major type and geographical area of both the Corps of Engineers and the private industrial dredging fleets.

Table 2.3: U.S. Dredge Fleet Inventory: Regional Distribution

Corps of Engineers						
Location	**Hopper**	**Sidecaster**	**Cutterhead**	**Dustpan**	**Clamshell**	**Dipper**
West Coast	3	0	4	0	1	0
Gulf Coast	5	0	2	0	0	0
Interior Waterways	0	0	4	8	2	2
Great Lakes	4	0	1	0	0	0
East Coast	3	3	0	0	0	0
Total	15	3	11	8	3	2

Private Fleets					
Location	**Cutterhead**	**Clamshell**	**Dipper**	**Plain Suction**	**Hopper**
West Coast	38	53	1	0	1*
Gulf Coast	75	23	0	0	0
Interior Waterways	31	33	5	19	0
Great Lakes	23	28	4	0	3*
East Coast	97	24	3	0	0
Total	264	161	13	19	4

*One on the West Coast and one on the Great Lakes are trailer hoppers and two on the Great Lakes are suction dredges mounted on hopper barges.

Source: CG-D-56-78

The availability of either Corps of Engineers or private dredges on short notice to the Coast Guard or other agencies responsible for hazardous material recovery is difficult to assess. Corps dredges are generally fully committed and though dredge-owning Corps Districts would be cooperative to the extent possible, it is doubtful that a rapid response in terms of days could be mounted. The private dredging industry, on the other hand, is generally under-committed and presumably would eagerly respond in the event of a spill that required dredging. However, the need for special preparations, precise predredging surveys, environmental impact statements, and contract negotiations may delay any actual response up to 2 months if conventional dredging practices are any indication. Moreover, private dredging industry officials may not have a full appreciation of the need for special precision, care, and precaution when dealing with hazardous materials, and thus extra supervision and coordination would be required.

Advances in Dredging Technology

As brought out earlier, the United States does not possess a state-of-the-art dredging industry, and, therefore, the advanced techniques discussed here, many of which could enhance the effectiveness of dredging for hazardous material recovery, would not be generally available. Until economic impetus or other considerations force the United States to employ on a large scale the innovative, technologically advanced methods and equipment found in western Europe and Japan, the ability to perform accurate, efficient dredging with minimum environmental damage will be severely limited.

Advanced dredging technologies are generally directed toward one or more of the following areas of improvement: greater depth capability; greater precision, accuracy, and control over the dredging process; higher production efficiency; and decreased environmental harm. Following are brief descriptions of the major innovations in production dredging that might be used to advantage in hazardous material recovery operations:

- Ladder-mounted, submerged pumps for higher production at depths up to 200 ft.
- Injection of buoyant material into pipe near dredging head (air and kerosene used to date) to provide increased lift and thus higher production and greater depth capability; similar in principle to airlift pumps.
- Improved designs of dredging heads to minimize material resuspension.
- Use of spud barges (aft of the dredge) to extend hull length and increase dredge swing; will increase production efficiency of cutterhead dredges; in limited use in the U.S. today.
- Longer ladders, connected further aft on the dredge hull to increase depth and permit greater control.
- Tandem pump systems for greater production efficiency and reliability.
- Articulated ladder designs to maintain constant dredging head bottom contact; will allow use without damage in rougher waters.
- Better hull designs, equipped with liquid stabilizing systems (swell compensators to allow use in heavier seas).
- Improved production instrumentation to monitor flow rates, cumulative production, etc.
- Improved navigation, positioning, and bottom profiling instrumentation; state-of-the-art includes advanced laser, electronic, and acoustical systems.
- Closed bucket modifications to reduce loss of fines and liquid in clamshell dredges.
- Depth and swing indicators for mechanical dredges.
- Clamshell to dipper convertible dredges.
- Use of silt curtains during dredging, as well as open-water disposal, to restrict turbidity plumes, and, in the case of hazardous materials, limit the added dispersion due to dredging. State-of-the-art is in U.S., but silt curtains are not in general use.

In addition to the improvements in conventional dredging discussed above, several other miscellaneous techniques and specific items of equipment deserve mention:

- The Amphidredge, manufactured by IHC-Holland, a major European dredge builder, is an amphibious, self-propelled vehicle that features backhoe and clamshell configurations as well as a shrouded cutterhead, similar to the Mudcat. The Amphidredge could prove to be a most useful and versatile device against spills in small, nonnavigable streams and marshy areas.

- A system (discussed earlier) has been developed by Takenaka Komuten Co., Ltd., of Osaka, Japan, for the dredging, treatment, chemical fixation, and land disposal of highly contaminated harbor sediments (organics, heavy metals). In this system, proprietarily named TST, a pneumatic Oozer dredge pumps the sediments to a sedimentation basin. The supernatant is charcoal-filtered and returned to the harbor while the sludge is mixed with a proprietary portland cement-based additive and pumped to a disposal site where it sets. The result is claimed to be a chemically inert landfill with excellent stability and mechanical strength, capable of supporting heavy construction. This system could probably be made available in the U.S. given a sufficient lead time. Though it cannot be considered a response system, it does address all phases of the hazardous material cleanup problem.
- A special dredging head named Cleanup has been developed in Japan to be used with standard hydraulic suction dredges for cleaning up highly contaminated sediments with a minimum of turbidity and hazardous material resuspension and optimum dredging accuracy. It consists of an articulated box that completely encloses the suction head and allows water to be entrained and mixed from one direction only. It is highly instrumented in order to ensure a constant cutting depth and slurry density, and it features a trap to capture noxious gas bubbles that are released as the sediment is disturbed. Results of demonstrations have shown that turbidity and chemical oxygen demand (COD) in close proximity to the cleanup head are virtually the same as that of the undisturbed ambient water.

Most of the conceptual thinking taking place in the dredging field continues to address the problem of increasing efficiency, production rates, and effective dredging depths along with considerable new interest in developing ways to mine the deep oceans for manganese nodules. Little thought, other than what has been discussed, seems to be directed toward hazardous material recovery. Mitigation of the adverse environmental effects of dredging and particularly of dredged material disposal continues to be of interest to the environmentally aware public and to the Corps of Engineers through the DMRP, but few engineering advances in dredging equipment and technology are on the horizon.

For purposes of enhancing the potential of dredging as a response to spills of hazardous materials that sink, the following areas of inquiry are likely candidates that need thought, basic research, development, or demonstration.

- Designs of dredging heads and improved techniques for better and more efficient control of sediment resuspension; study and demonstration of existing cleanup devices.
- Study and characterization of interaction of selected spilled hazardous materials with various sediments in various aquatic and marine environments.

- Dispersion behavior in various bottom environments of various solid and liquid chemicals that sink.
- Devices and instruments for in situ detection of various types of contaminants in bottom sediment, possibly on board recovery dredges.
- Development of a floating storage/treatment system for contaminated dredged material.

BURIAL

Burial of a spilled chemical, as the sole measure taken, will seldom be the best or most desirable response. It must be recognized that sooner or later, despite all efforts to prevent it, any covering layer may be reexcavated naturally by scouring, biological action, or both, especially in settings where storms, floods, unusual tides, etc., can produce extreme currents and turbulence. It must also be noted that burial would be a patently inappropriate response for a chemical spilled in a shipping channel that must be routinely dredged to remain open. For most chemicals in most circumstances, the best response is removal.

Nonetheless, burial as a response can fulfill a number of appropriate roles, and a feasibility assessment of the various techniques will be of value. Appropriate roles of the burial response include:

- As a temporary mitigating measure to retard the spread or reduce the hazard of a spill until a recovery (dredging) operation can be mounted.
- As a final step, following the recovery of most of a spill, to isolate any residual contamination of the sediments.
- As the sole response when recovery cannot be accomplished or in cases where the spilled chemical is harmless or nearly so.

Chemicals Amenable to Burial

Early in this program, it was decided that liquids were generally not suited to covering or burial. Even dense, viscous liquids such as PCBs can logically be expected to disperse when impacted by materials intended to bury them. The most likely candidates for burial were therefore limited to solid chemicals; however, it should be kept in mind that any hazardous chemical on the bottom of a water body will disperse to some extent. At the periphery of the dispersion pattern, the spilled substance will have no physical character of its own, either liquid or solid; its presence in the bottom sediment will be definable only by chemical analysis. Such perimeter areas, whether resulting from a liquid or solid chemical spill, may be amenable to burial even though the center of the spill must be treated otherwise. In that sense, burial should be considered at least a partial option for any spilled hazardous chemical that sinks. This is the rationale for the second role of burial stated above.

Only those chemicals that are essentially harmless or whose reaction products after burial with a chemically active material are harmless should be considered as candidates for burial as a first-choice response. Of the 70 materials listed in Table 1.18, only asphalts, sulfur, and barium carbonate can be considered to meet this condition fully. Several additional solid materials, which are moderately hazardous in an aquatic environment, could be rendered significantly less so by proper burial when removal, the best alternative, is not possible. These chemicals are aluminum fluoride, calcium fluoride, calcium oxide, calcium hydroxide, calcium carbide, and lead arsenate. Finally, there are two materials, red and white phosphorus, that are very hazardous to aquatic life if left in place, but possibly even more hazardous to human life if their recovery were attempted. Depending on specific circumstances, burial with a reactive material may be the best response. (See discussion below.) Several other solid materials would be extremely hazardous to dredge or recover due to their explosive nature (e.g., 2,4-dinitroaniline; 2,4-dinitrophenol), but it is felt by the authors that they are sufficiently hazardous to the environment that recovery is nevertheless mandatory. All remaining solids on the list are quite toxic or otherwise hazardous, but offer no chemical barrier to recovery and, therefore, should be removed.

The following paragraphs briefly characterize those chemicals cited above as being amenable to burial, either as a first choice response or as a second or last resort.

Asphalts: The asphalt fraction that sinks should be buried with an inert covering to prevent its being moved around by currents or tides. The covering will also help to retard the slow microbiological attack of the asphalt mass. Asphalt is an ideal candidate for burial, because its toxicity is not great and because it would be difficult to recover by dredging.

Sulfur: The primary hazard caused by a sulfur spill is related to the spill event itself, when molten sulfur enters the water. After sinking, plastic sulfur may be slowly acted upon by bacteria to release sulfuric acid. An inactive covering material will, as in the case of asphalt, retard this phenomenon. A basic material may be added to neutralize any acid produced.

Barium Carbonate: Barium ions are quite toxic. However, barium sulfate is virtually insoluble in water. Hence, it is relatively nontoxic. An active covering material which contains sulfate (e.g., calcium sulfate, ferric sulfate) should be used to cover the barium carbonate spill. Seawater may contain enough sulfate to precipitate barium. An inactive covering material would retard diffusion of barium from the spilled mass.

Aluminum Fluoride and Calcium Fluoride: The toxicological problem with each of these compounds is with the release of fluoride. Because these two substances are among the most insoluble of the fluoride compounds, covering with an active covering that would chemically convert the spilled mass to a more insoluble state is unlikely. An anion exchanger is a possibility, but not in seawater where the chloride would compete with the fluoride. An inert covering material could be used to retard the diffusion of fluoride from the spilled mass to the water column. Each of these compounds would tend not to change chemical form in the aquatic environment.

Calcium Hydroxide, Calcium Oxide, and Calcium Carbide: When calcium oxide or calcium carbide are spilled into water, they form calcium hydroxide exothermally. (The calcium carbide reaction also releases highly flammable acetylene gas.) Calcium hydroxide is quite caustic and, since a saturated solution of this compound has a pH of 12.4, it is also quite toxic. An active covering agent should be able to neutralize the caustic material (i.e., the agent should be acidic).

Also, if iron or aluminum ions were available, they would react with the hydroxide to form hydrous oxides. An inert cover could be used to retard diffusion of hydroxide from the spilled mass. However, this will not mitigate the caustic mixing zone created by the actual spill. Calcium hydroxide will react with sulfates and carbonates (carbon dioxide) which are common to most aquatic environments.

Lead Arsenate: Both the lead and arsenic portions of this compound are poisonous. This material should generally be recovered to prevent its toxic release. Because the compound is soluble in acid conditions, an active covering material, which has basic properties, would be desirable if the covering alternative is used.

Phosphorus, White: This is an extremely dangerous material since it self-ignites in moist air at 30°C. The principal manufacturers of elemental phosphorus are Monsanto and FMC Corporation. They maintain their own response capability in case of a spill and prefer to be called in to handle any and all response efforts. They normally attempt to recover the material under water. If burial is necessitated, an active agent, gently placed, should be selected to convert the phosphorus from its elemental form as quickly as possible. These might include sulfur, pyrites, and solid oxidizing agents such as potassium permanganate to help oxidize the phosphorus in situ. Inert covering should not be used because the phosphorus is so toxic. It is important to avoid stirring up the spilled phosphorus in a colloidal form.

A phosphorus spill, as in the case of most of the organics, will cause a fish kill. As long as the elemental phosphorus remains in the aquatic environment, organisms will continue to die. While phosphorus is very hazardous to the aquatic environment and very hazardous to recover, balancing the hazards must be done on a case-by-case basis; at times covering will be the response of choice because of hazards to personnel involved in recovery efforts.

Phosphorus, Red: Very little of this material is shipped. It may explode with friction. An active covering material may be useful; response techniques should generally be similar to those for white phosphorus.

Burial Materials

Several materials, both naturally occurring and man-made, that could be used to cover submerged spills of hazardous chemicals have been identified and divided into four categories for discussion: inert, chemically active, chemical additives, and sealing agents.

Characterization of covering materials will include discussions of the chemical nature of the material (where applicable); the ability of the material to retard leaching into the water column; susceptibility of the cover to scour; and potential impacts on biota. In this regard, grain size will be an important parameter as it has a significant effect on both leaching retardation and susceptibility to scour and resuspension.

Inert Covering Materials: Previous work on the covering of in-place pollutants has frequently alluded to the use of inert substances in experimental covering. Depths of 10 cm were considered by Jernelov (16) to be capable of inhibiting certain biological activity which might contribute to leaching into the water column. Pratt and O'Connor (17) suggest benthic activity to be a limiting factor to consider in determining covering depths, with some species surviving a 20-cm burial. Covering depths, then, should probably be in this range (equivalent to a 4- to 8-inch cover) to inhibit leaching due to biological activity. Cover thickness must be considered site specific, however, depending on the local benthos.

A list of potential inert covering materials has been developed and is presented here as Table 2.4. Two broad categories of inert materials are identified: coarse-grained and fine-grained. Coarse-grained materials include sand, gravel, crushed stone, etc. Fine-grained materials include commercially available clays (often used as soil sealants), diatomaceous earth, and other materials. It is recognized that transition materials of silt-like grain size exist and will combine many of the properties associated with fine-grained and coarse-grained sediments. Dredged material was included as a separate category because of both its variability in predominant grain size from location to location and the ease with which it can be obtained. Grain-size distribution and pollutant loads will be site-specific, but nearby availability of in situ covering material is assumed regardless of the spill location.

Table 2.4: Generalized List of Inert Covering Materials

Coarse-Grained	Fine-Grained	Mixtures
Gravel	Kaolin clay	Any combination of these materials, including dredged material and deliberate mixtures
Sand	Bentonite clay	
Crushed stone	Fuller's earth clay	
Crushed glass	Ball clay	
	Fire clay	
	Miscellaneous clays (local)	
	Diatomaceous earth/filter aid	

Source: CG-D-56-78

Susceptibility to Scour and Resuspension – Erosion of the bottom of a waterway with spilled hazardous chemicals on the bottom may be of concern for two reasons:

- Covering materials that have been placed on the spilled chemicals may be removed, thereby frustrating attempts at burial; and
- The spilled chemical itself, or sediments contaminated by it, may be eroded and dispersed.

The second effect would, of course, not occur if the covering material can be prevented from scouring. Therefore, it is the erosion of the burial material that is of primary concern and the subject of the following discussion.

Susceptibility of a sediment to scour will depend on: the particles (size, uniformity, shape, size distribution, texture, etc.); the dynamics of flow; slope of the bottom; angle of repose of the particles; and the degree of cohesiveness of a particular sediment. Thus, predicting scour and resuspension tendencies of such covering materials cannot be made in a general sense. Unfortunately, most of the available literature applicable to this subject is concerned only with natural materials (sand, silt, and clay). Application of erodibility considerations to the many man-made or man-emplaced burial materials discussed in this report is consequently based on theory and extrapolations.

From the viewpoint of erodibility, sediment used for covering may be generally classified as cohesive or noncohesive. As the term implies, noncohesive sediment consists of discrete particles the movement of which, for given erosive forces, depends only on particle properties, such as shape, size, and density, and on the relative position of the particle with respect to surrounding particles. For cohesive sediments, the resistance to any movement or erosion depends on those factors cited above for noncohesive sediment as well as on the strength of the cohesive bond between particles. This latter resisting force may be much more important than the influence of the characteristics of the individual particles.

Because the water currents required to erode or scour cohesive sediments are generally greater than those for noncohesive sediment with approximately the same grain size, the rate and extent of scour depends on this property (cohesive strength) rather than on the properties of the individual particles. Once the cohesive bond has been broken, the individual particles behave as noncohesive particles for which deposition, scour, and transport become functions of the properties of the separate particles or small groups of particles.

Additionally, once resuspended, the hydrodynamic behavior of cohesive sediments is complicated by the effects of flocculation. Floc size distribution depends not only on the physicochemical properties of the particles, but also on the flow conditions themselves. This dual dependence makes the processes of erosion, transport, and deposition of fine (cohesive) sediment fundamentally different from and considerably more complex than similar processes for noncohesive sediment.

Investigations (18)-(21) have also shown that for fine-grained sediments in particular the degree of consolidation, inversely proportional to the interstitial water content, has a significant effect on the ease with which they will erode. Whereas recently deposited (unconsolidated) clays would be among the most easily resuspended sediments, aged (well-consolidated) clays may be among the most resistant materials to scour, approaching the stability of coarse gravel. The time required for essentially complete consolidation to take place can range from minutes for coarse-grained deposited materials to hundreds of years for fine clays.

The subject of scour and resuspension of sediments defies generalization. However, for guidance of those who must base important spill response decisions on available information, Table 2.5 is presented as a qualitative guide to erosion and resuspension potential of various possible inert cover materials. In the subsequent section on active covering materials, similar considerations are presented.

Table 2.5: Tendencies of Natural Inert Sediments to Erode

Material	Approximate Range of Erosion Velocities (fps)	Mixtures
Coarse sand/gravel	1.3-9.8	Depends on grain-size distribution and mineralogy of fines, but generally more stable than covers of uniform grain size
Medium sand	0.7-1.3	
Fine sand	0.7	

(continued)

Table 2.5: (continued)

Material	Approximate Range of Erosion Velocities (fps)	Mixtures
Silt		
Consolidated	1.0-2.3	
Unconsolidated	0.3-0.7	Depends on grain-size distribution and mineralogy of fines, but generally more stable than covers of uniform grain size
Clay		
Consolidated	1.6-9.8	
Unconsolidated	0.2-0.5	

Source: CG-D-56-78

The need for accurate data on currents at the spill site cannot be overemphasized. For example, if consolidation of a cover is desired, knowledge of seasonal or storm-induced high currents is necessary to assess the likelihood of scour during the consolidation period.

Ability to Retard Leaching – Leaching of pollutants through cover material is expected to be directly related to grain size; the larger the grain size, the more leaching that will occur.

Permeability of materials of various grain sizes can be taken as an indicator of ability to retard leaching (the two will be considered inversely proportional). The coarser the material, the more permeable it is. Permeabilities reported in Lambe and Whitman (21) show clays being classified as generally being practically impermeable or having very low permeability. Sands are generally classified as of high permeability. Silts would vary between these two classifications. Permeability is also dependent on such factors as void ratio and composition.

Clays and materials with clay-sized particles such as bentonite find application as liners in ponds, lagoons, and similar bodies of water as an impermeable layer designed to retard leaching and aid in fluid retention. Similarly, clays would be the best type of inert cover for retarding leaching of spilled chemicals on a waterway bottom, if other factors are also favorable (e.g., scour resistance and ease of emplacement).

In addition to the effect of grain size on permeability and therefore leaching, the ability of a cover material to adsorb certain chemicals and its ion exchange capacity are important considerations in assessing its ability to retard leaching. Adsorptive and ion exchange capacities apply mainly to clays and are very site-specific, depending on the aquatic chemistry, the particular clay minerals present in the cover, the grain sizes, and the chemical that is being covered. In general, the ability of clays to physically/chemically tie up certain substances (mainly organics and metals) must be considered an additional advantage in their suitability as a covering agent.

Impact on Biota – There are two areas of concern regarding biotic effects: immediate impact of the dump and long-term effects of the bottom cover. It must be anticipated that a hazardous chemical spill that is to be covered will have already created a very impoverished benthic community, and therefore the immediate biotic effects of deploying the cover materials should be negligible. Indeed, any covering layer should be of sufficient thickness to ensure destruction of any surviving benthic organisms in order to preclude reexposure of the spilled chemical through escape holes. Covering depths for this purpose depend on the ability of the indigenous organisms to escape. This will vary from 10 cm for *Anodonta* (16), 20 cm for *Nephtys incisa* and *Mulinia lateralis* (22), and up to 50 cm for some species of deep burrowing bivalves.

For long-term effects, spill response coordinators should consider the bottom characteristics indigenous to the site. The biota of the area likely to recolonize the covered site will be attracted by material characteristics (especially grain-size distribution) compatible with the natural bottom sediments. In many cases, it may be desirable to select an incompatible cover material, one that will discourage recolonization and thereby preclude reexposure of the contaminant by organisms burrowing from the top. If recolonization occurs, depths should be greater than the burrowing ability of the new benthic organisms; this depth will generally be less than the depth through which the same organisms can escape.

Turbidity and suspended solids can also create biotic impacts in the water column during and after a covering operation. Fine-grained materials are generally more likely to cause these problems than coarse-grained materials in the short term. Bentonite, for example, has been known to cause fish kills due to gill-clogging during aquatic applications. After consolidation, however, fine-grained materials will have a lower tendency to be eroded and resuspended, which will lower their impact on biota.

Active Covering Materials: One covering strategy mentioned earlier involves the emplacement of a second chemical compound on the spilled hazardous chemical. This second compound would be chemically active (i.e., it would readily react with the spilled compound to neutralize or otherwise decrease its inherent toxicity). However, the active covering strategy differs from the inactive covering strategy because each spilled compound must be dealt with on a case-by-case basis. There are no universal active covering agents.

While active covering agents must be dealt with individually, some general conclusions can be drawn as to the importance of specific in-place properties such as chemical compositions, leaching, etc. In this regard, it is useful to draw some general parallels between the active and inert covering agents, followed by discussions specific to a number of obvious covering possibilities applicable to the hazardous materials under study.

Susceptibility to Scour and Erosion: There are some important differences between the active and inactive materials in terms of their in-place properties. First, with respect to susceptibility to scour and resuspension, the purpose of an active cover is to react with the spill, not solely to cover it. It is important only that the active cover remain in place long enough for this reaction to take place.

Since most of the active materials are fine-grained, many of the considerations (turbidity generation, in-place erodibility) discussed with the clays will be pertinent here. To alleviate the possibility of scour or erosion prior to inactivation of the spilled material, the active covering material could either be used alone or in concert with an inert stabilizer. In this case, the active material could be mixed into an inert base, the inert base acting as an antierosion vehicle. Mixing could occur in a barge, on land, or perhaps through slurry injection into a pipeline discharge. Another option would be to cover the active covering layer with a scour-resistant inert layer.

It should be noted that some active materials (gypsum, limestone) are pozzolanic in nature: that is, they tend to form cements. In this respect, these active covering materials may also be considered as a counter-erosion measure. Mixture with sand or a similar inert material could form a thick cement-like cover over the spill.

Ability to Retard Leaching – Another major difference between the active and inert covering materials is that the ability to retard leaching is not considered the key to the successful application of an active material. Most of the materials under consideration here are fine-grained, and might have some similarities to clays and silts in this regard, though a generalization is hard to draw. Should retardation of leaching become important, a mixture with or subsequent layer of a suitable inert material could likely be used to reduce permeability.

Chemical Characteristics/Impact on Biota – While the inert covers have little or no chemically related impact on biota, the active covering agents do. Indeed, impacts on biota may result in the early elimination of some potential covering materials from consideration for some spill scenarios. While accuracy of placement of inert material is important only from the point of view of cost effectiveness, the misplacing of active materials could be harmful to some organisms that had not been in the hazard zone of the spill itself.

Selection of Covering Materials – In the literature cited at the end of this chapter the authors have suggested a number of materials which have been considered for use as active covering agents. These considerations have always been made on a case-by-case basis, i.e., a specific active compound to react with a specific contaminant or spill under a specific set of environmental conditions (e.g., work related to covering of mercury-contaminated sediments). It is not sufficient to provide a fixed list of candidates for active covering materials as has been done with the inactive materials. Each spilled compound must be separately evaluated by a chemist familiar with both the compound and its chemistry and the environmental conditions (salinity, pollutant loads, etc.) at the spill site. Only with these factors fully understood can an informed selection of an active covering material be made.

A typical list of active covering materials, developed through reference to the hazardous material literature, is presented in Table 2.6. Many of these are discussed in later chapters concerning chemical methods or sorption and related techniques. Recommendations for specific chemicals are given in Table 2.7.

Table 2.6: Active Covering Materials

Material	Active Agent(s) or Mechanism
Scrap iron	Iron oxide absorption
Sulfide ores–pyrite	Iron, sulfide
Clays	Adsorption
Diatomaceous earth	Adsorption
Manganese dioxide	Adsorption

(continued)

Table 2.6: (continued)

Material	Active Agent(s) or Mechanism
Proteinaceous wastes	Sulfide
Wool	
Chicken feathers	
Xanthates	
Carbon compounds	Adsorption (especially organics)
Activated carbon	
Lamp black	
Bone char	
Charcoal	
Calcium carbonates	Carbonate, acid neutralization
Limestone	
Lime	
Chalk	
Stucco	
Spent tannery lime	
Gypsum (calcium sulfate)	Sulfate
Sulfur	Sulfur
Potassium Permanganate	Oxidizing agent
Alum (aluminum sulfate)	Aluminum, sulfate
Alumina (aluminum oxide)	Adsorption
Ferric sulfate	Iron, sulfate
Commercial ion exchangers	Ion exchange

Source: CG-D-56-78

Table 2.7: Recommended Covering Materials for Specific Chemicals

Spilled Chemical	Recommended Cover
Aluminum fluoride	Inert*
Calcium fluoride	Alumina
Asphalt	Inert
Barium carbonate	Inert
	Calcium sulfate (gypsum)
	Ferric sulfate
Calcium hydroxide	Inert
	Sulfates
	Carbonates–limestone
Lead arsenate	Inert
	Basic materials–limestone
Phosphorus, red or white	Sulfur
	Pyrite
	Solid oxidizing agents (e.g., potassium permanganate)
Sulfur	Inert
	Basic materials–limestone
Organic chemicals	Inert materials with low permeability and high adsorptive capacity (clays)

*Preferably naturally available materials similar to bottom sediments at the spill site

Source: CG-D-56-78

Chemical Additives: These agents are chemicals which can be added in some manner to a covering material in order to modify bottom chemistry and retard the entrance of spilled materials into the water column. Such additives will not be universally applicable to spills of the materials under study, though preliminary investigation reveals that the effects of spills of barium carbonate and lead arsenate could be lessened through such techniques. Due to the fact that application of these chemicals is not generally through simple burial techniques, discussion will be reserved for the chapter on chemical methods.

Sealing Agents: Sealing of the surface of a hazardous material spill is another alternative to inert covering. Such sealing could be done using concrete and grouts, gelling agents, or polymer films. In the case of concrete and grouts, it is possible to conceive of sealing being accomplished either in conjunction with in situ bottom material or as an entirely separate process using material pumped from the surface. Gelling agent use will require some advancement in technology from that available at this time. Most efforts to date have been conceived with land and/or surface water gel applications as discussed in a later chapter. Polymer films have been laboratory tested, but have not undergone full-scale field tests, so their technology remains somewhat underdeveloped.

Cement-Forming Mixtures – A Japanese firm (23) has done work in dredged material stabilization and deep-mixing of sediments using cement-forming compounds. Inorganics such as portland cement, gypsum, lime, and water glass have been used to solidify both sludges and in-place sediments. Organic materials such as chemical grouts, asphalts, and resins have been dismissed by the Japanese on the basis of cost. Deep-mixing has been accomplished with success. Methods using portland cement and lime appear to have particular significance for this project.

TRW (24) has studied the stabilization of wastes using both inorganic and organic cementation techniques. (Organic cementation techniques require heating and are not considered further for the purposes of this report.) Inorganic cements were classified as those materials which, when mixed with water, will form pastes that will harden in air or under water. Examples given include portland cement, lime, plaster of Paris, and calcium aluminate cements. Inorganic cement-forming mixtures were classified as mixtures, not necessarily cementitious in themselves, that would react with each other to form hardening compounds. Lime-clay, lime-pozzolan, and calcium aluminate-calcium sulfate were suggested. (Pozzolans include such material as diatomaceous earth, fly ash, and pumicites.) The lime-clay mixture suggests an immediate application of hardening agents in muddy harbor bottoms and, in fact, has been represented in the Japanese method outlined above.

The unique feature of all of these is that, when placed on top of or mixed in some manner with bottom sediments, they will tend to harden and form a crust, preventing erosion and resuspension of spilled material. They are primarily applicable in this regard to spilled solids. A study conducted by Walley (25) on the erosion resistance of freshly injected grout (cement-like in nature) to flowing water (about 0.3 fps) revealed that, as a general rule, short gel or set times were crucial to successful placement of grout. This applied whether the grout was a typical suspension grout (e.g., cement-sand-water mixture) or any of the chemical grouts available. Experiments showed that the fast-setting grouts were the most resistant to erosion and dilution. Grout mixtures involving Wyoming bentonite (a thixotropic or expanding variety) were also noted as showing a marked resistance to erosion. The bentonite-containing grouts were noted to be immobile after placement due to rapid gelling. This finding is not expected to hold under turbulent flow conditions.

Information on in-place properties of the grouts, especially the properties related to impact on the environment, appears to be sparse. Sales literature from a number of companies involved in grouting tends to center on engineering properties of treated soil, permeability, and the like. Low levels of permeability have been obtained, though permeability will vary with the specific grout, soil conditions, and application technique. In-place grout is generally inert and is not severely affected by acids, alkalis, salts, or similar substances. However, the unset grouting mixtures themselves, especially chemical grouts, can be toxic and require special handling (26).

There are a variety of potential grout/pozzolan delivery systems available for use. Pozzolans in their basic form could possibly be spread over the surface of the water (perhaps using a form of broadcast spreader) and allowed to sink in order to cover the bottom of a spill site. Hydration of the material prior to emplacement, followed by pressure injection, is also a possibility. It should be noted that the use of certain pozzolanic materials can impact the spill both as a chemically active cover (e.g., to neutralize acid) as well as an impermeable barrier. Limestone has a definite beneficial impact on spills of lead arsenate while gypsum is suggested as an active covering agent for barium carbonate. Information is scarce in the areas of spreading techniques, optimum grain size of material, limitations, and other topics necessary to devise a placement plan, even using this simple method.

Helicopter and truck transportable grouting units have been developed and these could be used in a strike-force configuration if required. These units are generally pressurized grouting systems. In the case of helicopter deployable units, mobilization could occur in less than a day. Units are available (27) for offshore oil applications which permit the addition of carefully measured liquid additives in order to control such characteristics of the grouting mixture as density, setting time, and the friction factor of the grout, permitting use of the same basic mixture under a wide range of conditions. This would be useful in the event that selected chemicals were to be added to the grout in order to react with those spilled on the bottom or in order to configure the high-density, fast-setting grouts necessary for placement in the flowing water.

It remains necessary to devise a diffuser for use on the bottom in conjunction with pressurized grouting systems. It does not seem feasible to inject the grout into the bottom sediments, as that might tend to resuspend both sediments and spilled material. Placement on top of the material seems most feasible. A diffuser device, which would lay the grout down in even bands, would be most useful. This would increase the chance of providing an even, unbroken cover over the spilled area. Placement of the diffuser would be difficult though it could be checked by remote television or perhaps by divers, depending on water clarity and on the material being covered. A hydraulic crane, or other rigid positioning system, might prove able to provide accurate placement and control of the diffusion head.

Grouting would probably find its best use as a barrier in low current areas. The most useful application would likely be in conjunction with chemical additives designed to neutralize in some manner those solid sinking hazardous materials trapped below it. It is not anticipated that the system would prove useful for liquid spills unless the liquid migrates completely into the sediment.

Grouting would probably be semipermanent after hardening. It would probably completely sterilize the bottom on which it was placed. It is possible that once the chemical underneath it was neutralized or otherwise rendered innocuous, the grout could be removed using a cutterhead dredge. If permanent coverage was desired, the grout could be covered with perhaps a 1-ft-thick layer, or more, of indigenous bottom material to allow recolonization by resident biota. As a result of the slowness of application and semipermanent nature of certain grout-cement mixtures, they must be rated as having moderate to high overall potential as a cover.

Soil Sealants – Commercially available soil sealants of the Wyoming or expanding bentonite variety may prove useful in the covering of spills and sinking materials. Their primary use is expected to be with spilled chemicals that are solid and sink or that will bond temporarily to sediments and have a tendency to leach into the water column with time. These sealants result in very low permeabilities in the sealed material. The technique could prove to be useful, though it probably would not prove to be efficient in an emergency situation due to limited supplies of the covering material at any particular spill site.

Two commercially available soil sealants for use in contaminated water or water containing significant concentrations of salt have been identified: Dowell Soil Sealant, manufactured by the Dowell Division of the Dow Chemical Company, Tulsa, Oklahoma (28), and Volclay Saline Seal-100, manufactured by the American Colloid Company, Skokie, Illinois (29). Both are Wyoming-grade (expanding) bentonite based. Their primary use is in the sealing of sanitary landfills, sewage and process plant lagoons, and similar applications. These salt-resistant grades of soil sealant can be successfully used in salt water.

Traditionally, soil sealants are best worked into the impoundment bottom prior to filling. In emergency situations, however, this will not be possible and it will be necessary to apply the mixture from the water surface at the spill site. In salt water or contaminated fresh water, it will be necessary to prehydrate the material into a slurry. (It has been applied dry by shovel from a rowboat in seal repair on freshwater reservoirs, but it is not expected that conditions favorable to this application method will occur in many emergency situations.) Mixing could be done with mechanical mixers, recirculation mixers, or portable air-driven or electrical-driven mixers using 55-gal drums. Application rates would be slow: perhaps 20 to 30 gal of mix every ½ hour. For higher application rates, it might be necessary to adapt a grouting mixer and pump-down device.

These self-contained units can achieve up to 70 cu ft/min of coverage, as was reported in a previous section. Essentially the same rig as used for grouting, including spreader bar, could be adapted for use with the bentonite slurry. A source of fresh water for prehydration would have to be supplied. (It should be noted that bentonite is often used in the grouts and drilling muds used in the offshore oil industry.) Hand mixing for small spills is possible, though it would be very slow.

Soil sealants would be best used in quiet waters, areas such as drainage ditches, coves, and off channel areas of ports and harbors seeing relatively little tide and current action. Wading and burrowing animals may also prove a problem, though in areas where the material is to be ultimately dredged this will not prove to be serious. Where long-term use of the bentonite sealant is anticipated, consideration should be given to the addition of coarse-grained material to the slurry to inhibit erosion. Information concerning impact on biota is not available. However, it should be noted that no special handling procedures for either material were noted by either the Dow Chemical Company (28) or the American Colloid Company (29). Toxicity of Dowell Soil Sealant was low, and a 1 mg/l concentration in potable water is not considered toxic to humans (28).

Saline Seal-100 and Dowell Soil Sealants are generally available in truckload quantities (such as might be used on a spill) on 7 days notice. Due to the long lead times involved, stockpiling might be necessary.

In summary, these materials will likely find their best use in protected waters. This would include nonnavigable waters and quiescent areas of rivers, ports, and harbors. Use in open waters, due largely to environmental considerations, is not recommended. Pumps and grout injecting systems are capable of pumping up to distances of several thousand feet, so water depth will not prove a significant problem. It should be noted, however, that problems in locating the spill and of access by support divers will increase with depth. Use of surface-applied sealants will likely be confined to very shallow water and drainage ditch spills. Considering the limitations, soil sealant usage can be considered only of moderate potential as a response technique, due largely to cost and erosion consideration.

Polymer Covers – Early work on coatings, originally intended for ocean bottom stabilization during salvage work, was reported by Epstein and Widman (30)(31) and Roe et al (32). Epstein and Widman (30), in discussing work sponsored by the U.S. Navy, stated that an effective ocean bottom stabilization system must operate under a wide range of environmental conditions, including a current of up to 3 knots and depth of up to 850 ft. It should also be designed to be effective in both fresh and salt water. Laboratory examinations showed that a sodium alginate system was found to produce potentially useful gels, though it has problems such as shrinkage and nonbondability. It was also subject to deterioration due to the mineral content of seawater.

Roe et al (32) reported on concurrent work at the Naval Civil Engineering Laboratory and stated that a polymer film overlay system would best achieve the goal of bottom stabilization. Bonding of adjacent strips was a desirable feature of this method. Flocculating agents and soil stabilization were also considered. Flocculating agents did not keep sediment from resuspending if disturbed and stabilization of soil required injections of large amounts of binding materials into the sediment in what was felt to be a semipermanent procedure. It should be noted, with reference to this study, that permanence is not necessarily desired in a bottom coating, as many spilled materials could be safely removed after initial control of the spill is obtained.

Extension of the polymer film cover concept to mercury pollution control was reported in the work by Widman and Epstein (33). In this report, the sodium alginate films were found to be highly biodegradable and, due to their high water content, were not considered to be good pollution barriers. It was from this work that the concept of using preformed commercially available films was formulated. Erosion-control capabilities of a polymer overlay could be useful in combined usage with, for example, a fine-grained soil sealant such as one of the bentonite-related materials.

Widman and Epstein (33) report that the most effective film material was preformed nylon 6 (polycaprolactum), shown to be an effective barrier to both inorganic and organic mercury compounds. Other films tested [high-density polyethylene, low-density polyethylene, polyvinyl chloride, alcohol-soluble nylon, and poly(ethylene-vinyl acetate) copolymer hot melts] were equally satisfactory against inorganic mercury compounds. They varied in performance against organic mercury compounds, none being as effective as nylon 6.

There is an indication that the preformed film overlay system would be deployable in water from 25 to 30 ft deep (33). This would make it usable in most nonchannel areas of harbors and ports, as well as rivers. Nonnavigable water could not be traversed by the application equipment, while open waters would possibly be too deep for effective application. Difficulties with its use include the necessity of vent holes to allow trapped gases to escape, weighting, and the possibility of puncturing. The materials are not expected to cause environmental problems, other than the obvious smothering effects on benthos.

There are several current candidates for use as covers. A number of polymeric liner materials (including some tested by Widman and Epstein) have been reported as being used for sanitary landfills (34), among them: polyethylene, polyvinyl chloride, butyl rubber, Dupont Hypalon (a synthetic rubber), ethylene propylene diene monomer, and chlorinated polyethylene. Polyethylene, the least expensive, has good chemical but poor puncture resistance. It might be used on smooth bottoms or as a temporary measure. Polyvinyl chloride is tolerant of a number of chemicals, but becomes stiff in cold weather. Butyl rubber is not recommended for cover of hydrocarbons and solvents, but is resistant to water-based inorganic salts, sewage, oxidizing chemicals, and animal and vegetable oils and fats. Hypalon suffers largely from high cost, but is also low in tensile strength. Ethylene propylene diene monomer (EPDM) is not recommended for solvents or hydrocarbons, but is resistant to mild concentrations of acids, caustics, and other chemicals. Chlorinated polyethylene has low resistance to chemicals, acids, and caustics and is not considered a good prospect.

Polymeric covers should be considered of moderate potential, due primarily to the capital equipment required to place them.

Emplacement and Erosion Control Measures

Several methods for the emplacement of covering materials have been investigated. Initial attention was paid to existing methods of dredged material placement: point dumping from scows, barges, and hopper dredges and open-pipe discharges from hydraulic dredges. These methods, readily available and capable of placing a cover, were found to have a number of drawbacks requiring further attention (uneven cover, turbulent impact on bottom and scouring). Several alternatives to these techniques, such as hopper dredge pump-down or sand-spray systems, hydraulic dredge pump-down, and submerged discharge apparatus for hydraulic dredges, have potential for emplacement, although at an increase in cost for new equipment. Further attention was given to special-process apparatus (film roller systems, deep chemical mixing apparatus) as well as low-potential techniques such as plowing and the use of grab-buckets for emplacement.

Silt curtains, borrow pits, and submerged dikes have been mentioned in the literature as possible turbidity or dispersion control methods for dredging operations and could be used in conjunction with the covering methods outlined above. Control of the covering material during emplacement is a concern from two points of view. First, the covering material may tend to disperse during emplacement. This concern would especially apply to fine-grained materials, which may form a mud-flow capable of travelling great distances under water. This kind of dispersion must be avoided to minimize loss of covering material. Second, turbidity caused by the covering action would have a negative effect on biota in the area. Thus, dispersion should be controlled for both economic and environmental reasons.

Erosion control methods, such as artificial seaweeds, submerged dikes, and artificial breakwaters, may be necessary in high current areas. Scour velocities will vary with the covering material. Cases of erosion of the covering layer could occur, re-exposing the spilled hazardous material. Erosion control measures could reduce these effects. The possibility also exists of downstream or down-current trenches being excavated to catch any cover and hazardous material that might erode away. These methods are beyond the scope of this section on burial techniques, however.

BARRIERS, DIKES AND TRENCHES

An Overview

> The information in this section is based on *Manual for the Control of Hazardous Material Spills. Volume 1: Spill Assessment and Water Treatment Techniques*, EPA-600/2-77-227 prepared by K.R. Huibregtse, R.C. Scholz, R.E. Wullschleger, J.H. Moser, E.R. Bollinger and C.A. Hansen of Envirex for the Environmental Protection Agency.

Figures 2.1 through 2.5 provide a graphic illustration of the types of containment measures considered in this chapter.

Figure 2.1: Containment of Spills on Land

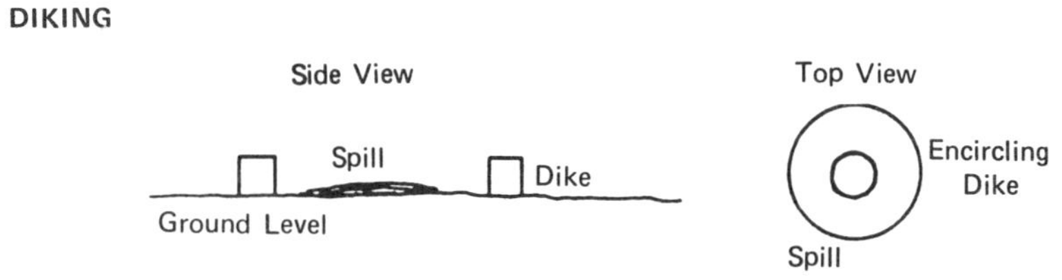

(continued)

Figure 2.1: (continued)

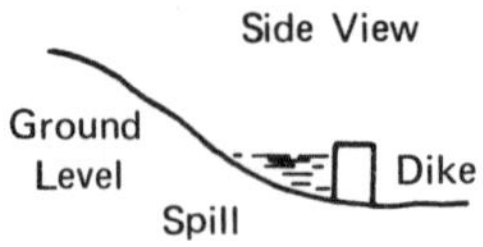

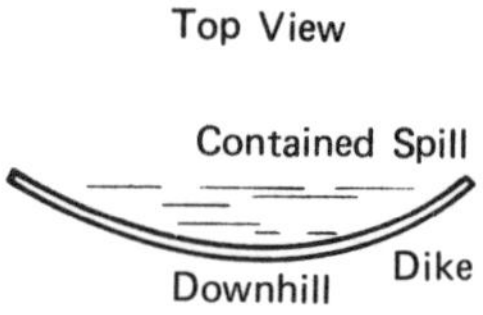

1. Suitable dike materials must be available, either natural soil, sand bags or foam.
2. Proper earth-moving equipment or specialized foam dike equipment must be obtainable.
3. An advantageous site must be available.
4. This procedure may not eliminate percolation of the spill through the soil.

EXCAVATIONS

1. Equipment and land must be available to accommodate the excavation.
2. In certain areas, soil or subsoil nature may render excavation impossible or ineffective.

Source: EPA-600/2-77-227

Figure 2.2: Containment of Spills Heavier than Water

EXCAVATION AND DIKING

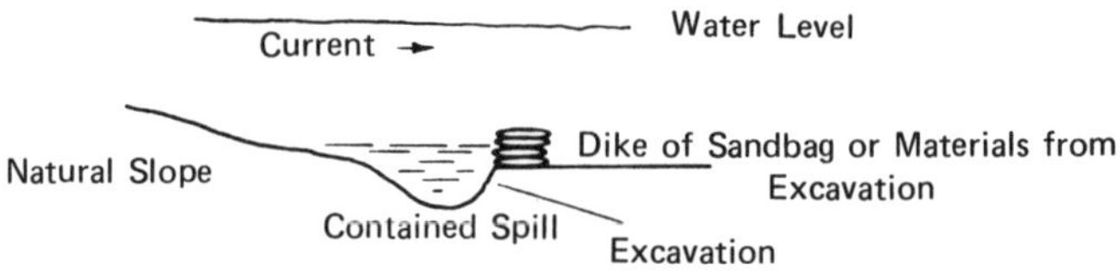

1. Difficulties may be encountered when attempting excavations under water.
2. Natural forces such as currents or slope of the bottom may be utilized advantageously.
3. Increased turbidity may hamper the activity of scuba divers.
4. Material needs may include divers, pumps, earth-moving equipment, dredges, and diking material.

Source: EPA-600/2-77-227

Figure 2.3: Containment of an Entire Water Mass

DIVERSIONS

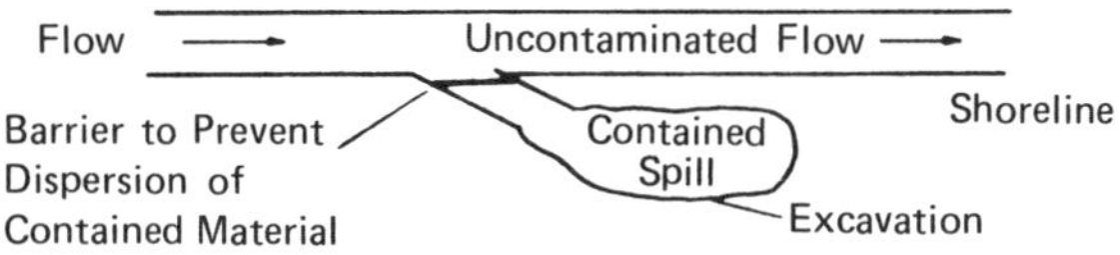

1. The equipment and suitable land areas must be available for a large excavation upon short notice.
2. An effective method of back-filling the excavation must be available.
3. It is possible to use pumps for stream diversion.

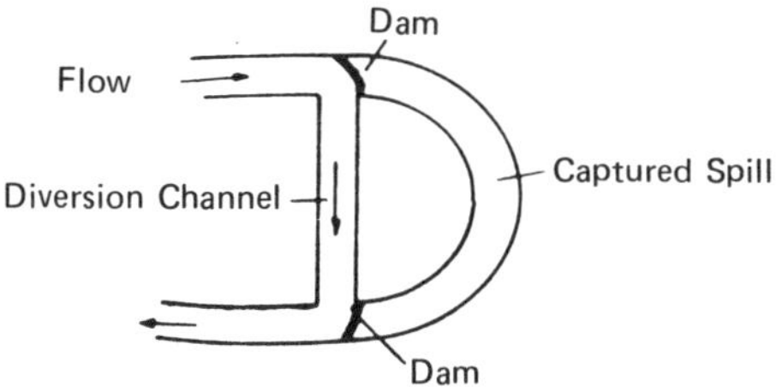

1. Equipment and suitable land area must be available for a large excavation upon short notice.
2. An effective method of enclosing the spill with dams must be available.

GELLING

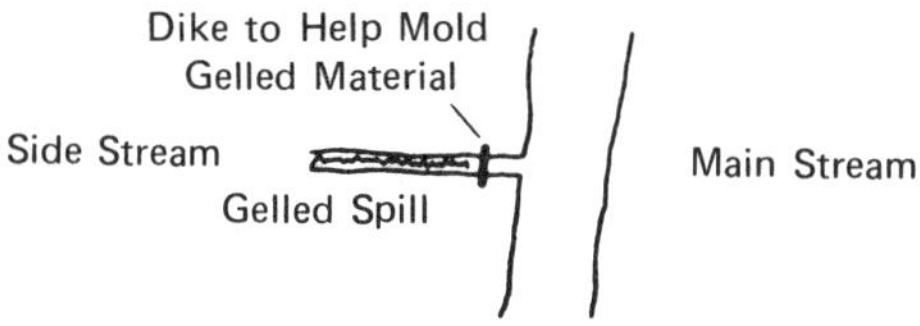

1. Gelling is useful when a limited volume of waste is to be treated.
2. Treatment and/or disposal of the entire gelled mass is necessary.
3. Application of gelling agents must be implemented a short time after the spill occurs.
4. Trained personnel and specialized equipment must be available.

(continued)

Figure 2.3: (continued)

CONTAINING AN ENTIRE WATER BODY

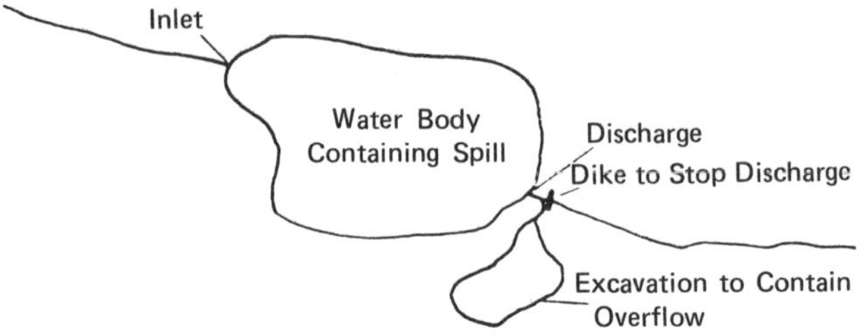

1. Voluminous overflows may be difficult to retain.
2. Earthmoving equipment must be obtainable.
3. When more than one overflow originates from a water body, all overflows must be contained.

SEALED BOOMS

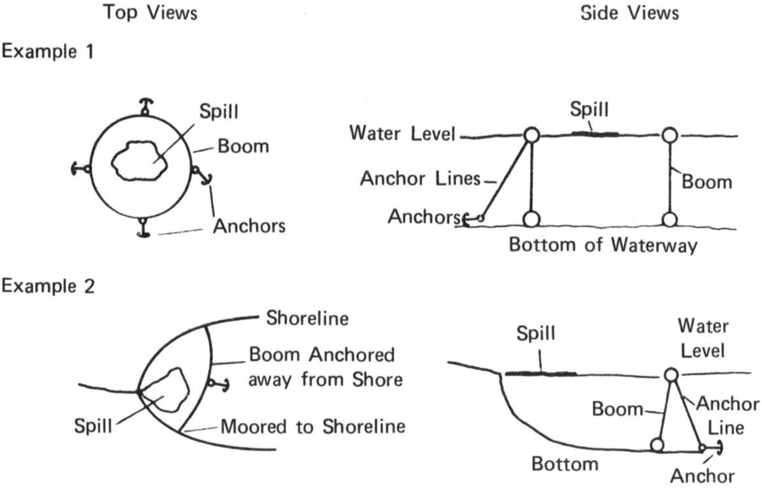

1. The spill must be of limited volume.
2. Wind or current may render this containment method ineffective.
3. Difficulties in obtaining a sealed boom system may be encountered.
4. This containment method is applicable in shallow water only due to bottom seal and anchoring difficulties.

Source: EPA-600/2-77-227

Figure 2.4: Containment of Floating Spills

WEIRS

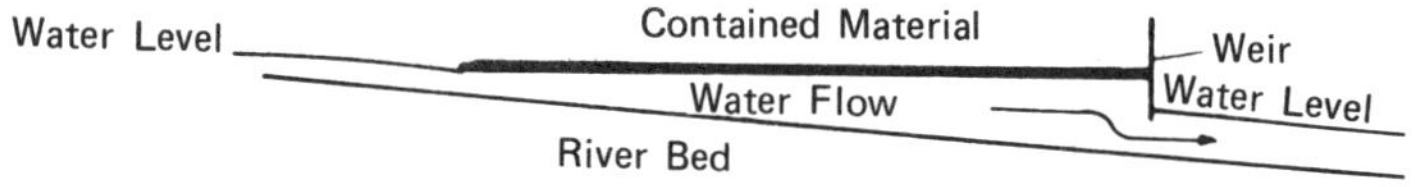

1. Weirs may be difficult to deploy properly.
2. Turbulence behind the weir may render this containment method ineffective.
3. To be effective, the spilled material must be in the upper layer of water.

USING FIREHOSES OR PROPWASH TO DIRECT AND CONTAIN SPILLED MATERIALS

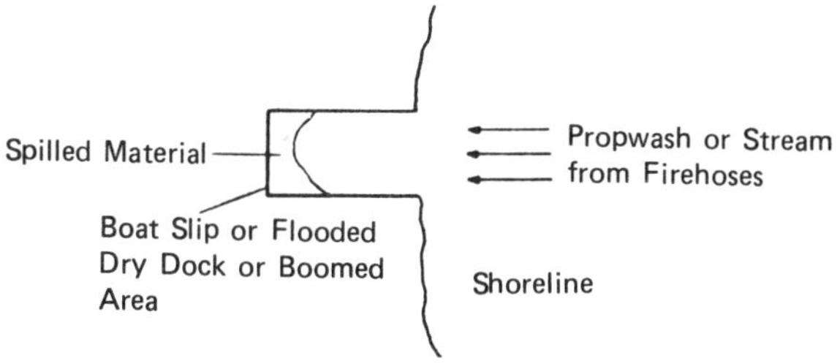

1. Adverse winds or currents may render this method ineffective.
2. Firehoses or ships must be obtainable, and have access to the spill site.
3. A suitable containment area must be available.
4. Winds and currents must be taken into account.
5. Impact water with fire stream at least 6.1 to 9.2 m (20 to 30 ft) away from spill.

PNEUMATIC BARRIERS

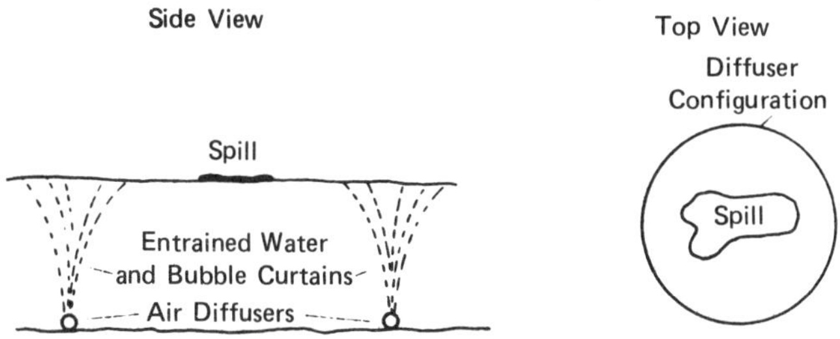

1. Wind or current may render this containment method ineffective.
2. Obtaining proper diffusers and compressors may be difficult.
3. Deep water requires suspension of diffusers in the water column to increase efficiency of unit.
4. Rarely used.

(continued)

Figure 2.4: (continued)

DIVERSIONARY AND CONTAINMENT CONFIGURATIONS

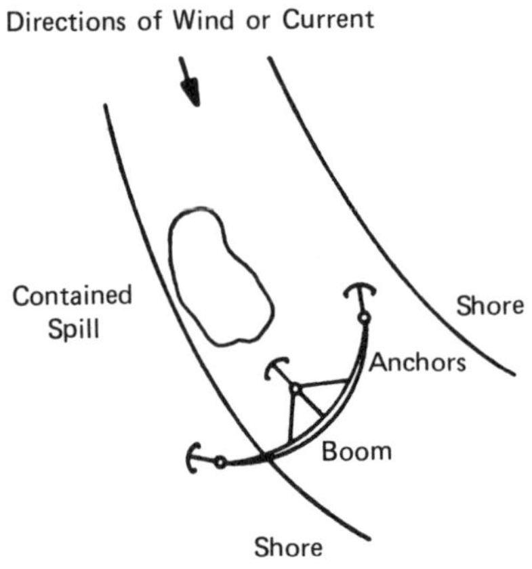

1. Whenever possible use with outside bend and sufficient clearance to reach site.
2. An intermediate tie is needed to keep bucket from forming.
3. Technique is limited to currents up to 2.4 to 3.1 mps (8 to 10 fps).
4. Proper booms and deployment systems may be difficult to obtain.

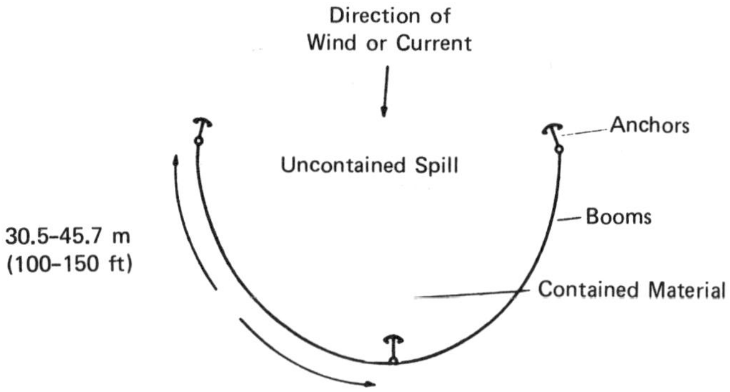

1. Heavy seas may render this containment method ineffective.
2. Wind or current shifts may render this containment method ineffective.
3. The proper booms and deployment devices may be difficult to obtain.
4. This technique is limited to currents up to 0.3 to 0.61 mps (1 to 2 fps).

(continued)

Figure 2.4: (continued)

ENCIRCLING BOOM

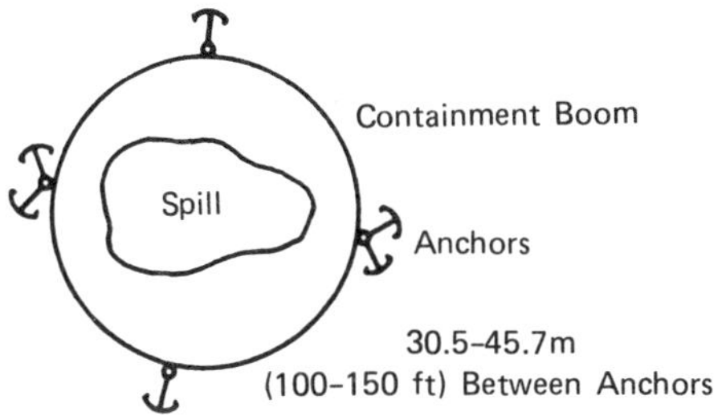

1. Heavy seas may render containment with this method ineffective.
2. Wind or current shifts may render this containment method ineffective.
3. The proper booms and deployment systems may be difficult to obtain.

COLLECTION AND TOWING OF A SPILLED MATERIAL WITH A BOOM

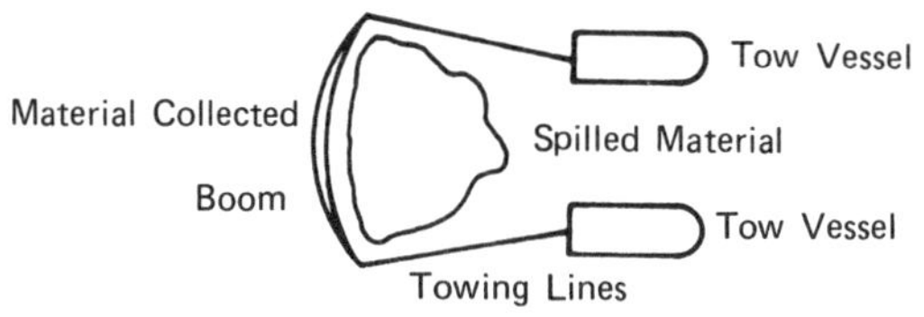

1. Heavy seas may render containment with this method ineffective.
2. Wind or current shifts may render this containment method ineffective.
3. The proper booms and deployment systems may be difficult to obtain.
4. Towing speed may be limited to 2.8 km/hr (1.5 knots) or less.

Source: EPA-600/2-77-227

Figure 2.5: Suppression of Air Spills

MISTING TO REMOVE CONTAMINANTS

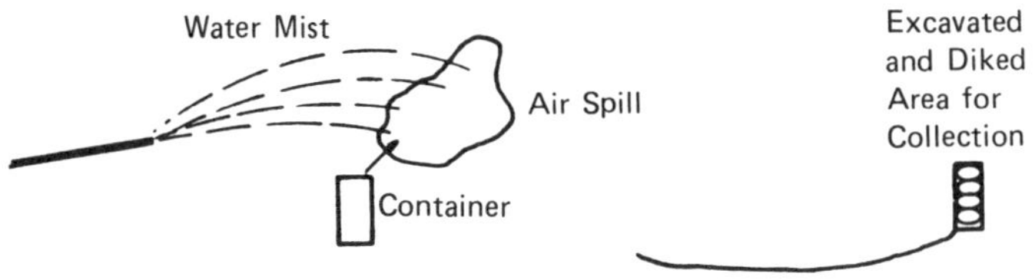

1. Not all materials will be removed in this manner.
2. Water source must be available.
3. Require large area for containment of resulting water.

Source: EPA-600/2-77-227

In-Water and Submerged Dikes

The information in this and the following section is based on *A Feasibility Study of Response Techniques for Discharges of Hazardous Chemicals that Sink,* CG-D-56-78 prepared by T.D. Hand, A.W. Ford, P.G. Malone, D.W. Thompson and R.B. Mercer of the U.S. Army Engineer Waterways Experiment Station for the U.S. Coast Guard.

Dikes could be used to retain spills of sinking hazardous materials, especially those liquids that might have a tendency to spread by bottom flow. The literature does not contain information on the application of dikes to spilled hazardous materials.

Retaining dikes would likely be confined to nearshore, shallow-water areas, though the use of dikes for open-water island construction has been practiced by the Dutch (35). In-water dike construction experience is limited and has generally been conducted to depths of no more than 30 ft at mean low water. Retaining dikes have historically been sloped earth embankments constructed either in the water or directly adjacent in bordering lowland areas and islands. Earth-filled cellular and double sheet pile retaining walls have also been used, but are expensive (36).

One advantage of earth dikes is their potential for construction by existing dredging equipment and earth-moving equipment such as bulldozers. There is a good chance that such equipment would be available near a spill, especially if it occurs in a major metropolitan area. Materials for dike construction (preferably gravels or coarse sands) would probably be available near the site, depending again on spill location.

The feasibility of dikes for spill control decreases as the spill gets larger. Very small spills, in generally confined or nonnavigable waters, could be contained quickly and readily with no more requirement than men with shovels, a backhoe, or similar crude response techniques. As the water environment becomes more difficult to work in and the spill becomes larger, mobilization and construction times will increase, making the diking operation at some point infeasible as an emergency response procedure. The diking operation itself could begin to pose a greater problem than the spill, through its impacts on the environment, obstruction to free use of waterways, and similar considerations. Open-water dike construction, while feasible, could fall into the above category.

Trenches

Trenches could act as a trap for sinking chemicals, particularly in the case of rivers or other bodies of water with predictable currents. Liquid sinking chemicals are often noted for their "pooling" tendencies in bottom depressions, and trench construction in selected locations could take advantage of this tendency. Some solid sinking chemicals might also be carried along the bottom and collect in bottom depressions. The liquid or solid could then be removed from these trenches by dredging.

Experiments have been performed (37) to assess the feasibility of gravity collection of immiscible heavier-than-water chemicals in streams. These experiments, performed using ethylene dichloride (liquid), indicate that gravity separation using trenches is a feasible method of controlling hazardous material spills. A slow-moving stream was simulated with continuous flow from hoses down a 2% grade. A 55-gal spill of ethylene dichloride was used in the simulation. It was noted, that the solvent formed beads that moved with the flowing water and collected in a trench. Approximately 85% of the original volume of ethylene dichloride was recovered from the trench. Attempts at simulating fast flows revealed no essential change in the collection process. The experimental configuration is illustrated in Figure 2.6, taken from the reference cited.

Figure 2.6: Cross Section of the Ditch as Modified to Trap Heavier-than-Water Spills

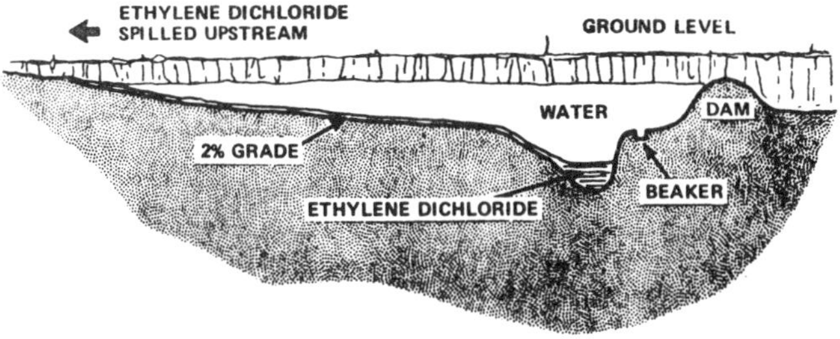

Source: EPA-670/2-75-042

The trenching method could be accomplished in the field through use of hydraulic dredges, land-based or water-based clamshell dredges, or similar equipment. An adaptation of a 50-ton plow used for placing pipelines offshore could possibly be used (38). The 50-ton plow would require 80 tons of pull, and the availability of vessels with that capability would be a factor, as well as transportability. The scaling down of such a concept could yield acceptable results. The choice of a trenching device would be dictated by the locale and characteristics of the spill area. Spills in flowing, nonnavigable waters could be handled by a land-based clamshell, a handheld suction dredge, or even by hand shovel if the spill were small enough. In deeper water and with larger spill areas, hydraulic dredges of increasing size could be used, making repeated passes over a trenching site to develop a trench large enough to collect all of the spilled material.

Trenches would be most useful in areas of predictable current, such as rivers, drainage ditches, canals, etc. Trenches could be located downstream of the spill, perpendicular to the direction of flow. In areas of reversing tidal flow, upstream and downstream trenches could be used. In off-channel areas of ports and harbors, and in relatively shallow open waters (certainly up to 50 ft, the approximate practical limiting depth for hydraulic dredges), trenches could be dug completely surrounding the spill, such that movement of the spill in any direction could be arrested.

Offshore pipe burial with the 50-ton plow mentioned earlier has been conducted in waters up to 500 ft deep. However, once trenched, there are no U.S. dredges available for use in waters deeper than about 100 ft, rendering such a deep trench less effective since removal of collected material could not take place. Costs would be commensurate with those for mobilization of a dredge or other equipment to the site and operation for whatever period would be required to dig the trenches. Material dredged in digging the trenches must be redeposited, incurring further costs and delay.

It should be noted that natural depressions, particularly in streams, would probably not be successful as containment areas. It must be assumed that whatever forces created the depression would also remove the spilled material that might tend to collect there.

Soil Surface Sealing to Prevent Penetration of Spills

> The material in this section is based on a paper given at Conference 1978 by J.P. Lafornara of the U.S. Environmental Protection Agency and M.D. Marshall, M.J. McGoff and J.S. Greer of MSA Research Corporation. The title of the paper was "Soil Surface Sealing to Prevent Penetration of Hazardous Material Spills."

The problem of control of chemical spills on land has recently received a great deal of attention from the public, in general and the technical community, in particular, due at least in part to several incidents throughout the country where groundwater has been contaminated. Most of the scientific activity in this area has been in the monitoring and health effects areas, but little has been done to develop methods to alleviate or mitigate the problem of the contaminated groundwater. Presently, mitigating action involves either the physical removal of the soil from the spill-impacted area and disposing of it at a secure landfill or the drilling of "cone of influence" wells followed by the removal of the pollutant from the water by conventional treatment technology. These techniques are both time consuming and expensive.

The scenario for the use of a soil sealant requires that it be employed immediately after the spill, possibly in conjunction with diking materials so as to contain the material in the immediate vicinity of the accident. It is envisioned that railroad and trucking personnel or possibly public safety personnel from a nearby community will have to determine the probable course of flow of the spill, locate a suitable site for containment and initiate the application of the sealant and installation of dikes.

Site preparation for applying the sealant must be minimal. Conceivably large pieces of debris, sticks, stones, etc., might be removed, but in general and ideally, the sealant should be capable of handling these obstacles. Since accidents frequently happen in inclement weather, wet and frozen surfaces are major conditions that must be overcome

The design goals for the system are as follows. It must be:

(1) Portable and capable of operation by one man.
(2) Self-contained.
(3) Operable in a safe manner.
(4) Operable in a manner such that no permanent secondary environmental damage results.
(5) Capable of fabrication from commercially available components to the maximum extent possible.

(6) Capable of reducing soil penetration by 95% in a land area of 1200 ft^2 under 1 ft of liquid head.
(7) Capable of sealing many types of soils under varied terrain conditions.
(8) Capable of application in varied weather conditions.
(9) Capable of sealing when applied before or after spill front has passed an area.
(10) Capable of preventing the spilled material from seeping under the edge of the sealed area.
(11) Capable of being removed after the spill incident is over.

Selection of Challenge Agents and Sealants: Table 2.8 lists the thirty hazardous chemicals which are believed to best represent the total spectrum of chemicals that are bulk-transported and, if spilled, would be a significant threat to groundwater supplies. The list includes representatives from most chemical classifications.

Table 2.8: Primary Representative Hazardous Chemicals

Inorganic acid	Alcohol
Sulfuric	Isopropanol
Nitric	Cresols
Inorganic base	Ester
Sodium hydroxide	Ethyl acetate
Inorganic halide	Ethyl acrylate
Phosphorus trichloride	Amine
Inorganic oxidizers	Ethanolamine
Phosphorus oxychloride	Pyridine
Inorganic sulfur	Ketone
Carbon disulfide	Acetone
Organic acid	Methyl ethyl ketone
Acetic	Mixed solvent
Acrylic	Naphtha
Organic halide	Turpentine
Methylene chloride	Lacquer solvent
Trichloroethylene	Mixed hydrocarbons
Aldehyde	Gasoline
Acetaldehyde	Kerosene
Benzaldehyde	Aromatic
Acrolein	Toluene
Plastic monomer	Xylene
Styrene	
Vinyl acetate	

Source: Conference 1978

Soil surface sealant candidates can be grouped into three general classes: reactive, nonreactive and surface chemical. This classification is a function of how the sealant is formed chemically and the interaction of the sealant with soil surface.

Nonreactive sealants are those which have been previously polymerized and are either dispersed or dissolved in either an aqueous or solvent system. Such materials are primarily thermoplastic in nature and include such materials as: Bitumastic, rubber, acrylic, cellulosic, fluoroplastic, phenolic, polyester, polystyrene and polyvinyl chloride.

Some of these materials can be used in aqueous media. An aqueous media has no particular application hazard associated with it but will freeze at low temperatures and film formation depends upon the evaporation of the water which is a function of both temperature and air velocity. At low temperatures, the rate of film formation could be slow. A typical example of such a temperature sensitive system is a household latex paint.

Solvent systems depend upon solvent evaporation for film formation. Again, temperature is an important factor. In addition, many solvents are offensive or flammable and may constitute a potential health or fire hazard.

All nonreactive aqueous or organic solvent systems can be subject to pinhole formation. When applied over gravel or stones, the drainage could be excessive and the film formed could contain defects. When applied to a very porous substrate, film formation would be very difficult and might require several coats. When coating rough surfaces, it is necessary to cover or seal cracks and to set all sides of any large particle, and this could be very difficult to accomplish.

Reactive sealants are usually two-component systems in which one material is either reacted or catalyzed with a second material to yield a polymer. Such materials include: epoxy, unsaturated polyester, phenol/formaldehyde, urea/formaldehyde and urethane. These materials may or may not utilize an aqueous or organic medium and are more likely to form films under adverse cold and wet conditions, although they, too, have low temperature limitations. All would have to have the reactants at about 55°F or higher in order to have sprayable components. The substrate could be at a lower temperature, even below freezing, if the components are sufficiently warm to initiate reaction.

Reactive sealant candidates with the ability to foam in place are particularly attractive. The foaming ability helps to eliminate pinholes and "shadowing" effects around stones and debris as a result of the spraying operation. This "bridging" ability, along with the resultant resiliency of the foam, has made polyurethane foam the best performing antispill mine sealant known. Foamed materials are equally attractive because of weight limitations.

The third sealant type consists of repellent chemicals which, when applied to surfaces, modify the surface characteristics such that the surfaces are not penetrated.

Extensive repellent technology has been developed for four classes of materials: textiles, paper, leather and masonry. In each, there is a broad range of techniques and chemical systems; but only two classes of chemicals appear promising, silicones and fluorocarbons. These are used to provide the so-called durable finishes on textiles, those which are essentially unaffected by either laundering or dry cleaning.

The silicone systems are employed either alone or in combination with melamine or urea-formaldehyde crease-proofing resins. For optimum properties, these require some crosslinking, usually accomplished in the textile industry with a short-term elevated temperature (150°F) cure. The most widely used materials are perfluoro derivatives of the polyacrylates. Marketed as Scotchgard, Zepel, and the U.S. Army material Quarpel, they are available as emulsions and dissolved in organic solvents for spray application.

Samples of sealants were requested for a preliminary evaluation if, in the opinion of the authors or the supplier, the candidate could meet the following broad criteria:

(1) A proven ability to form a continuous film.
(2) Sprayable under conditions practical for a portable system, or could be made so with minor modification to the formulation.
(3) Able to, or having the property of being able to, form a protective skin on the film in a reasonable time period. Fifteen minutes at moderate conditions appeared reasonable.
(4) Would not pose any serious threat in itself to the operator or the environment.

Manufacturer's compatibility data were essentially ignored since most such data are based on, and intended for long-term exposures, whereas this intended use was short-term. Also conditions such as blistering or swelling, which may affect the sealant's use in other applications, are of little hindrance to its use as a barrier to penetration of the chemical into the substrate.

Sealant candidates were screened by either spraying or brushing the materials on a cardboard backing. Physical properties and set times were observed with the more promising candidates being exposed to a selection of chemicals.

The initial preliminary challenge chemicals were water, sulfuric acid, sodium hydroxide, trichloroethylene, methanol, naphtha and methyl ethyl ketone. Failure of the sealant generally was catastrophic with loss of liquid. For those cases where failure was observed, similar chemicals in the classification were tested for verification. Those considered for further testing generally passed all preliminary tests.

Viable sealant candidates were compared based on their sprayability, film-forming capability and compatibility with selected chemicals in testing designed to select three to five sealant candidates from the total field.

The status of the various sealant candidates as a result of the preliminary and subsequent sprayability and compatibility testing is shown in Table 2.9 with comments given as to their performance. The latex Polyco 2607 (Borden Chemical) forms a surface film within minutes and is compatible with most organics. The urethanes of Callery and Ashland Chemical also exhibit a rapid cure and good compatibility. Both of these foam slightly, giving films in the 15 to 20 pcf range at 70° to 80°F. This low foaming is undoubtedly due to the thin film applied and resultant heat loss. The types of portable systems considered to apply these coatings included hand pumps, propellant-pressurized systems and powered sprayers (battery or gasoline engine driven).

Hand Pumps: Hand pump equipment, such as used for insecticide spraying and painting, is available. The capacity of these spray systems is between 1½ to 3 gal. For a 3-gal size, filled weight would be of the order of 40 lb. The primary shortcomings for hand pump sprayers are the necessity for continuous hand pumping and pressure variations, since there is no pressure regulator. Spray patterns and rates would vary because of the unregulated pressure head. Although this type of system is attractive in terms of weight and simplicity, it would be difficult to maintain rates of 2.67 to 10 lb/min to apply 100 to 10 lb of sealant.

Table 2.9: Status of Candidate Sealants

Sealant Type	Company	Product Name	Performance	Rejection
Adhesive	Goodloe E. Moore	Tuff-Bond 35	Questionable	Limited compatibility
		Tuff-Bond 66	Questionable	Limited compatibility
		Quick Set	Questionable	Limited compatibility
	Oneida Electronics	Instant Weld	Poor	Limited compatibility
	Borden Chemical	Polyco 2445	Good	Long cure time
		Polyco 2186	Good	Long cure time
		Polyco 2607	Good	–
Acrylic Paint	Emerson & Cuming	Eccocoat AC-8	Questionable	Limited compatability
	Cole-Parmer	Nalgene	Questionable	Limited compatability
	Carboline	Versikote 54	Questionable	Limited compatability
Silicone	General Electric	85-1693 A/B	Poor	Limited compatibility
	Dow Corning	3-5000	Questionable	Limited compatibility
Urethane	Irathane Systems	Irathane 149-C/R	Questionable	Viscosity and compatibility
	Mameco International	Vulkem-205	Good	Viscosity excessive
	Randustrial	P-250	Questionable	Limited compatibility
	Metalcrete Mfg.	Vulcanox	Good	Viscosity excessive
	Callery Chemical	115	Good	–
	Ashland Chemical	6100/6300	Questionable	Poor filming
		EP65-86/88	Good	–
Surfactant	DuPont	Sample kit	Questionable	Limited compatibility
		Zonyl RP	Good	Poor sealing
	Clintwood Chemical	Clindrol 100CG	Poor	Limited compatibility
Polyester	Ceil-cote	Flakeline 252	Good	Viscosity excessive
	Koppers	1070 Resin	Questionable	Limited compatibility
Aqueous (two-component)	Pacific Anchor Division P.V.O. International	Chem-cure	Questionable	Limited compatibility

Source: Conference 1978

Propellant Pressurized Systems: Pressurized systems incorporate a pressurized cartridge or a pressurized storage tank to propel the reagent and blend, if necessary. Commercially available portable fire extinguishers offer a simple approach. One would replace the extinguishing agent with the sealant candidate. Another pressurized system type is the urethane foam pack. These are self-contained, portable, and weigh around 40 lb. All of these pressurized systems would be capable of maintaining 2.67 to 10.67 lb/min liquid flow rates. However, sufficient gas capacity would not be available for providing atomizing air if it were required; only enough capacity is available to propel the liquid coating.

The pressurized system will exceed the minimum 40 lb weight target if gas cylinders are employed. It is, however, more favorable than a hand pump system, although more complex. These would be intrinsically more easily maintained compared to a powered system.

Powered Sprayers: A powered unit requires, in addition to containers, pumps and a power supply, be it batteries or gasoline type engine, making this concept less attractive from weight, complexity and maintainability than other system approaches. The pumps would be required to propel coatings or to spray air. A proportioning type pump would be used to move liquid out of the tank. If atomizing gas is required, it would be preferable to use this same pressurized air source to propel the coating out of the tank as well.

Curtain Barriers for Control of Sinking Chemicals

> The information in this section is based on *A Feasibility Study of Response Techniques for Discharges of Hazardous Chemicals that Sink,* CG-D-56-78 prepared by T.D. Hand, A.W. Ford, P.G. Malone, D.W. Thompson and R.B. Mercer of the U.S. Army Engineer Waterways Experiment Station for the U.S. Coast Guard

Information has been gathered on two generic forms of curtain barriers: one type designed specifically for control of turbidity from dredged material discharge (the silt curtain), and the other specifically intended for use in containment of hazardous materials spills. The silt curtain is not considered promising for confinement of sinking hazardous materials, as it is not designed to reach the bottom of a body of water. This would allow the sinking chemical to disperse underneath the curtain barrier. The hazardous material barrier, though not without problems, has been specifically designed for use in cases such as those under consideration here. Further discussion will be centered on this device.

The hazardous material barrier was developed for the EPA. Individual barrier units (which may be joined together) consist of a 200-ft long section of flexible, fiber-reinforced plastic, with buoyancy provided by an inflatable flotation collar. There is also a water-inflated bottom seal, secured by explosive anchors. It is claimed by its manufacturer (Samson Ocean Systems, Boston, Mass.) to be effective in 1-knot currents, 40-knot winds, and 6-ft waves (39). There is a 25-ft depth limitation. The barrier was designed to be effective in spills of a number of hazardous chemicals. Among these are aldrin and DDT, which are both under consideration in this report. An artist's conception of a deployed barrier is shown in Figure 2.7.

Figure 2.7: Deployed Hazardous Material Barrier

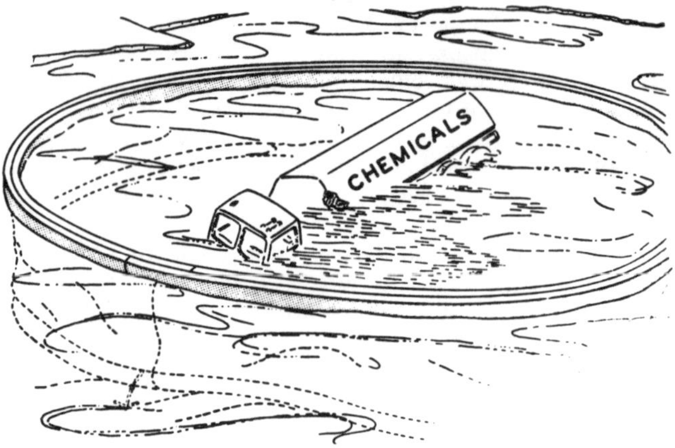

Source: CG-D-56-78

Deployment of the barrier would require a crew of a minimum of five men, portable air compressor(s) and integrated water pump, and one or two outboard powered boats or inflatable rubber rafts. The system was designed to be transported in a small truck, helicopter, or similar conveyance. A typical deployment sequence (Figure 2.8) follows:

- Transport barrier system to the site of spill.
- Inflate and activate boats.
- Survey site and locate source of pollutant. If barrier is submerged, position marker buoys.
- Make up a barrier length compatible with the size of the barrier and spill.
- Moor the barrier temporarily on the surface in the desired configuration.
- Emplant the vertical holding anchors and attach the pulling down lines to the barrier.
- Pull the barrier down to the bottom.
- Inflate the buoyancy chamber using the portable compressors.
- Fill water-inflated seal.

The manufacturer claims transportability of individual units by U.S. Coast Guard HH2 and HH3 helicopters.

Currents are apparently the most significant problem when considering use of this system. Difficulty of deployment in currents of three-quarter knot has been experienced, though training of personnel is felt by the manufacturer to have eliminated much of this difficulty (39). Currents remain a limitation to the maximum depth of the barrier, however, due largely to the tremendous sail area exhibited by the deployed barrier. Small increases of barrier depths, to perhaps 30 ft, are presently considered feasible.

Figure 2.8: Barrier Deployment Procedure

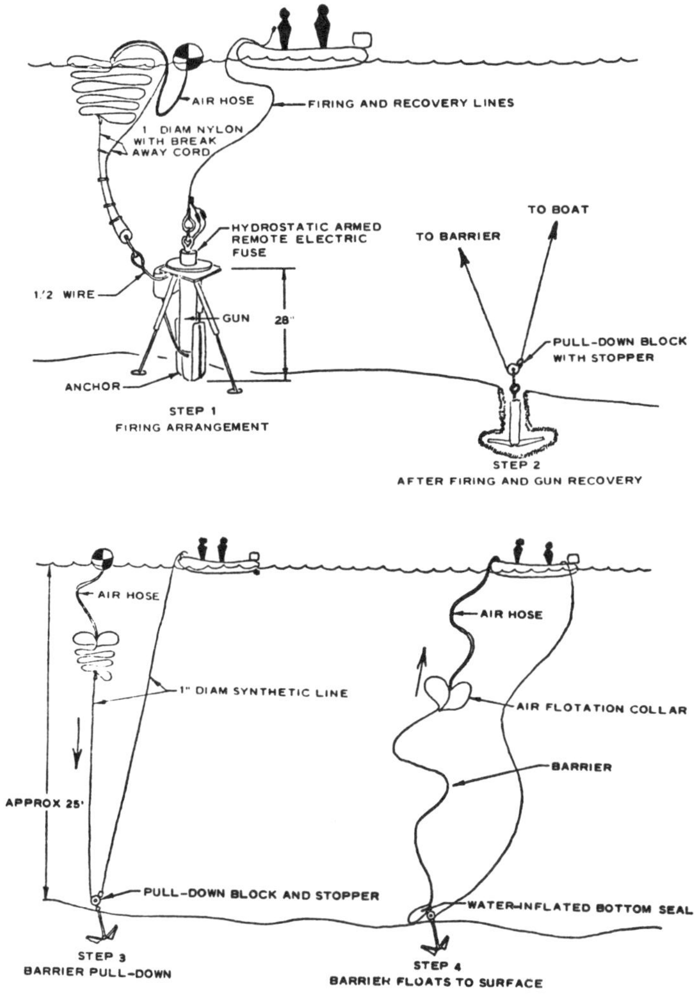

Source: CG-D-56-78

While the barrier has been designed for bottom to surface application, it might be possible to use it in deep water applications. It would then function as a "bottom-containment boom" in much the same manner as floating booms are used on oil spills today.

Usefulness of the system would probably be confined to spills in quiescent waters (nonnavigable waters; shallow coves; low current, nearshore areas of rivers; and shallow, low current areas of ports and harbors). It has not been tested for use in open waters. The system could be used as a part of a "strike team" concept, being purchased in advance and readied for rapid air deployment in the event of a spill. A significant time lag could occur before deployment and this would limit the barrier's usefulness.

Encapsulation

> The information in this and the following section is based on *Feasibility Study of Response Techniques for Discharges of Hazardous Chemicals that Float on Water,* CG-D-56-77 prepared by J.S. Greer of MSA Research Corporation for the U.S. Coast Guard.

Encapsulation is a direct means for localizing both the spill and vapor hazards arising from it. The equipment presently available has the ability to contain most hazardous chemicals for subsequent removal or recycle. Vapors are presently being contained at near 100% effectiveness over lagoons, where off-gases are collected. The effectiveness index of the encapsulation technique is adequate.

There appears to be no problem for acquiring membrane covers which are capable of encapsulating floating, hazardous chemicals. Several areas are uncertain, however, in the deployment over large areas of water and in joining the separate sheets into one impermeable cover. The efficiency index of this technique is conditionally acceptable.

The materials and equipment required for encapsulating hazardous chemical spills will not present an environmental burden, if specifically chosen for compatibility, but polyethylene, polyvinyl chloride and some of the other common plastics may be attacked and dissolved by some of the solvents included among the hazardous chemicals, creating more problems than they solve.

The basic encapsulation concept can be visualized as a floating boom enclosure, fitted with an elastomeric membrane cover (lid). Putting this lid over a floating hazardous chemical is a direct approach for containing the toxic vapors with nearly 100% efficiency. Accomplishing this feat quickly and safely over a typical chemical spill will severely test engineering acumen.

Floating cover assemblies are presently being fabricated by a number of suppliers for such purposes as protecting drinking water reservoirs, collecting methane gas from sewage waste lagoons, and sealing a broad range of liquid chemicals for atmospheric isolation. Minimal modifications to existing designs are anticipated for sealing floating hazardous chemicals.

A typical example of this technique is performed by Globe Linings, Inc. (Long Beach, CA) which manufactures a Hypalon membrane (0.8 to 1.1 mm thick) fitted with polyurethane floats for varied chemical service, including off-gas collection. Normal installation is by using 7.6 m (25 ft) wide sections, supplied in rolls, to build up the full area required. Covers up to 9,750 m^2 (105,000 ft^2) have been installed, typically by two men within two days. Skirting, similar to containment booms, is being currently evaluated for reservoir repair access service. Application potential for floating spill service is excellent.

Another producer, Gulf Seal Corp. (Houston, TS) provides a continuous polyethylene foam underlay to float the sealing membrane for water service. This supports heavy snow, ice, or personnel, including polyethylene work platforms.

Normal installation is by floating 3 to 30.5 m (10 to 100 ft) wide sections and tying them together at grommet provisions. Sizes in excess of 4.05 x 10^3 m^2 (1 acre) pose no problem. Material compatibility can be attained by substituting resistant foams and membranes. Portability and rapid deployment provisions need development since their continuous float arrangement may merely push the spill material ahead of the cover as it is pulled over the spill.

Polyethylene, vinyl and other common plastics are used as cover materials. Their light weight, availability and prices are major factors contributing to their use on these large installations. Polyurethane is an alternative cover material and composite or layered materials could also be considered.

The selection of encapsulating materials and method of deployment must be made using factors such as: probable spill size, material compatibility, weight, cost, portability, availability, reuse or disposal requirements, environmental compatibility and interfacing with spill removal systems.

Polyethylene film is stocked in clear, black or reinforced versions, in 0.1 to 0.25 mm (4 to 10 mils) thicknesses, up to 12 m x 30 m (40 ft x 100 ft) size rolls, weighing 34 to 91 kg per 373 m^2 roll (75 to 200 lb per 4,000 ft roll), and in an economical cost range. It can easily be heat sealed to form larger shapes, flotation chambers, or to repair leaks. Compatibility with most (solvent type) spill materials is generally limited. Polyethylene membranes may serve for a single response against many of the hazardous chemicals.

Carlisle Tire and Rubber routinely fabricates large elastomeric membranes for liquid containment or exclusion service in ponds or reservoirs as liners and/or covers. These materials are compatible with the floating, hazardous chemicals.

Depending upon chemical resistance requirements, a variety of synthetic rubbers (butyl, EPDM, neoprene, Hypalon, etc.) can be supplied. Standard sheets are commonly 0.8 to 3.3 mm (0.03 to 0.13 in) thick, up to 12 m x 30 m (40 ft x 100 ft) size rolls and weighing 227 to 1,000 kg per 372 m^2 roll (500 to 2,200 lb/4,000 ft^2 roll). Sheets can be provided with reinforcing fabrics for added strength. Larger lengths are feasible, but larger sections are usually made up by appropriate joining methods (vulcanizing, gluing, mechanically). When used for cover service, suitable flotation or mechanical support provisions must be obtained to keep these materials from sinking.

Deploying the thin covers over a hazardous chemical spilled on water presents a more difficult problem than covering sewage lagoons, reservoirs and other landlocked waters. Combining the operations of deploying the cover and a boom appear to be the best approach. The cover would be slid over the boom or unfurled as the boom was being set in place. This operation would be done from the upwind side of the spill to avoid the vapor hazard.

The difficult conditions encountered for deploying these membrane covers over spills of floating, hazardous chemicals may be assumed to double normal installation costs.

Close-Packed Particles as a Spill Response Technique

The distribution of light particulates over a liquid surface is another means for preparing a physical barrier to encapsulate the liquid and reduce evaporation losses. This technique is presently used for open storage tanks, ponds and reaction vessels, where evaporation losses create air polution or increase process costs. Evaporation is reduced by the presence of the densely packed layer of particles and its ability to reduce convection currents and insulate the liquid surface.

Plastic Systems, Inc. of Santa Ana, California has developed an evaporation control system which incorporates the geometrical ordering of dodecahedrons to produce this close-packed array. The producers say that one thousand particles will cover an area of 0.38 m^2 (4.12 ft^2).

Solid particulates are spread over the surface by gravity, surface tension and convection forces. The dodecahedron geometry permits a single layer to cover more than 99% of a surface, when in the close-packed arrangement.

Polypropylene foam lacks the chemical resistance necessary for responding to spills of floating, hazardous chemicals. Some research and development would be required to improve compatibility or develop a similar process for forming this geometry with other materials. Polyurethane appears to be the most promising material at the present time.

This technique would require the dispersal of approximately 10^7 particles to cover a spill of 38 m (10,000 gal). Auxiliary equipment for storage and dispersal would be similar to that used for sorbent materials. Dispersal equipment would have to be specially selected to avoid damage to the particle geometry.

Plastic Barriers for Plugging Leaking Chemical Containers

> The material in this section is based on a paper given at Conference 1972 by R.C. Mitchell, M. Kirsch, C.L. Hamermesh and J.E. Sinor of Rocketdyne, North American Rockwell. The title of the paper was "Methods for Plugging Leaking Chemical Containers."

In so highly an industrialized society as ours, tremendous quantities of chemicals are continually produced, usually at one location, and then shipped to another site for use in some application. Many of these chemicals are quite hazardous if introduced into natural waters. The likelihood of accidental release to the watercourses is obviously increased by the many sequences involved in the production, transfer, shipping, unloading, and ultimate utilization of the chemicals. Thus, it is not at all surprising that spills do occur.

Such spills can occur in a variety of ways. For example, the most dramatic and generally most serious type occurs when a container is violently ruptured (sometimes accompanied by fire or explosion), and large quantities of the hazardous material are spilled almost instantaneously. A less catastrophic spill results when the container maintains its integrity, but suffers enough damage to cause leakage of the hazardous material at a moderate rate. The leaking chemical can enter the watercourse either directly (for example, as the result of a barge accident or a land-based container falling into the water) or by flowing or being washed into a drainage channel or percolating into the groundwater supply.

Countermeasures which neutralize or treat hazardous chemicals which are already mixed and in the waterway may require hours or days to be initiated and involve the very difficult problem of handling large volumes of water. Therefore, a vital need is for a system which can prevent further dissemination of the hazardous chemical by stopping the leak regardless of whether the container is on land or under water.

The original goal of the program was to demonstrate the feasibility of plugging and controlling spills of hazardous materials by application of suitable plastic barriers. Particular emphasis has been on methods for plugging leaks in damaged chemical containers.

The original approach to plastic barriers which would seal leaks and prevent further dissemination of the chemical into a watercourse was based on the use of chemical systems such as polyurethanes which would be forced into the holes in the container, and, by expansion, seal these leaks. As the program developed, it was concluded that an equally valid approach is the use of sealants which do not expand when forced into the holes (as do the polyurethanes), but rather are already of the proper size and provide suitable adhesion to plug these leaks. The foam approach has the advantage that the increase in volume on foaming allows for a tighter seal. However, the chemical reaction of the foaming may be inhibited or retarded by the ingredients in the container. The true "sealant" of the second approach is unaffected by this latter problem.

Another way of classifying sealants depends on specifying the number of components in the final sealant. For example, one-component sealants include polysulfide rubber, polystyrene Instant Foam, and swelled Neoprene W; two-component sealants include many of the urethane foams and the epoxy systems. Many of the two-component systems adhere to a variety of surfaces including painted metal, whereas the one-component sealants often produce a mechanical plug to seal the leak.

One-component sealants have the advantage of being ready to apply at the moment a leak is discovered, whereas the need for mixing two or more components entails a delay during which the hazardous chemical continues to contaminate the environment. However, one-component sealants do not necessarily form a permanent seal, and they must be reinforced by applying a more permanent seal such as that formed by a suitable epoxy or polyurethane sealant.

Because of early success, initial screening tests were soon expanded to include most of the hazardous chemicals that rank among the top twenty on the priority list (40). Chlorosulfonic acid and phosphorus pentasulfide are probably too reactive with water or the sealants and would probably require a special sealant system. On the other hand, many of the other top twenty hazardous chemicals can probably be sealed with one or more of the sealant candidates that are effective with either benzene or phenol. For example, sealants useful for benzene leaks would also likely be effective against the other hydrocarbons, styrene, isoprene, xylenes. Sealants useful for phenol would probably also seal acrylonitrile and acetone cyanohydrin. The range of hazardous chemicals sealed or likely to be sealed with sealants already tested is shown in Table 2.10.

Table 2.10: Leak Sealing Tests

Hazard Ranking*	Hazardous Material	. . Sealed by Available Sealants . . From Tests Completed	Expected from Tests to be Made
1	Phenol	–	**
2	Methyl alcohol	x	–
3, 8, 14, 18	Insecticides, rodenticides	x (DDT)	x (others)
4	Acrylonitrile	–	–
5	Chlorosulfonic acid+	–	–
6	Benzene	x	–
9	Phosphorus pentasulfide+	–	–
10	Styrene	–	x
11	Acetone cyanohydrin	x	–
13	Nonyl phenol	x	–
15	Isoprene	–	x
16	Xylene	–	x
17	Nitrophenol	x	–

Note: x is polyurethane or epoxy sealants and + is not sealed with available sealants.

*1 is most hazardous.

**Butyl, polysulfide and neoprene rubber sealants.

Source: Conference 1972

The epoxies and urethanes tested are all base-catalyzed; acid interfered seriously with their polymerization and resulted in breakdown of already polymerized material. In general, these sealants, when set up, retain nonpolar material, but are unsuitable for phenol and acrylonitrile. Sea Goin' Poxy Putty and the metal-filled Devcon putties set up in a relatively short time, but needed some mechanical assistance (e.g., solid material patch) to support the epoxy while it hardened.

The three different types of rubber tested are inert to phenol and swell or dissolve in aromatic hydrocarbons. Uncompounded Neoprene W was obtained from DuPont, and was allowed to swell in either carbon tetrachloride or benzene, forming a relatively soft, highly pituitous mass before testing as a sealant. In this condition, it was stuffed into a two-inch diameter hole near the bottom of a five-gallon drum full of water. The rubber elasticity of the swelled Neoprene allowed the pressure of the water to produce normal stresses that contributed to expansion of the mass on the inside of the container. The material still behaved as a viscous liquid and so slowly extruded through the hole; a permanent seal is required to prevent the leak from redeveloping.

Another two-inch diameter hole was permanently sealed by first applying a plug of 3M polysuflide rubber putty, followed by a patch of the Sea Goin' Poxy Putty on a strip of nylon cloth which encircled the container. The cloth girdle held the epoxy in place while it hardened, and in about two hours the entire seal was hard enough to allow any usual handling of the five-gallon container.

Instant foams supplied by Monsanto were contained in pressurized cylinders. Upon opening a valve, solid foam is delivered. These materials may be useful in sealing leaks of compatible hazardous chemicals by forming a mechanical plug.

Additional evaluation tests and scale-up investigations were made with some of the most promising sealants. These included investigation of the effects of: barrier type (e.g., patch, outside plug, plug expanded outside and inside), thickness of barrier, speed of application, type of applicator, technique of application, and temporary support during the curing stage. The primary emphasis of the scale-up work has been to increase the hole sizes and hydraulic heads against which leaks can be plugged.

Earlier testing with various available plastic systems indicated the need for having a flexible laboratory mixing/delivery system in order to perform effective evaluation tests. No available mixing system was found which could perform the necessary functions, so a system was designed and constructed. It uses an in-line "static" mixer and can meter the components to supply various mixtures over a range of flow rates. It has been used for tests with urethane foams, and can be modified to permit its use with more viscous systems, such as some of the epoxy formulations.

A promising technique for sealing large holes with urethane foam against appreciable hydrostatic heads is to expand it both inside and outside the hole. This results in a structural bridge. It is important to apply the foam at a point in its curing cycle where it will adhere to the container surface. This requires some type of temporary support. This has led to the concept of either a special applicator design (large cross section, long residence time), or a two-stage application technique. In the latter concept, the foam is mixed externally and allowed to cure for a few seconds until it reaches the right stage of tackiness. It is then applied to the leak with some device which provides temporary support for an additional few seconds until sufficient rigidity has developed.

An important key to successful operation with this concept is that the applicator be designed in such a way as to provide some compression of the tacky foam while forcing it through the leaking hole. Thus, the foam will expand on the other side of the hole to provide a structural bridge, but have sufficient internal cohesion while expanding to avoid being washed away.

A variety of applicator concepts can be envisioned for practical usage. An extremely simple one has been tested with considerable success. A large suction cup "plumber's helper" is filled with foam and allowed to set for a few seconds until the proper tackiness is reached. The cup is then placed against the leaking stream, and some of the foam is forced through the hole with the plunger. The sides of the cup provide enough restraint so that the necesssary compression of the foam cells can be achieved, resulting in expansion on the inside of the hole.

It is not necessary that the cup exactly conform to the surface of the container. This technique has been used successfully to seal a 1½-inch hole in a 55-gallon drum with 3 ft of head. Tests with larger holes and greater liquid heads are scheduled. The hydraulic force of the stream from such a hole would immediately wash away any sealant material if it were not applied with special application system and in sufficient quantities to bridge the hole.

OTHER PHYSICAL METHODS

Solvent Extraction

The information in this section is based on *A Feasibility Study of Response Techniques for Discharges of Hazardous Chemicals that Disperse through the Water Column,* CG-D-16-77 prepared by E. Drake, D. Shooter, W. Lyman and L. Davidson of Arthur D. Little, Inc., for the U.S. Coast Guard.

For the purposes of this report, solvent extraction can be defined as mixing water containing a hazardous chemical with a liquid which is immiscible with water and then allowing the two liquids to separate again. At this point, the hazardous chemical is distributed between the two phases; its concentration in the water has been reduced.

Practical Limitations of Solvent Extraction: For purposes of ameliorating spills of hazardous chemicals in water, practical limitations restrict the types of systems considered. Constraints are imposed by the chemical components of the system, particularly solvent and spill chemical, which determine whether adequate extraction and separation of the two phases can occur. Contacting equipment must be selected to insure that the procedure can be carried out on site in a reasonable period of time and in a manner consistent with Coast Guard clean-up operations.

Phase Separation – The ease with which the two phases, solvent and water, separate after an appropriate mixing interval is an important criteria for successful solvent extraction. The two critical physical properties are the difference in densities between the two phases and the difference in interfacial tension between them. The ease of separation is a compound function of these two properties. The interfacial tension may, in practice, be a critical property because the presence of small amounts of impurities, particularly those with detergent properties, can modify the surface film severely and prevent efficient separation.

Solubility of Solvent in Water – Although any useful solvent must be essentially insoluble in water, all solvents will have a finite solubility (in addition to possible losses by entrainment of droplets). This loss of solvent will occur after every contacting operation and should be as low as possible, certainly less than 1% by weight of the solvent. If the solvent solubility were higher than this figure, then a second extraction (with another solvent) would be required to remove the first solvent from the water. The complexities and disadvantages of such a two-step extraction process would appear to outweigh the potential limitations of a single-step extraction (where the solvent may be less efficient in extracting the spill chemical but is not lost by solution in water). An equally important criteria is that the equilibrium concentration of solvent in water at the solubility limit must be lower than the toxicity limit for that mineral. Otherwise the addition of solvent compounds the problem by introducing a second material at a toxic level.

Solvent Recovery – Recovery and reuse of solvent during the clean-up operation is an essential part of the solvent extraction system; without solvent recovery any multistage operation would require such large volumes of solvent as to be impractical for any but the smallest spills. To simplify this portion of the equipment, the relative volatility of the spilled chemical solvent system should significantly exceed unity so that the chemical can be removed overhead without the necessity of evaporating and condensing the solvent.

Partition Coefficient – The partition coefficient of the spill chemical (weight fraction in solvent divided by weight fraction in water, following extraction) should be greater than or equal to one. This means that a single extraction (using equal volumes) would remove 50% of the spill chemical; five separate extractions would give 97% recovery, assuming 100% theoretical efficiency in each extraction stage. Appropriate equipment is limited to about five extraction stages in a single operation.

Selection of Equipment: There are two major types of equipment for solvent extraction.

Single-Stage Equipment – The fluids are mixed, extraction occurs, and the immiscible liquids are allowed to settle and separate. A cascade of such stages may be arranged. A single-stage must provide facilities for mixing the liquids and for settling and decanting the emulsion or dispersion which results. In batch operation, mixing, settling and decanting may take place in the same or in separate vessels. In continuous operation, different vessels are required.

Multistage Equipment – The equivalent of many stages may be incorporated into a single device or apparatus. When the liquids flow countercurrently through a single piece of equipment, the contacting may be equivalent to as many stages as desired. In such devices, countercurrent flow is induced by virtue of the different densities of the liquids; with few exceptions, the equipment takes the form of a vertical tower which may or may not contain internal devices to influence the flow pattern. Other forms include centrifuges, rotating discs, and rotating buckets. Depending upon the nature of the internal structure, the contact may be either stage-wise or continuous.

Examples of various types of contacting are given in Table 2.11 below.

Table 2.11: Classification of Contacting Equipment

Type of Agitation	Discrete Stage Contact	Continuous Contact
None	–	Spray column
–	–	Baffle plate column
–	–	Packed column
–	–	Sieve-tray column
Rotary devices	Holley-Mott	Schiebel column
–	Simple mixer-settler	Oldshue-Rushton column
–	Pump-mix mixer-settler	Rotating disc contactor
–	Individual stage centrifuges	Multistage mixer column
–	Stacked-stage mixer-settler	Podbielniak extractor
–	–	DeLeval extractor
–	–	Luwesta
–	–	Graesser
–	–	Westfalia
Pulsed	Pulsed mixer-settler	Pulsed packed column
–	–	Pulsed sieve-plate column

Source: CG-D-16-77

For the amelioration of chemical spills, the most important considerations are the number of stages required (about five for the singly substituted hydrocarbons); the need for a high throughput per unit volume of contactor (so that a small portable unit can be used); and, the need to separate two liquid phases having small density differences.

Selection of Solvent: All organic chemicals which have a low solubility in water are potential solvents in an extraction process. Therefore, it is necessary to develop criteria which will rapidly eliminate less desirable solvents from consideration, in order to concentrate on those which show the best properties with respect to clean-up of the listed spill chemicals. Criteria are listed in Table 2.12; 1 is considered most important, 2, 3, and 4 about equally important, and 5 and 6 less important.

Table 2.12: Criteria for Choice of Solvent

(1) Solubility in water less than toxic limit
(2) Low solvent losses
(3) High partition coefficient (⩾1)
(4) Easy separation of solvent and water
(5) Ease of handling
(6) Availability and cost

Source: CG-D-16-77

Toxicity Limitations — Twenty-two solvents were selected on the basis of availability of data (for NAS hazard rating, solubility in water and partition coefficients, etc.) and also to represent the various classes of solvents. These solvents were first evaluated to decide whether their solubilities in water were less than their acceptable toxic limits according to the NAS rating. On this basis, only three solvents were clearly acceptable: heptane, vegetable oils, and oleyl alcohol. Octanol is of borderline acceptability because of its appreciable solubility and its moderate toxicity. Heptane was eliminated from further consideration because of its relatively low boiling point and appreciable fire hazard and because of the low partition coefficients for the chemicals of interest in this solvent.

A literature search revealed that partition coefficients between these solvents and the spill chemicals reached 1 only in the most favorable cases; for many spill chemicals, they were much less than 1. Octanol shows the most favorable range of partition coefficients, with values as high as 250, although some compounds still had coefficients much less than 1. In addition to potential toxicity, the low interfacial tension for octanol would make separation somewhat more difficult. Oleyl alcohol, with a higher interfacial tension, higher boiling point and very similar range of partition coefficients, looks more attractive than octanol, although the available data on use of oleyl alcohol as a solvent is much more limited.

Vegetable oils appear to be the most attractive compounds as solvents because of their low toxicity. The density difference between vegetable oil and water is only about 10%, but this should be acceptable because of the high interfacial tension. The partition coefficients are acceptably high for many of the spill chemicals; however, some of the list would not be readily extractable with vegetable oils.

Soybean oil is a typical vegetable oil which is produced in high volume (greater than 7.5 billion lb annually) and is widely available at acceptable cost. Vegetable oil is readily stored in closed containers (to prevent oxidation and bacterial attack) and is stable to heat, with a flashpoint above 280°C.

Extractability of Spill Chemicals: The range of partition coefficients encountered with all of the promising solvents suggest that some spill chemicals may be difficult to extract with any solvent. In addition, data are not available for many compounds and a method of estimation is desirable. Therefore, an examination was made of the different chemical classes of compounds to evaluate the potential for solvent extraction with vegetable oil. Data show that it is clearly possible to separate the list into two types of compounds, i.e., those with only one hydrophilic group (monosubstituted hydrocarbons) and those with more than one hydrophilic group (polysubstituted hydrocarbons).

Available data for the polysubstituted hydrocarbons indicate that these compounds in general have a very low partition coefficient for vegetable oils (and for other promising solvents); they therefore are not readily extractable because they have such a great affinity for water compared to any organic solvent. The data for monosubstituted hydrocarbons are more promising for many of the chemical classes. All the monosubstituted classes except acids show some partition coefficients of one or higher; acids run an order of magnitude lower and therefore would not be readily extractable with vegetable oils.

Data from the literature were also evaluated for different chemical classes to see the effect of carbon chain lengths on the partition coefficient. The partition coefficient increases with increasing carbon chain length in each chemical class; the large molecules are more extractable than small molecules. Even the acids, normally the most difficult class to extract, have sufficiently large partition coefficients in compounds with more than about 6 carbon atoms.

Each compound in Table 2.13 has been evaluated in terms of aquatic toxicity and extractability and classified in one of four groups; extractable with vegetable oil (T), not extractable with vegetable oil (NX), no treatment required (NT), and no data available (ND). Those compounds classified as extractable have partition coefficients ≥ 1. Those compounds classified as not extractable have partition coefficients <1. "No treatment required" indicates low toxicity (aquatic toxicity rating of 1 or 0). "No data available" means both insufficient data available to estimate the partition coefficient and appreciable toxicity (aquatic toxicity 2 or 3). Only about 20% of the chemicals listed are extractable with vegetable oil; an additional 30% require no treatment because the spill chemicals have a relatively low toxicity.

Table 2.13: Chemicals Subject to Solvent Extraction

Chemical	Comments
Acetone cyanohydrin	NX
Allyl alcohol	NX
Aminoethanolamine	NT
n-Amyl alcohol	T
1,4-Butanediol	NX
1,4-Butenediol	NX

(continued)

Table 2.13: (continued)

Chemical	Comments
n-Butyl acetate	T
n-Butyl alcohol	T
sec-Butyl alcohol	NT
tert-Butyl alcohol	NT
1,4-Butynediol	NX
Carbon bisulfide	T
Chloroform	T
Chlorohydrins (crude)	NX
Corn syrup	NT
Dextrose solution	NT
Diacetone alcohol	NT
Dichloromethane	T
Diethanolamine	NT
Diethylene glycol	NT
Diethylene glycol dimethyl ether	NX
Diethylene glycol monoethyl ether	NX
Diethylene glycol monomethyl ether	NX
Diisopropanolamine	NX
Dimethylsulfate	ND
Dimethylsulfoxide	NT
1,4-Dioxane	NX
Dipropylene glycol	ND
Epichlorohydrin	ND
Ethoxytriglycol	NT
Ethoxylated dodecanol	NT
Ethoxylated pentadecanol	NT
Ethoxylated tetradecanol	NT
Ethoxylated tridecanol	NT
Ethyl acetate	T
Ethyl acrylate	ND
Ethyl alcohol	NT
Ethylene cyanohydrin	NX
Ethylene glycol	NT
Ethylene glycol monobutyl ether	NX
Ethylene glycol monoethyl ether	NX
Ethylene glycol monomethyl ether acetate	NX
Ethylene glycol monomethyl ether	NX
Ethyleneimine (monoethanolamine)	NX
Formaldehyde solution	T
Glycerin	NT
Hexylene glycol	NX
Isoamyl alcohol	T
Isobutyl alcohol	NT
Isopropyl acetate	NT
Isopropyl alcohol	NX

(continued)

Table 2.13: (continued)

Chemical	Comments
Methanearsonic acid, sodium salts	ND
Methyl acrylate	T
Methyl alcohol	NT
Methyl amyl alcohol	T
Methyl isobutyl carbinol	T
Methyl methacrylate	T
Monoethanolamine	NT
Monoisopropanolamine	NT
Morpholine	ND
Paraformaldehyde	T
Polypropylene glycol methyl ether	NX
n-Propyl acetate	T
n-Propyl alcohol	NX
Propylene glycol	NT
Propylene glycol methyl ether	NX
Propylene oxide	NT
Sodium alkylbenzene sulfonates	T
Sodium alkyl sulfates	T
Sorbitol	NT
Sulfolane	ND
Tetrahydrofuran	ND
Triethanolamine	NX
Triethylene glycol	NT
Vinyl acetate	T

Source: CG-D-16-77

Preferred Methods of Solvent Extraction: The use of vegetable oil as the solvent would seem to allow a relatively simple system design. It would consist of the following basic components: initial settling tank and filter, differential centrifugal extractor, boiler tank with propane heater, condenser, temporary storage for extracted hazardous chemicals, surge storage for the solvent, possibly a heat exchanger, pumps, valves, flow meters and associated piping, and motor-generator.

The whole system, except for the motor generator is shown schematically in Figure 2.9. The actual size of the system would depend on the limitations of the components (principally the centrifugal extractor), the availability of funds, and the size and weight limitations for a mobile unit.

The data in this report show a number of chemicals are theoretically amenable to solvent extraction treatment. However, except for contained spills or spills into small stagnant water bodies, many practical problems are encountered. First, the equipment is more complex, costly and cumbersome than that typically required for neutralization. The treatment skid is estimated to weigh about ten to twenty tons. Because of its weight, these skids would have to be designed for transport by truck or C130 aircraft. For marine application, the skids would be moved onto a barge using a construction crane or a dockside marine loading crane. It will be more difficult to follow a spill in moving water with a barge.

Figure 2.9: Schematic Diagram of Extraction System

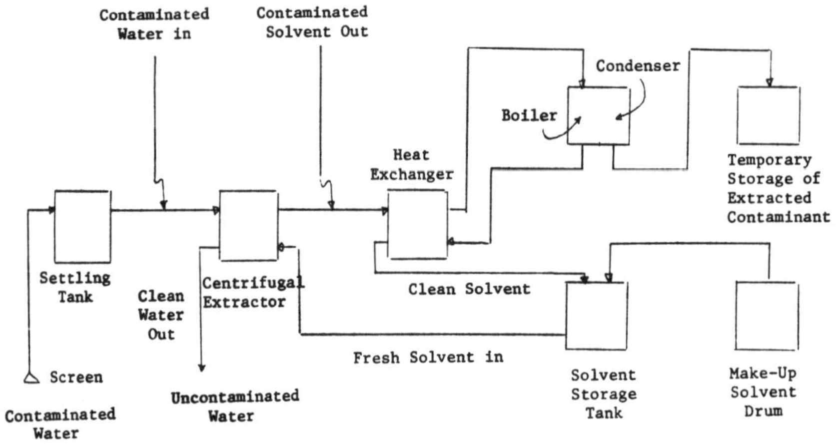

Source: CG-D-16-77

Another problem is that treated water must be discharged beyond the contaminated zone. Otherwise, dilution of contaminated intake water would make the extraction process much less effective.

An interesting possibility seems to be the development of a multiskid system. One skid, with an inlet pump, settling tank and filter would be designed to accomplish a precipitation response in addition to its use as part of the solvent extraction scheme. Another skid with the solvent extraction portion of the system may also provide a means for recovery of certain chelating agents.

At present, the usefulness of a solvent extraction spill response concept seems limited; however, this method may be the only way of coping with some particularly severe spill hazard.

Environmental Effects: Several environmental effects may result from the discharge of vegetable oils.

Formation of a Surface Film – This is likely only if the vegetable oil is discharged in excess of its effective solubility. Such a surface film could (a) interfere with the natural processes of reaeration and photosynthesis, (b) adversely affect waterfowl in a manner similar to that of petroleum oils, and (c) create adverse aesthetic effects such as the fouling of shorelines and beaches.

Reduction in Level of Dissolved Oxygen – The relatively high BOD of vegetable oils could lead to a serious depletion in dissolved oxygen if the discharged (i.e., treated) waters were not sufficiently diluted by the receiving waters. Assuming a discharge with 0.1 g oil per liter and a BOD_5 of 0.39 g O_2/g oil, the resulting 5-day oxygen demand is 0.039 g O_2/l. Since the receiving waters may have around 4 to 8 ppm dissolved oxygen (0.001 to 0.002 g O_2/l), a dilution factor of at least 50 would seem to be desirable in waters where the level of dissolved oxygen was low and the rate of reaeration was small. Without such dilution or reaeration, the level of dissolved oxygen could temporarily fall below a point necessary for the survival of many species of aquatic life.

Direct Toxic Action – Vegetable oils may be considered nontoxic to humans. The FDA stipulates that vegetable oils used in human food contain less than 2% free fatty acids. There does not appear to be any data relating to the aquatic toxicity of vegetable oils. One recent EPA paper indicates that the aquatic toxicity of vegetable oils depends upon the content of fatty acids, and that solubility calculations indicate that fatty acids are sufficiently soluble to exceed the toxicity threshold for fish (42). The oil does not, of course, have to be in a dissolved state in order to be ingested by aquatic organisms. In spite of the lack of any real data, the level of aquatic toxicity may be considered quite low.

Indirect Toxic Action – Free oil and emulsions, if present in sufficiently high concentrations, could accumulate on sensitive portions of aquatic organisms (e.g., gills) in such a manner as to seriously affect them.

Tainting of Flavor – Oil ingested by fish or other aquatic life may cause a tainting of the flavor, though the effect would be expected to be small.

Cryogens

> The information in this and the next two sections is based on *Feasibility Study of Response Techniques for Discharges of Hazardous Chemicals that Float on Water,* CG-D-56-77 prepared by J.S. Greer of MSA Research Corporation for the U.S. Coast Guard.

Cryogens can be used to reduce the vaporization rate of all floating hazardous chemicals. Vapor pressure or equilibrium vapor concentrations can be reduced by 40 to 95% if the final temperature is maintained at the ice point.

The quantities of cryogen and rates of dispersal required for adequate securing of a hazardous chemical spill create problems for both storage and delivery. These will require some modifications to presently available equipment or to the ships carrying them to respond to a spill.

Selection of the cryogenic material determines the environmental burden of this response technique and the compatibility index. Most refrigerants must be excluded on the basis of their toxicity; carbon dioxide would change the pH of water, when used in the quantities required. Liquid nitrogen (LN_2) and solid water (ice) are considered to be the only materials which can be used in this response technique to give an acceptable rating for the compatibility index.

The efficiency of cryogenic cooling to ameliorate spills of floating, hazardous chemicals is controlled by four factors: delivery, physical properties, freezing point of water and spilled chemical and environmental factors.

Delivery of the cryogenic material affects the ultimate efficiency of this technique much the same as the other techniques. It is imperative that the cryogen be spread evenly, over as large an area as possible, very quickly. Vaporization rate reductions will be directly proportional to the efficiency with which a homogeneous cooling blanket of cryogen is placed over the spill.

The presence of a tremendous excess of water limits the equilibrium temperature to 0°C (32°F) although the evaporation rate reductions are favored by lower temperatures, the freezing point of water is a more realistic temperature limit.

Environmental factors, such as rain, wind and water turbulence, are just as important in this spill response technique as in others. Warm days with bright sunshine will impose a higher surface heat load, while rain, wind and waves will increase the heat load by mixing.

Cryogenic material selection for spill cooling may well be limited to liquid CO_2, as indicated by Dr. S.M. Mathews of AIRCO (43). Liquid CO_2 was recommended over LN_2 based on:

- Loss-free storage, since CO_2 tanks are equipped with refrigeration cycles to reliquefy boil off.
- CO_2 storage tanks are single wall, in contrast to the vacuum tanks used for LN_2. This makes CO_2 tanks less costly and more resistant to the "high-g" stresses experienced aboard ships.
- CO_2 costs less than LN_2 based on equivalent "effective" regrigeration effects expected during spill disposal. Spraying would vaporize all LN_2 within a few feet from the nozzle. Pouring LN_2 on the spill surface may spread slowly by flowing over local ice plates, initially formed. The LN_2 would lay and boil off these plates with excessive cooling at the point of application.
- CO_2 dispersal is easier to visualize since "snow horns" (described below) are available for local application. Modifications for large application areas may be feasible using such methods as: positioning booms for standard snow horns; a snow lance at the perimeter to create a snowstorm at some loss in transfer efficiency; blowers to broadcast the snow horn feed directly, or intermediately using a water spray heat transfer media to broadcast a mixture of wet/dry ice.

It must be noted, however, that in addition to the deleterious environmental reactions noted previously for carbon dioxide, this response technique would also face the problem of generating uninhabitable gas mixtures near the surface of the spill where response personnel must function.

Liquid CO_2 stored at 300 psi will yield about 43 cal/g (77 Btu/lb) of cooling in the spill material, after allowing for transfer inefficiencies using snow horns directed over the spill area. To meet the cooling requirements of a maximum size spill, 12,000 kg (13 tons) are needed for initial cooling, followed by 3,600 kg/hr (3.9 ton/hr) for maintaining the lower temperature.

Liquid carbon dioxide storage facilities would be similar to those recently prepared by AIRCO Industrial Gases Company for use on barges. This facility consisted of a 27,000 kg (30 ton) capacity tank, fitted with a heat loss refrigerator installed on an ISO frame, occupying 77 m^3 (2,770 ft^3), weighing 18,000 kg (20 tons) empty.

Expansion nozzles known as snow horns are available from AIRCO converting the liquid CO_2 to solid CO_2 snow at high (42 to 45% by weight) efficiencies. Largest standard horns have a capacity of 19 kg/min (42 lb/min) of snow, using 45 kg/min (100 lb/min) of liquid CO_2 supply. To apply initial cooling within one-half hour, about 10 snow horns are required. Equipment required for dispersing the snow over the spill area as discussed, will add to the system cost. This could be a conventional snow blower, modified for cryogenic service, using stainless steel contacting parts. Operation would require 2 to 3 men. Electrical power (20 hp) is required for storage refrigeration. Standby maintenance is minimal. A 10-year useful life would be expected for hazardous spill response service.

Alternative spill refrigerant systems could be based on using either solid CO_2 (dry) ice or solid water (wet) ice. The solids would be crushed and pneumatically broadcast over the spill using a snow thrower or mulch spreader. Shipboard storage of ice would include high losses and considerable labor and mechanical handling equipment. Special materials of construction would be required for the cryogenic temperature of dry ice. Offsetting advantages would include higher "effective" refrigeration effects, smaller quantities of refrigerant required, and lower cost per Btu content. Use of water/ice system could prove particularly attractive, since it can be generated "in situ" for maintaining an adequate supply.

Both these coolants could be dispersed by equipment similar to the model R-300-76 snow blower, manufactured by American Snowblast Corporation.

The quantities of cryogen and rates of dispersal required in response to these spills presuppose the use of the largest blowers and material handling facilities. Response to a spill would be seriously impeded by the necessity to carry large storage vessels; however, incorporating cryogenic regeneration units into the response equipment significantly increases the equipment costs.

Surfactant Films

This technique, as developed by Moran et al (44), utilizes fluorochemical surfactants in water solution to form a thin film over the surface of a spilled chemical. The surfactant/water film should be between 10 and 20 μ (3.3×10^{-5} and 6.6×10^{-5} in) thick to seal the surface of the hazardous chemical and reduce evaporation.

Two distinct methods may be used to form the surfactant/water film. An aerosol or mist of the surfactant/water solution may be formed to settle out over the surface of the spill and spread to form the thin film. An alternative procedure is to use the fluorochemical surfactant in a special foam formulation (aqueous film-forming foam) to be generated and spread over the surface of the spill. The foam is formulated for rapid collapse, to form the surfactant/water film.

The amount of solution required for this response technique can be calculated from the typical size and volume estimates established for spills of floating, hazardous chemicals and the film thickness. The volume of solution required for responding to the typical spill of 100 to 10,000 gallons would be between 4 and 76 liters (1 to 20 gallons).

In the alternative method for preparing surfactant/water films the fluorochemical is incorporated into the foam. This aqueous solution may be used with low expansion foam generators, to form a foam having a nominal expansion ratio of 10 to 1. The foam has been formulated to collapse rapidly and form an aqueous film, which seals the spilled chemical.

Some assumptions must be used in calculating application rates for aqueous film-forming foams. The amount of surfactant/water solution required to form the film is constant. The film should be only slightly affected by the presence of foaming agents, extenders and stabilizers, since they are also present at low concentrations. Fire-fighting experience with foams will influence the actual use of spill response techniques, however, and significantly increase the amount of foam used.

Normal fire-fighting applications require between 120 and 200 times as much foam as theoretically necessary to form the surfactant/water film. The excess liquid would form "lenses" in the film and increase the probability for securing film to drop through the spilled chemical layer. Using approximately 100 times the minimum surfactant/water solution required for film formation should permit visual verification of the foam cover and a realistic safety factor for establishing the film upon collapse of the foam.

Either hydraulic or pneumatic equipment could be used to deploy the surfactant/water solution over the surface of the spill. The small volume required for sealing these spills favors hydraulic equipment, but pneumatic systems could cover equal or larger areas without relying on prevailing winds to obtain particulate drift.

Typical hydraulic equipment might be a Model FFF Fog Foam Nozzle, manufactured by Rockwood Manufacturing. This nozzle could be used to generate either aerosol or foam.

Hydraulic equipment for preparing the surfactant/water films would require a minimum of deck space. These units are approximately 1.75 m^2 (19 ft^2) and may be attached to the ship's water system.

The "Eagle" model blower, manufactured by Finn Equipment Company, is a typical pneumatic dispersing system. This equipment would deploy the surfactant/water solution as a mist, using misting nozzles and other optional equipment for this operation.

This sytem would occupy approximately 10 m^2 (109 ft^2). It could also be atatached to the ship's water supply or have a separate hydraulic pump to supply the water solution.

Either hydraulic or pneumatic equipment would provide a long service life (10 to to 20 years). Normal operation and maintenance would be similar also for these marine installations. It is estimated that either hydraulic or pneumatic systems could be operated by one man in response to spills of hazardous chemicals.

The foam generators used in the alternative method of preparing surfactant films would use between 400 and 7,570 liters (100 to 2,000 gal) of the foam solution to cover a typical hazardous chemical spill. This foam would have a nominal 10:1 expansion ratio to give a maximum cover of approximately 0.7 cm (0.27 in). A foam generator, similar to the Model FFF Fog Foam Turret Nozzle, would be used to generate the foam. The equipment for this response technique would include a storage tank, pumping unit and auxiliary hardware for transporting and metering, in addition to the foam generating nozzle. Such a system would have a capacity for generating approximately 380 l/min (100 gal/min) of low expansion foam.

A low expansion foam generating system would weigh approximately 910 kg (2,000 lb) and occupy about 3.7 m^2 (40 ft^2) of deck space. Additional equipment should include a gasoline motor, to make the system completely self-sufficient. The foam response technique requires one man to operate the equipment and direct the foam over the spill. The service life of foam generating systems would be expected to be in the range of 10 to 20 years, with proper operation and maintenance.

Aerosols

There are two different applications for aerosols in the response to spills of floating hazardous chemicals. Fast-settling aerosols can be used as an alternative procedure for the preparations of surfactant films. Slow-settling aerosols can be used as flame inhibitors or arrestors. The dispersed particulates used in response to spills of hazardous chemicals must be inert towards response personnel, while actively suppressing flames.

Aerosols may be prepared from water, or other liquid suppressants, and from powdered solids, usually common salts or minerals. The effectiveness of this technique can be modified by proper selections of the inhibitor material, particle size, surface area and particulate loading. Inhibitor materials can be chosen which have no toxicity, but flame inhibition requires particulate loadings in the range of 2×10^4 to 3×10^4 mg/m^3 (1.2×10^{-3} to 1.8×10^{-3} lb/ft^3), which are well into the range of nuisance dusts. The accepted TLV for most nuisance dusts is approximately 10 mg/m^3 (6×10^{-7} lb/ft^3) which indicates the necessity for having response personnel wear particulate filter masks.

Small particles of inorganic salts may be used as aerosols or mixed in the spill to suppress flammability. The amount required to secure a spill would depend upon the surface area of the particles (size), the rate of settling, density of the particles and chemical interaction between the spilled chemical and particulates.

Several manufacturers supply equipment capable of preparing aerosols to the specifications required for responding to spills of floating, hazardous chemicals. The maximum particulate generation is estimated at 5,700 kg/min or 12,500 lb/min, which corresponds to approximately 3.4×10^5 kg/hr or 375 tons/hr. The problems of storing and delivering these quantities of solids could be overcome, but may be used to emphasize the greater praticality of water aerosols. These aerosol specifications may be translated to liquid flow rates of approximately 5,700 l/min or 1,500 gal/min.

Pneumatic systems can be used to generate and distribute aerosols. Commercial equipment, such as that used for orchard spraying or fertilizer spreading, is adaptable for aquatic spill response. Snow blowers and mulch spreaders appear to be particularly well suited for the required specifications. Typical examples are manufactured by Finn Equipment Company, Kool Brothers and American Snowblast Corporation. Power requirements for these units are between 50 and 100 hp and they are fabricated as separate units, occupying 3.7 to 9.3 m^2 (40 to 100 ft^2).

The capacity of most commercial pneumatic systems is limited to between one-quarter and one-third of the calculated maximum application rates. Multiples of three or four units could easily be used without interference and possibly obtain a more uniform aerosol distribution.

The expected life of these systems, in use against spills of hazardous chemicals, would be in the range of five to twenty years. Proper maintenance and storage are assumed in estimating this operating lifetime. Exposure to salt water during operation and the rigors of deck exposure would make frequent maintenance chekcs necessary, but these would require minimal time and effort. Operation would require one to two men for startup; only a minimum amount of control or adjustment would be necessary during long term spill response applications.

Hydraulic equipment could also be used to generate liquid particulate aerosols. This equipment is manufactured for fire fighting and for spraying insecticides or fertilizers. Rockwood Systems Corporation, Vandermolen, Inc., Fire Control Engineering Company and FMC Corporation are representative of the companies manufacturing hydraulic systems which could be adapted for spill response.

A typical example of this approach is the Rockwood Systems Corporation's Fog Foam Turret Nozzle, which is capable of handling water flows to 1,900 l/min (500 gal/min). The fog nozzle could be permanently mounted or a similar portable model used in conjunction with a pumping unit sized to meet the capacity. Such a unit would occupy approximately 5 m^2 (54 ft^2) and require a 75 hp motor and pump assembly for 500 gal/min at 150 psi. Electric, gasoline or diesel power could be used.

Normal capacities of commercial hydraulic systems are 1,900 l/min (500 gal/min) maximum. This would mean that three or four units would be necessary to secure the largest spill area.

The aerosol technique has been considered practical only when water is used as the dispersed phase. The amount of water required to reduce the vapor flammability of hazardous chemicals will vary with the reactivity of the chemical but this technique can be used against all hazardous chemicals. The necessity for using large amounts of nonpolluting material eliminates all other alternatives.

Aerosols (except cryogens) have negligible effects on the vapor concentrations of hazardous chemicals and will not ameliorate the hazards associated with the toxicity of vapors.

Several problem areas may be envisioned during the removal of the hazardous chemical in the presence of aerosols. Some experimental verification tests could be done to determine which recovery devices and procedures could be used without disrupting the layer of floating chemical, visibility and safety of response personnel.

REFERENCES

(1) *United States Coast Guard, Chemical Hazards Response Information System (CHRIS),* 4 Vol, CG-446, Office of Marine Environmental Systems, Washington, DC (January 1974).

(2) McCullough, J., Schram, Inc., West Chester, PA, Private Communication.

(3) Dawson, G.W., Shuckrow, A.J. and Mercer, B.W., "Strategy for Treatment of Waters Contaminated by Hazardous Materials," *Proceedings of the 1972 National Conference on the Control of Hazardous Material Spills,* PB-228 736, Houston, TS (March 1972).

(4) Srinivasan, S., Thomas, T., Genzen, A., Coyle, A. and Kimm, C., *Influence of Environmental Factors on Selected Amelioration Techniques for Discharges of Hazardous Chemicals,* Final Report No. CG-D-81-75, Columbus Laboratories, Battelle Memorial Institute, Columbus, OH (February 1975).

(5) Swisher, R.D., *Surfactant Biodegradation,* Marcel Dekker, Inc., NY (1970).

(6) U.S. Army Engineer District, Seattle, CE, "PCB Clean Up Operations—Duwamish Waterway," Seattle, WA.

(7) Willmann, J.C., "PCB Transformer Spill, Seattle, WA," U.S. Environmental Protection Agency, Region X, Seattle, WA; Parts of this article published in *Proceedings of the 1976 National Conference on Control of Hazardous Material Spills,* New Orleans, LA (April 25-28, 1976).

(8) Takamura, Y., et al, "Hedoro Dredging in the Osaka Port Area," reprinted from the Italian journal *Inguinamento (Pollution),* pp 2-7, translated by Joint Publications Research Service (November 1973).

(9) "TST System," Takenaka Komuten Co., Ltd, Tokyo, Japan.

(10) Sato, E., "Application of Dredging Techniques for Environmental Problems," *Dredging: Environmental Effects and Technology,* Proceedings of WODCON VII, World Dredging Conference, San Pedro, CA (1976).

(11) Nawrocki, M.A., *Removal and Separation of Spilled Hazardous Materials from Impoundment Bottoms,* EPA-600/2-76-245, U.S. Environmental Protection Agency, National Environmental Research Center, Cincinnati, OH (September 1976).

(12) LaFornara, J.P. and Wilder, I., "Solution of the Hazardous Material Spill Problem in the Little Menomonee River," *Proceedings of the 1974 National Conference on Control of Hazardous Material Spills,* San Francisco, CA (August 25-28, 1974).

(13) Faldi, G., "Portable Dredge Pump Proves Adaptability," *World Dredging,* Vol 12, No. 9, pp 19-22 (August 1976).

(14) Acres Consulting Service, *Evaluation of Procedures for Removing and Decontaminating Bottom Sediments in the Lower Great Lakes,* Canada Center for Inland Waters, Department of the Environment (August 1972).

(15) Murden, W.R. and Goodier, J.L., "The National Dredging Study," *Dredging: Environmental Effects and Technology,* Proceedings of WODCON VII, World Dredging Conference, San Pedro, CA (1976).

(16) Jernelov, A., "Release of Methyl Mercury from Sediments with Deposits of Inorganic Mercury at Different Depths," prepublication correspondence with JBF Scientific Corp. (1971).

(17) Pratt, S.B. and O'Connor, T.P., "Burial of Dredge Spoil in Long Island Sound," Marine Experiment Station, University of Rhode Island, Kingston, RI (March 1973).

(18) Graf, W.H., *Hydraulics of Sediment Transport,* McGraw-Hill, NY (1971).

(19) Postma, H., "Sediment Transport and Sedimentation in the Estaurine Environment," *Estuaries,* American Association for the Advancement of Science, Washington, DC, pp 158-179 (1967).

(20) Partheniades, E., "Erosion and Deposition of Cohesive Soils," *Proceedings of the American Society of Civil Engineers,* Vol 91, HY1, pp 105-139 (January 1965).

(21) Lambe, T.W. and Whitman, R.V., *Soil Mechanics,* John Wiley and Sons, New York, NY (1969).
(22) Saila, S.B. and Pratt, S.D., "Dredge Spoil Disposal in Rhode Island Sound," *Marine Technical Report No. 2,* University of Rhode Island, Kingston, RI (1972).
(23) Takenaka Komuten, Ltd., "Recent Developments in Dredged Material Stabilization and Deep Chemical Mixing in Japan," promotional brochure, Tokyo, Japan (June 1976).
(24) TRW Systems Group, *Recommended Methods of Reduction, Neutralization, Recovery, and Disposal of Hazardous Waste,* prepared for U.S. Environmental Protection Agency, National Environmental Research Center, Cincinnati, OH (August 1973).
(25) Walley, D.M., *Investigation of the Resistance of Freshly Injected Grout to Erosion and Dilution by Flowing Water,* Miscellaneous Paper C-76-4, U.S. Army Engineer Waterways Experiment Station, CE, Vicksburg, MS (June 1976).
(26) American Cyanamid Company, unpublished technical and promotional data on "AM-9" Grout, Wayne, NJ.
(27) Dodd, D., B.J. Hughes, Inc., Houston, TS, personal communication.
(28) Unpublished Technical and Promotional Data, Dowell Division of the Dow Chemical Company, Tulsa, OK.
(29) Unpublished Technical and Promotional Data, American Colloid Company, Skokie, IL.
(30) Epstein, M. and Widman, M., *Coatings for Ocean Bottom Stabilization,* paper presented at 158th Meeting, American Chemical Society, New York, NY. (1969).
(31) Epstein, M. and Widman, M., "The Application of a Polymer Film Formed Underwater," *Marine Technology Society Journal,* Vol 4, No. 6, (November-December 1970).
(32) Roe, J., et al, *Chemical Overlays for Sea Floor Sediments,* presented at The Second Offshore Technology Conference, Houston, TX, Paper No. OTC 1170 (May 1970).
(33) Widman, M. and Epstein, M. *Polymer Film Overlay System for Mercury Contaminated Sludge—Phase I,* No. 16080 HTZ, U.S. Environmental Protection Agency Water Pollution Control Research Series (May 1972).
(34) Geswein, A.J., *Liners for Land Disposal Sites—An Assessment,* Report No. SW-137, U.S. Environmental Protection Agency, Office of Solid Waste Management Programs (March 1975).
(35) Biggs, R.B., "Offshore Industrial Port Islands," *Oceanus,* Vol 19, No. 1 (Fall 1975).
(36) Murphy, W.L. and Ziegler, T.W., *Practices and Problems in the Confinement of Dredged Material in Corps of Engineers Projects,* Technical Report D-74-1. U.S. Army Engineer Waterways Experiment Station, CE, Vicksburg, MS (May 1974).
(37) Pilie, R.J., Baier, R.E., Ziegler, R.C., Leonard, R.P., Michalovic, J.G., Peck, S.L., and Bock, D.H., *Methods to Treat, Control, and Monitor Spilled Hazardous Materials,* EPA-670/2-75-042, U.S. Environmental Protection Agency, National Environmental Research Center, Cincinnati, OH (June 1975).
(38) "Underwater Plow Prepares Trench for Stratfjord Loading Pipeline," *Ocean Industry,* Vol 12, No. 8 (August 1977).
(39) Brockman, D.B., Samson Ocean Systems, Inc., 99 High Street, Boston, MA, personal communication.
(40) Wilder, I., and Lafornara, J., "Control of Hazardous Material Spills in the Water Environment: An Overview," presented before the Division of Water, Air and Waste Chemistry, American Chemical Society, Washington, DC (September 1971).
(41) Lagsdail, P.H., and Lowes, L., "Industrial Contacting Equipment," *Recent Advances in Liquid–Liquid Extraction,* C. Hanson, ed., Pergamon Press, New York (1971).
(42) Crump-Wiesner, H.J., and Jennings, A.L., "Properties and Effects of Nonpetroleum Oils," 1975 Conference on Prevention and Control of Oil Pollution, San Francisco, CA (March 25-27, 1975).
(43) Matthews, S.M., AIRCO Industrial Gases, Private Communication.
(44) Moran, H.E., Bernett, J.C., and Leonard, J.T., *Suppression of Fuel Evaporation by Aqueous Films of Fluorocarbon Surfactant Solutions,* NRL Report 7247 AD 723 189 (April 1971).

CHEMICAL METHODS

GENERAL GUIDELINES

The information in this section is based on *Manual for the Control of Hazardous Material Spills. Volume 1: Spill Assessment and Water Treatment Techniques*, EPA-600/2-77-227 prepared by K.R. Huibregtse, R.C. Scholz, R.E. Wullschleger, J.H. Moser, E.R. Bollinger and C.A. Hansen of Envirex for the Environmental Protection Agency.

Safety Precautions

It is stressed that many treatment chemicals are hazardous themselves and must be treated as such. All operators handling chemicals should wear goggles and in most cases chemically resistant aprons and gloves should be used. All chemicals on site should be stored in an isolated area so that unauthorized personnel will not come in contact with them. The handling of chemicals on site must eliminate compounding the hazard that already exists.

Troubleshooting

The most common problem in treating spills with chemicals would be that the end point may be overshot. This would most likely occur at the beginning of the operation before the kinetics of the reaction are familiar to the operators. When this situation occurs, the following procedure can be used: (1) Take a sample of the overtreated water; (2) Back titrate to the end point using contaminated water; (3) Note the volume of wastewater needed to reach the desired end point; and (4) Scale up the results to determine the increased volume of wastewater to be added. Other possible problems involved in chemical reactions include: (1) Concentration gradients in tank. If this is suspected, check jet mixing systems to be sure they are operating properly; and (2) Incomplete chemical reactions after addition of entire chemical amount. Repeat bench tests and recalculate required volumes.

NEUTRALIZATION

Water Quality Criteria for pH

> The information in this and the next four sections is based on *A Feasibility Study of Response Techniques for Discharges of Hazardous Chemicals that Disperse Through the Water Column*, CG-D-16-77 prepared by E. Drake, D. Shooter, W. Lyman and L. Davidson of Arthur D. Little, Inc., for the U.S. Coast Guard.

The pH Range in Natural Waters: The acidity-alkalinity range of terrestrial waters varies over nearly the entire pH range. The waters of Kata-numa, a volcanic lake in Japan, are essentially fairly concentrated sulfuric acid with a pH of 1.7. Another crater lake, Kawah Idjen, in Indonesia, may even have a pH of less than 0.7. Such highly acid waters are by no means restricted to volcanic regions. Biological activity and the decay of dead vegetable material can produce high acidity. For example, the sphagnum-mat-edged pools around Mud Lake, Michigan, can have a pH as low as 3.3. Biological acidification of fresh water can be greatly aggravated by human activity, as in the well-known case of acid mine drainage. Waters in the bituminous coal region near Elkins, West Virginia, for example, have pH values close to 3.

On the other extreme, highly alkaline waters are also formed in nature, especially in arid regions such as the American Southwest. Lake Nakura, an alkaline lake in Kenya, Africa, has a pH of 12. But such highly acid and alkaline waters are the exception rather than the rule and "normal" terrestrial waters have pH values usually within 1 unit of neutral, pH 7. Lake Erie, for example, is slightly alkaline with pH of 7.7. A series of streams reviewed by Stumm and Morgan (1) fell in the pH 6.6 to 8.0 range, and a series of ground waters, pH 6.8 to 8.0.

In contrast to terrestrial waters, the pH of seawater is restricted to a narrower range. Surface values in the Western Pacific, for example, stay pretty much within the range pH 8.0 to 8.4. Within this range, however, the deep-sea pH profile can exhibit some significant features. As in terrestrial waters, biological activity, notably photosynthesis, can affect the pH of seawater, giving rise to interesting diurnal and seasonal fluctuations. For example, in a shallow Texas bay, Park et al (2) observed that on a summer day the pH rose from about pH 8.2 in the morning to pH 8.9 in the afternoon, and that in winter the pH ranged about 0.5 unit less. In biologically rich tropical water even greater diurnal ranges of pH 7.3 to 9.5 are not uncommon.

The oceans of earth are a highly buffered aquatic system and this controls the pH to within narrow limits. The carbon dioxide-carbonate system is believed to be responsible for this control; however, in recent years the importance of the contributions of ion-exchange and other equilibria, especially involving silicate minerals, has gained growing recognition. The salt content of the seas is in fact the consequence of a gigantic titration or neutralization taking place throughout geological time of alkaline crustal material mobilized by weathering and acidic outgassing of the earth's crust. This global neutralization is also occurring locally in terrestrial waters but, because of their transitory nature, the end point is often never approached.

Change of pH is a product of more critical biological stress than a sustained pH level. Natural rates of change in pH are probably fairly slow. In the case of the shallow Texas bay cited above, the pH changed by about 0.7 within a 12 hour period, due to photosynthetic activity; this probably represents a fairly drastic change in a natural environment. Solar warming could also contribute to the afternoon peak in pH since the solubility of atmospheric CO_2 in water decreases with increasing temperature.

Water Quality Standards for pH: The Committee on Water Quality Criteria of the Environmental Studies Board of the National Academy of Sciences, National Academy of Engineering (3), has recommended that, for marine waters, (1) "the normal range (presumably of the site in question) of pH in either direction should not be extended by more than 0.2 (pH) units," that (2) "within the normal range the pH should not vary by more than 0.5 units," and that (3) "addition of a foreign material should not drop the pH below 6.5 or raise it above 8.5." The 0.5 unit change in recommendation (2) is, as has been seen, less than the range of rapid natural changes but may nevertheless be reasonable. In recommendation (3), the lower limit of 6.5 may be much too low, especially when one remembers that the synergistic effects of high acidity may be considerable.

Recommendations for terrestrial waters are in terms of level of protection (Table 3.1). Comparison with the ranges in natural waters suggests that the recommendations for a "high level" of protection are reasonable and also seem to correspond roughly to the same degree of protection that is recommended for marine waters. Whether terrestrial waters need the same, a greater, or lesser level of protection than marine waters is subject to debate.

Table 3.1: Recommendations for Terrestrial Waters

Level of Protection	pH Range	pH Change
Nearly maximum	6.5-8.5	0.5
High level	6.0-9.0	0.5
Moderate level	6.0-9.0	1.0
Low level	5.5-9.5	1.5

Source: CG-D-16-77

Some uses are less restrictive than others. A pH range of 4.5 to 9.0 has been proposed as acceptable for irrigation and 5.0 to 9.0 for public drinking water, as compared to 6.0 to 9.0 for fresh water and 6.5 to 8.5 for marine aquatic life (4).

The pH Limits for Aquatic Life: Organisms have adapted to environments that are astonishingly extreme with respect to pH. Acid pools and alkaline lakes can teem with life. pH 2 to 10 appear to be the extreme limits for living organisms. The cells of a stomach are undamaged by the gastric juices (a hydrochloric acid solution) and acid-mine drainage, it might be noted, is the consequence of the activity of microorganisms. Natural, unbuffered terrestrial waters with pH 4 derived from carbon dioxide and natural organic acids can support fluorishing biocommunities, and waters with a pH as high as 9.5 can support fish but tend not to be highly productive.

Not only do different species exhibit different tolerances to pH, but also the tolerance of a given species can be different for different stages in its life cycle. Reproduction, for example, can be seriously reduced by pHs which do not appear to adversely affect the adult organisms. Inasmuch as terrestrial waters, being relatively unbuffered, generally exhibit greater pH ranges and fluctuations than seawater, fresh water organisms have adapted correspondingly and tend to be more tolerant of pH change than marine organisms. Oysters appear happiest in brackish waters of about 7.0, but at only half a pH unit lower, at pH 6.5, their activity is very significantly reduced. Not unexpectedly, organisms can tolerate relatively large pH changes within the normal ambient range, whereas relatively small changes at the limits of these ranges can be harmful.

A spill of an acidic substance is not a single impact; it is a barrage of environmental stresses presenting, in the case where the spill is combated by neutralization, at least 10 critical aspects. There is first the natural pH level and the natural fluctuation of the pH level which define the pH range to which the impacted biota are accustomed. Both the change of pH resulting from the spill and the rate of this change are important. Rapid changes in the ambient environment tend to be especially traumatic and one might expect a gradual movement, even amounting to a greater pH increment, to be less disruptive than a rapid but somewhat smaller increment. Similarly, the duration of exposure to the stress may be as important as the magnitude of the stress. Very little appears to be known about the effect of exposure duration.

However, while some species may have some capability to adapt to an extended exposure stress, generally one expects the larger the exposure the greater the damage. The neutralization of the spill is another traumatic environmental impact which could in itself be very damaging even if there is no overshoot of the neutralization. This raises a question which needs a great deal more reflection. Will a second rapid traumatic event, the neutralization, do more damage than persistent, but slow diminution by dilution of the original impact? Even after the pH range of the waters has been restored to its former value, damage, for example, in the form of the altered biopopulation distributions, may persist.

Response to a pH-altering spill must also take into consideration the mixing and flow characteristics of the impacted water to minimize biological damage. If the water is quiescent and the spill has destroyed biota in a water volume, the expansion of lethality (by dilution) can be slow and of diminishing intensity. However, in the case of a river, a moving lethal slug will travel downstream exposing further aquatic environment to damage. In the latter case, neutralization may spare more organisms from destruction than the former.

The Environmental Impact of pH on Aquatic Systems: It is probably safe to assert that most of chemical equilibria in terrestrial and marine waters are pH dependent. Crustal weathering (with consequent erosion and increased silting), cation exchange with mineral species, precipitation and colloid chemistry, and of course, the $CO_2/CO_3^=$ system are only a few examples of major pH-dependent geochemical processes.

The same may also be said of biochemical reactions. Many of the fundamental life processes are so pH dependent that the pH of blood and other body fluids is strictly regulated by the organism.

Chemical speciation of many dangerous pollutants is pH dependent. The solubility of toxic heavy metals often increases with decreasing pH; the toxicity of pulp and paper mill effluents is strongly pH dependent, as are also metal plating and finishing industry effluents, especially cyanides. Sulfides are found in industrial effluents and are also formed in poorly aerated anoxic waters, both marine and fresh, and again the equilibria involved are pH dependent. Thus, a spill which causes a change in pH, particularly an acid spill, may solubilize other toxic components in the water body with far-reaching effects. To complicate matters still further, the response of organisms to other environmental stresses such as oxygen level, temperatures, and tolerance to a host of chemical species encountered in natural waters can be pH dependent.

Finally, it should be noted that, in general, because of the strong dependence of aquatic organisms on chemical sensation for predator and prey detection, for environmental identification, and for intra-species identification and breeding, chemical pollution of waters can be a particularly disruptive form of environmental stress. In addition to damage to an individual organism's health, chemical pollution by confusing intra-species identification can (as in the case of catfish) destroy community relationships and structure, and by the confusion of environmental identification can disrupt migratory habits. The role of pH changes in aggravating these stresses is largely unknown.

Required Reduction of Contaminant: From the evidence available and the proposed water quality criteria, it seems that any treatment agent should act so as to return the pH to within the range of 6.0 to 9.0. Because of the potential adverse effects caused by rapid changes in pH, the problem of pH changes caused by misapplication of the neutralization agent or excess of the neutralization agent must be carefully evaluated.

Factors Influencing the Choice of Neutralizing Agent

The ideal neutralizing agent should have the following characteristics:

- It should return the pH to the range of 6 to 9 without a large overshoot (in pH) if a small excess of the agent is added by mistake or it is applied outside the area of the spill.
- It should be relatively nontoxic to aquatic life, and not form more toxic compounds in the neutralization reaction.
- Its biological oxygen demand (BOD) should be small.
- It should be safe to use by Coast Guard personnel (i.e., have low flammability, high TLV, etc.).
- It should be easy to handle and store.
- It should be commonly available in bulk quantities and be low in cost.

Addition of Excess Neutralizing Agent: In many instances the exact amount of acid or base spilled will not be known; also the concentration in different parts of a water body will vary due to dispersion. This is particularly true where actual conditions may differ considerably from those predicted by a simple model. Under some circumstances it may be desirable to add an excess of reagent to ensure that the spill has been effectively neutralized. However, it is also important that the excess reagent does not cause the pH to move out of the range pH 6 to 9 in the opposite direction.

Calculations showing the pH obtained after the addition of 20% excess neutralizing agent are given in Table 3.2 for different neutralizing agents. Strong acids and bases (e.g., sulfuric acid, sodium hydroxide) are definitely unacceptable because excess neutralizing agent will produce extreme changes in pH. Even acetic acid is only marginally acceptable since the final pH would be well below pH 6. For neutralization of bases, the best choice appears to be sodium dihydrogen phosphate (boric acid and sodium bicarbonate are unacceptable because of the large quantity of neutralizing agent required). Similarly, the most promising agents for neutralization of acids are sodium carbonate and sodium bicarbonate.

Table 3.2: Consequences of Overshoot or Misapplication

Neutralizing Agent	M	pH After Adding 20% Excess Past pH = 9	pH for Miss of the Neutralizing Agent
Neutralization of 0.1 M Sodium Hydroxide			
Sulfuric acid	1.0	0.85	0.3
Acetic acid	1.0	5.46	2.4
Sodium dihydrogen phosphate	1.0	7.87	4.5
Boric acid	0.1	8.87	5.2
Sodium bicarbonate	1.0	9.0	8.3
Neutralization of 0.1 M Hydrochloric Acid			
Sodium hydroxide	1.0	12.2	14.0
Lime (saturated solution)	0.024	11.8	12.4
Sodium carbonate	1.0	7.1	12.1
Sodium bicarbonate	1.0	6.2	8.3
Urea	1.0	6.1	7.1
Limestone (saturated solution)	0.00015	–	9.4
Borax	0.1	–	9.2

Source: CG-D-16-77

Misapplication of Neutralizing Agent: If it is necessary to keep the pH in the range pH 6 to 9 when the neutralizing agent is accidentally applied in the wrong place, then the choice becomes more restricted. Table 3.2 shows that sodium dihydrogen phosphate is still the best reagent for bases; in the event of misapplication the pH could reach pH 4 to 5. Sodium bicarbonate is still acceptable for neutralization of acids, but misapplication of sodium bicarbonate could result in a pH of about 12.

Methods of Application

Viable methods to apply a neutralizing chemical must recognize that the hazard zones cannot be well defined a priori. Thus, application cannot be planned on the basis of calculations; disregard of this fact can easily lead to situations where the spilled acid or base is followed (or preceded) by a zone containing the ameliorating chemical; therefore the spill must be tracked with surface craft to allow one to delineate the extent of the hazardous zone before any neutralizing chemical is applied. In the discussion to follow, it is assumed that this vital information is available.

Neutralizing chemicals (sodium bicarbonate for acid spills and sodium dihydrogen phosphate for alkaline spills) are assumed to be available in the immediate vicinity of a spill from cooperating chemical companies. Normally such neutralizing chemicals are stored as dry powder (or granules) in 50 to 100 lb bags. This assumption is important because no special power equipment is needed to move the material from storage to the waterfront. If storage is in bulk or large barrels, then it is unlikely that men could carry out the transfer without prior planning and ready availability of bulk carriers, power loaders, etc. Not only must the neutralizing chemical be packaged in sizes convenient for one (or two) men, but there must be a sufficient amount stored in hazardous areas to be useful for the largest credible accident. For example, for each ton of 100% hydrochloric acid spilled, 3.3 tons of sodium bicarbonate are required to neutralize the spill. Thus, for multiton spills, it is apparent that a large quantity of neutralizing chemical must be available.

Assuming that a sufficient supply of the desired neutralizing chemical is available and can be transported to the spill site and that means are available to track the spill downstream, techniques must be found to apply the neutralizing chemical to the hazardous area at the appropriate location and in the proper amount. A simple analysis indicates that application can be made in three ways: (A) from a mobile marine ship, (B) from an aircraft, and (C) from a convenient shore or bridge location. Further, application can be made: (1) with the dry powder (as received), (2) with a slurry of powder in water, and (3) with a solution of the powder in water.

First, let us consider (A), (B), and (C). In case (A), appropriate ship(s) must be available and have the ability to load and store sufficient quantities of the neutralizing chemicals on deck or in an accessible hold with protection from the weather. The crews must have had some prior training in simulated spill scenarios in order to exercise good judgment and to cooperate with the local Coast Guard officer in charge of the application. Communications with other ships involved in the operation must be available, especially with those craft which are responsible for monitoring the local pH values. Strict adherence to orders relating to the quantity of neutralizing chemical applied and the time of application is necessary.

For case (B), appropriate fixed-wing aircraft or helicopters must be available. An advantage such craft have over mobile marine units is that they can be moved rapidly to the scene of the spill. Also, it is possible that with helicopters, the transportation of the chemical for storage to the waterside may be avoided. Aircraft have, however, the serious disadvantage that they have a low carrying capacity relative to marine units and multiple trips would probably be required. Also, precise dumping in hazardous areas is obviously much more difficult from an aircraft.

In case (C), there would appear to be few advantages. At a fixed position on the shore or on a bridge, there is only one opportunity to apply the appropriate amount of neutralizing chemical as the hazardous zone passes by. From shore locations, it is difficult to achieve any more than local neutralization of the water near the shore line. Calculations have indicated that, even with dispersion, the downstream hazardous zone is still very localized and moves as a wave with approximate stream velocity. To neutralize this wave, even for a convenient bridge crossing, would require detailed planning and cooperation between monitoring ships and those on the bridge. The critical time period to add the neutralizing chemical is short. Except in very special situations, case (C) does not seem to be a practical method and it will not be discussed further.

Of cases (1), (2), and (3), the first is obviously the simplest. One could visualize the operation as follows:

- The monitoring craft locates an area of pH imbalance.
- Free-floating buoys are set to indicate this zone.
- The cooperating marine unit or aircraft are alerted and are told how much neutralizing chemical to disperse in the marked zone.
- After application, the monitoring craft measures the pH (surface and lower layers) and indicates if more chemical is desired.

Note that excellent communications are required between the monitoring and application craft, although it is possible both could be in the same vessel. If the dry powder is simply dumped overboard, little mixing will take place and the bulk (dense) powder will sink. There are several ways this application method can be improved, although each requires more equipment.

- A device can be used to distribute the powder over a wider area. Mechanical "slingers" are a possibility as are compressed-air powder sprays. By breaking up the bulk powder, there is less tendency for it to "clump" and sink and solution is achieved more rapidly.
- The powder can be mixed with water to form either a slurry (case 2) or a true solution (case 3). For either of these cases, mixing and pumping equipment must be available. The more neutralizing chemical dissolved in water, the more rapid is the effect in the hazardous area. Also, there is little tendency for sinking if a well-dispersed slurry or a true solution is distributed over a wide area.
- The powder may be added carefully in the region of maximum propeller turbulence. This should promote more rapid solution and neutralization.

It is difficult to visualize an aircraft carrying any equipment to prepare or dispense slurries or solutions. Even the concept of mechanical slingers is not reasonable because of wind drift problems. Thus, for aircraft, simple powder dumps are the only feasible alternative. Some dispersion in the wind or propeller (rotor) wash will occur but not in a very controlled manner and this will increase the difficulties of accurate placement of powder. Therefore, the preliminary analysis indicated that only the following application methods are candidates for further study:

- Apply powdered chemical with mechanical slinger or air jet from a mobile marine unit.
- Prepare slurries or solutions in mixing tanks on board and spray on the lee side of a mobile marine unit.
- Possibly, use aircraft with powder dumps for localized spills when the time element is critical.

A preliminary design was carried out for four potential methods: (1) Manual dump of powdered agents into prop wash of marine craft, (2) mechanical slinger dispensing powder over side of marine craft, (3) water pumped into mixing tank where powdered agent is added; mixing solution or slurry overflows overboard into spill area, and (4) water pumped, powder injected from hopper, slurry passed through static mixer and discharged overboard.

In order to compare alternatives, a spill size of 10,000 gallons was selected as representative of the largest size spill that might occur and still be amenable to full treatment over practical response periods of 6 to 24 hours. Smaller spills could be treated more quickly with this same equipment. Should a larger spill occur, partial treatment would still allow some amelioration of the hazard. Since extremely large spills occur so infrequently, it seems economically and logistically unreasonable to design response equipment for these situations, especially since the response equipment would then be cumbersome to transport and there might be less flexibility as to the type of deployment craft that might be used.

As a preliminary design basis, treatment of a 10,000 gallon spill of 37% hydrochloric acid was considered. Six hours after a sudden release of this quantity of acid into a wide river, for example, unacceptable acidity would cover about 20,000 ft^2 of river surface (and would extend from the water surface to the river bottom) according to the assessment models used. The affected zone would be roughly circular, with a diameter of about 160 ft. It is evident that finding a hazardous zone of this size after a several hour transit downstream will require detection equipment in addition to the basic treating equipment considered in this study.

For the assumed spill, it was estimated that a minimum of 90 tons of sodium bicarbonate would be needed to bring the affected zone back within acceptable pH limits. Since the treating agent for neutralization is a fairly common chemical, available in powder form in 100 lb bags, the required amount of treating agent probably could be obtained and transported to the vicinity of the spill location using C130 aircraft. Although the bags could be handled manually, for a spill of this magnitude some 1,800 bags would be required. Several 130C aircraft or a shuttle operation would be needed for the transport of 90 tons of treating agent.

In the basic response scheme, treating chemical would be supplied to marine craft which would search, locate and follow the hazard zone. The types of marine craft that are likely to be available for a response effort at a random site probably will be able to carry loads of 1 to 10 tons at a time so several craft and/or a shuttling of treatment chemical to the hazard zone will be required. The weight of chemical involved for treating a spill of this size mitigates against deployment from a helicopter (in addition to wind drift problems).

Coast Guard sea-going cutters in the buoy tender class (WLBs) and the medium- and high-endurance classes are candidates for the water transport task, as are other public vessels and those commercial vessels certified for offshore operations. The principle water-carrier design parameters are the cargo-boom-handling capacity, the deck space available, the habitability of the deck space during transit, and the on-scene cargo-handling capability. Typical buoy tenders have adequate deck space, normal load capacities in the 2 to 10 ton range and hydraulic hoists capable of lifting 1 to 2 ton weights.

At the contaminated zone, the powdered neutralizing agent must be introduced into the water. If the bags of chemical were dropped overboard intact, they would sink to the bottom before much chemical could dissolve and the spill in moving water would pass by. Therefore, as a minimum, the bags would have to be slit open and the contents dumped. Again, depending on water depth and turbulence, the treating agent may not fully dissolve. Some advantage might be taken of the extra mixing available in the vessel's wake.

Manual dumping of hundreds of bags of powder, while not impossible, involves much heavy labor and is not a very good method for distributing the treatment powder throughout the water. Some skid-mounted equipment could greatly facilitate the operation. To treat the assumed design spill in a 16 hour period, sodium bicarbonate would have to be applied at a rate just under 6 tph. If the powder were dissolved in water, the solution would have to be applied at a rate of about 325 gpm. If the powder were applied in slurry form, lower water-pumping rates could be used.

For the three equipment alternatives listed earlier, a mechanical slinger system to dispense powder at a rate of 185 lb/min would weigh about ½ ton. The pump and mixing tank system and the pump-powder injection system both would weigh about one ton. The latter concept is the more attractive since the rate of powder injection is controlled mechanically. It might be desirable to add another system for conveying and dumping powder into the supply hopper to reduce the manual labor requirements.

The skid-mounted treatment equipment is light enough to be transported by helicopter to the spill location and could be conveniently used on boats of the 45 ft buoy boat type which usually have load capacities of 5 to 10 boats and deck space for equipment and chemicals.

A more sophisticated system incorporates a metered slurry mixture with a convenient automatic bag-opening device. This concept provides for the proper percentage of slurry mix by prior adjustment of the auger screw feed into the eductor. Furthermore, with the addition of the bag opener not only is the manual labor reduced but the dust level is also controlled by the enclosure. A bag disposer may also be added. The entire unit including skid-mounting pump and drive would weigh around 3,000 lb.

Results of Neutralization Treatment

Physio-Chemical Consequences: The major physio-chemical consequences of a spill of an acid or base into water is, of course, a significant change in the pH (assuming a relatively strong acid or base and a relatively large spill). Physio-chemical consequences expected to be relatively minor are associated with:

- an increase in the ionic strength of the water;
- a slight increase in water temperatures—for most, but not all, spills—due to the heat of solution;
- a possible increase in the Chemical Oxygen Demand (COD) resulting in lower levels of dissolved oxygen;
- the resolubilization of heavy metals from a lowering of pH; and
- other changes in water chemistry.

Biological Consequences: The severity of the biological consequences will, as mentioned previously, depend on several factors, e.g.:

- amount of spill; resulting rate and extent of pH change;

- duration of spill effects (before neutralization);
- identity of the chemical; its specific toxicity, BOD, COD, etc.;
- the nature of the affected water; fresh versus salt, naturally high or low pH, buffering capacity, rate of mixing; and
- the season; this will be a determining factor in the number and types of aquatic organisms present.

Ultimate Fate of the Spilled Chemicals: In all cases, dilution will eventually reduce the concentration of the spilled chemical to essentially harmless levels. If the spill is into a water body (or stream) of adequate size, then dilution will be the primary factor in the natural amelioration of the spill. Various other factors will also be involved depending on the chemical and the affected waters:

- Adsorption onto colloidal particles and sediments may be significant, especially for the weaker electrolytes;
- Cation exchange may take place;
- Hydrolysis–this may be important for some organic species;
- Biochemical oxidation–this may be important for some chemicals; and
- Evaporation–this may be a factor only for chemicals such as NH_3 and HCl.

Consequences of Optimum Treatment: Optimum treatment, i.e., the attainment of a safe pH of 6 to 9 in the spill site with no overshoots or misses, will essentially eliminate the worst aspect of the initial spill, the excessively high or low pH. This major benefit will be slightly offset by:

(1) A further increase in the ionic strength of the water, to which aquatic organisms will be variously sensitive;
(2) A slight increase in temperature, probably no more than 1°C, due to the heat of reaction; and
(3) An increase in nutrient concentration–primarily in the case of neutralization with NaH_2PO_4–which could lead to algal growth (and, thus, an increase in the oxygen demand), and a temporary increase in the rate of eutrophication in confined waters.

None of the offsetting factors mentioned are expected to pose a serious problem. The increase in ionic strength due to the addition of the neutralizing agent is not likely to make matters much worse, compared to the damage done initially by the spill. The temperature increase due to the heat of reaction is unlikely to exceed 1°C and, thus, should not result in any adverse effects. The addition of a nutrient, phosphate in the case of NaH_2PO_4, may or may not lead to temporary increase in algal growth. Phosphate is sometimes the biomass-determining nutrient, but it is expected that in a majority of cases the limiting nutrient will be some other compound or element (e.g., N, S, Ca, Mg, K, Fe, Mn, Zn, Mo, or B). Carbonate, like phosphate, is a nutrient, but it is rarely the limiting nutrient in water bodies.

The other components of the neutralizing agents (Na^+ for both acid and base neutralization and HCO_3^- for acid neutralization) should not have any significant adverse effects on natural waters aside from the ionic strength aspect mentioned above. Both components may be considered essentially nontoxic; both are constituents of unpolluted waters at modest levels.

Addition in the Wrong Place: If a recommended neutralizing agent were added in the wrong location the consequences (to aquatic life) would be negligible to moderate depending on the conditions.

A misapplication of $NaHCO_3$ to neutral, unbuffered waters could, at most, raise the pH to about 8.3, which is technically within the designated safe range. This pH is achieved only for $NaHCO_3$ concentrations near the ultimate solubility of about 1.0 M. The abrupt, but temporary, increase in ionic strength associated with a misapplication could affect aquatic life; ionic strength affects transport across biomembranes and their osmotic properties, and a sufficiently abrupt change in ionic strength may even result in cell rupture. This effect would be expected to be local and temporary.

A misapplication of NaH_2PO_4 to neutral, unbuffered waters could, at most, lower the pH to about 4.1. This pH is achieved by NaH_2PO_4 at a concentration of 1.0 M; the ultimate solubility is about 10 M. This pH lowering could have moderate, local impact on the aquatic life. The problem of ionic strength change would be present as would any problems relating to nutrient addition (where phosphate was the limiting nutrient).

Addition of Too Little or Too Much: Addition of an inadequate amount of neutralizing agent, such that the pH is not returned to the range of 6 to 9, will improve the situation. Partial neutralization is better than no neutralization since it will reduce the extent of damage due to the original spill.

Addition of too much neutralizing agent should, generally, have a negligible adverse effect. The recommended agents, $NaHCO_3$ and NaH_2PO_4, were chosen (in part) because they would not result in a serious overshoot problem. If neutralization is stopped at the limits of the safe pH range (e.g., at 6 when neutralizing acids and at 9 when neutralizing bases), then a significant excess can be added without having the pH leave the safe range.

Preferred Techniques for Water-Soluble Chemicals

Spill Chemicals Subject to Neutralization: The full list of chemicals considered for amelioration by neutralization techniques can be subdivided into the following four categories: (1) acids; (2) compounds that react with water to form acids; (3) bases; and (4) compounds that react with water to form bases. Chemicals falling into the first two categories are listed in Table 3.3 and into the second two categories in Table 3.4.

Table 3.3

Chemicals Classified as Acids (Total of 14)

Acetic acid	Hydrogen fluoride
Acrylic acid	Nitric acid
Formic acid	Oxalic acid
Hydrochloric acid	Phosphoric acid
Hydrofluoric acid	Propionic acid
Hydrogen chloride	Sulfuric acid
Hydrogen cyanide	Sulfuric acid (spent)

(continued)

Table 3.3: (continued)

Chemicals That React in Water to Give Acids (Total of 16)

Acetic anhydride → Acetic acid
Aluminum chloride → Hydrogen chloride (+ Aluminum hydroxide)*
Benzoyl chloride → Hydrogen chloride + Benzoic acid
Bromine → Hypobromous acid
Chlorosulfonic acid → Hydrogen chloride + Sulfuric acid
Maleic anhydride → Maleic acid
Nitrogen tetroxide → Nitric acid (+ Nitric oxide)*
Nitrosyl chloride → Hydrogen chloride + Nitrous acid
Oleum → Sulfuric acid (+ Sulfur trioxide)*
Phosphorus oxychloride → Hydrogen chloride + Phosphoric acid
Phosphorus pentasulfide → Hydrogen sulfide + Phosphoric acid
Phosphorus trichloride → Hydrogen chloride + Phosphorous acid
Polyphosphoric acid → Phosphoric acid
Sulfur monochloride → Hydrogen chloride + Sulfuric acid (+ Others)*
Sulfuryl chloride → Hydrogen chloride + Sulfuric acid
Titanium tetrachloride → Hydrogen chloride (+ Ti-hydroxychlorides)*

*Reaction products listed in parentheses are not acids.

Source: CG-D-16-77

Table 3.4

Chemicals Classified as Bases (Total of 28)

Aminoethanolamine	Hexamethylenetetramine
Ammonium hydroxide	Hydrazine
Aniline	Methylethylpyridine
Caustic potash solution	Monoethanolamine
Caustic soda solution	Monoisopropanolamine
Cyclohexylamine	Morpholine
Diethanolamine	Potassium hydroxide
Diethylamine	Pyridine
Diethylenetriamine	Sodium hydroxide
Diisopropanolamine	Triethanolamine
Dimethylformamide	Triethylamine
1,1-Dimethylhydrazine	Triethylenetetramine
Ethylenediamine	Trimethylamine
Hexamethylenediamine	Urea

Chemicals That React with Water to Give Bases (Total of 6)

Anhydrous ammonia → Ammonium hydroxide
Ethylene imine → Monoethanolamine
Lithium aluminum hydride → Lithium hydroxide (+ Hydrogen + Aluminum hydroxide)*
Sodium → Sodium hydroxide (+ Hydrogen)*
Sodium amide → Ammonia + Sodium hydroxide
Sodium hydride → Sodium hydroxide (+ Hydrogen)*

*Reaction products listed in parentheses are not bases.

Source: CG-D-16-77

Spill Chemicals Which Should Not Be Neutralized: For a number of the chemicals listed, neutralization should be approached carefully because of the hazards to personnel from toxic or irritant vapors which may be present in the vicinity of a spill (e.g., hydrogen chloride, ammonium hydroxide). However, if the reaction with water is complete these chemicals can be treated by neutralization. It is undesirable to treat the following compounds by neutralization for the reasons given. Acids: Hydrogen cyanide–high toxicity, vapor and liquid. Bases: Aniline, dimethyl formamide, hexamethylene tetramine, methyl ethyl pyridine, pyridine, and urea–very weak bases which will not violate pH criteria.

Preferred Techniques: For response to spills subject to neutralization, the evaluation of potential treating agents identified two preferred agents: sodium bicarbonate for treatment of acidic spills and sodium dihydrogen phosphate for treatment of basic spills. The choice of these agents was based on several criteria including considerations of availability, cost, efficacy, and safety. The primary consideration in selecting these two preferred agents, however, was that excessive application or misapplication would have a minimal effect on the water quality. If the preferred agents are not available, second choice alternatives are probably sodium carbonate (for acidic spills) and boric acid (for basic spills). Both these agents will produce more severe pH excursions in the water body if misapplied than will the preferred agents.

For practical purposes, spills in excess of about 10,000 gallons are rare and also will require tremendous quantities of treating agent. For this reason, it has been recommended that response equipment be designed to cope with spills smaller than about 10,000 gallons. The equipment still would be useful for partial treatment of very large spills.

In the preferred response scheme, treatment skids would be constructed and stored at Coast Guard spill response depots. Each skid would contain the following equipment: a conveyor and bag opener for the treating agent; a feed bin; a positive mixer for adding treating agent to water in the desired proportion; a pump; and a fishtail nozzle for dispersing neutralizing solution over the spill zone.

The skid weight is estimated at about 3,000 lb. In marshalling a response effort the skid could be carried by helicopter to the response location. Simultaneously, treating chemical would be obtained from the nearest source and trucked or air lifted to the scene. The response strike team would be at the spill scene to oversee the installation of the treatment skid on a boat of the buoy tender class. (For spills into confined areas such as drainage ditches, the skid could be truck mounted and hauled to the proper location.) Treatment chemical, available in 50 or 100 lb bags, would be shuttled to the skid for application by the strike team.

This response technique appears feasible for neutralization of hazardous water-soluble chemicals which result in unacceptable water pH levels. In stagnant water, long response times are acceptable; in flowing water, response within 6 to 12 hours appears to have considerable potential for spill hazard amelioration. In estuaries or tidal waters, response becomes less likely to be effective unless accomplished before the next tidal reversal.

An important factor in the overall feasibility is, obviously, the ability to locate the contaminated zone during the response period. This facet was not a part of the present study. Nevertheless, overall feasibility of the treatment by neutralization will depend on the ability to locate the contaminated region; also equipment for marking the spill and for monitoring in situ pH before and during treatment will be required in addition to the response skid equipment.

Since neutralization was considered as a general technique, some special properties of individual chemicals were not considered. For example, a dense soluble chemical might tend to sink before entering the water column and dispersing. Such factors would have to ultimately be worked into the response strategy data base available to the response team.

Neutralization of Chemicals That Sink

> The information in this section is based on *A Feasibility Study of Response Techniques for Discharges of Hazardous Chemicals that Sink*, CG-D-56-78 prepared by T.D. Hand, A.W. Ford, P.G. Malone, D.W. Thompson and R.B. Mercer of the U.S. Army Engineer Waterways Experiment Station for the U.S. Coast Guard.

A sinking hazardous chemical would mix with water to form a suspension with a density greater than 1.0. This suspension would spread laterally on the bottom due to gravitational forces. The process of eventual uniform distribution and dilution would be slow, requiring days for completion. In shallow and quiescent waters, the early part of spill amelioration could involve a containment procedure, such as hazardous chemical barriers. Use of such a barrier would allow for the effective application of a neutralizing agent followed by dredging or covering of the neutralized material. Application within the confined area would reduce the danger of damage to the environment from an excess of neutralizing agent.

The literature yields four primary candidates for neutralization of the chemicals related in Table 3.5. Neutralization of acidic materials can be achieved by calcium carbonate, sodium carbonate, and sodium bicarbonate. Sodium dihydrogen phosphate has been suggested for use on bases.

Table 3.5: Candidates for Neutralization

Acids	
Benzoic acid	
Bases	
Calcium Hydroxide	
Chemicals That React in Water to Give Acids	
Aluminum chloride	Hydrogen chloride + (Aluminum hydroxide)
Benzoyl chloride	Hydrogen chloride + Benzoic acid
Bromine	Hypobromous acid
Phthalic anhydride	Phthalic acid
Chemicals That React in Water to Give Bases	
Calcium carbide	Calcium hydroxide + (Acetylene)
Calcium oxide	Calcium hydroxide

Source: CG-D-56-78

Application can be accomplished in either solid or slurry form (powdered or fine grained) or as a solution. A number of techniques can be used for application of a solid, including hand shoveling, broadcast spreading, dumping from aircraft, or similar methods. Injection into a slurry of inert cover material could be used to form a cover that both retards leaching and neutralizes. A solution could be injected in the same manner or could be pumped directly to the bottom using an eductor with metering or direct pumping. This material could be applied either in open waters or quiescent or confined spill locations in non-navigable waters, rivers, ports, and harbors. A monitoring device (remote pH meter) could be used to locate pH imbalances and follow the neutralization reaction.

Calcium Carbonate: Calcium carbonate neutralizes acids by reaction with hydrogen ions to form carbon dioxide and water. The over-application of a solution of calcium carbonate could result in a pH of up to 9.4. This is close to the pH range (6 to 9) of natural waters.

In neutralizing hypobromous acid, formed from bromine spills, the carbonate ions act to form soluble bromide ions. Bromide is a natural constituent of seawater and has a low toxicity when compared to bromine. In neutralizing benzoic acid and phthalic acid with calcium carbonate, calcium benzoate and calcium phthalate are formed. Both compounds are slightly soluble and eventually will be taken into solution. The toxicity of the benzoate and phthalate anions may still present a problem. Neutralization should be followed by some other chemical reaction or by physical removal (i.e., dredging) of the precipitate.

Calcium carbonate neutralizer can be made from limestone or marble. The more finely ground the material, the faster the neutralization reaction. Several companies supply 325-plus mesh calcium carbonate in bulk quantities.

Sodium Carbonate: Sodium carbonate (soda ash) neutralizes acid spills in the same way as calcium carbonate. The over-addition of a sodium carbonate, however, can result in a pH of 12 or higher. Application is suggested by Drake et al (5) as a 1 M solution.

Sodium Bicarbonate: Like calcium carbonate, sodium bicarbonate neutralizes acids by forming carbon dioxide and water. Misapplication of sodium bicarbonate could result in a pH of up to 8.3. The sodium salts of benzoic acid and phthalic acid are very soluble and the use of sodium bicarbonate as a neutralizing agent for these chemicals would preclude dredging.

Unlike the preceding neutralizing agents, which are bases for the neutralization of acids, sodium bicarbonate can also neutralize bases by forming carbonate ions. In the neutralization of calcium hydroxide, calcium carbonate would be precipitated. Drake et al (5) found sodium bicarbonate unacceptable for base neutralization because of the large quantity required.

Sodium Dihydrogen Phosphate: Drake et al (5) found sodium dihydrogen phosphate to be the best choice for the neutralization of bases. Stronger acids were found to be unacceptable because excess neutralizing agent would cause extreme changes in pH. The misapplication of sodium dihydrogen phosphate could result in a pH as low as 4.5. In the neutralization of calcium hydroxide, calcium phosphate would precipitate. This compound is insoluble and would exert no toxic effect on the environment.

Results: The different chemical techniques discussed in this section may prove to be very useful in control of sinking hazardous chemical spills. Many of the techniques suffer from being chemical specific. Most of the techniques have not been tested in the field. These techniques appear to have a high potential as intermediate response techniques which would then be followed by dredging or burial. A great deal of development work is needed on most of the techniques as applied to hazardous chemicals that sink.

Neutralization of Representative CHRIS Chemicals

> The information in this section is based on *Agents, Methods and Devices for Amelioration of Discharges of Hazardous Chemicals on Water*, CG-D-38-76 prepared by W.H. Bauer, D.N Borton and J.J. Bulloff of Rensselaer Polytechnic Institute for the U.S. Coast Guard.

Neutralization was examined as a potential response to spills of the following representative CHRIS chemicals: chlorine, hydrogen fluoride, ammonia, bromine, concentrated sulfuric acid, sodium hydroxide, ethylene diamine, and sodium alkylbenzenesulfonates. In only the following four cases, however, was it determined to be a recommended procedure.

Neutralizers for Sodium Hydroxide: Carbon dioxide permits neutralization of sodium hydroxide without danger of acid overrun (6)(7) and sodium bicarbonate may be used for neutralizing sodium hydroxide solutions. Zeolite neutralization might be more advantageous for small confined spills (8).

Neutralization for Concentrated Sulfuric Acid: Ion exchange and precipitation methods can both effect neutralization. Lime, limestone, soda ash and ferrochromium slags have been discussed for amelioration. Calcined dolomite, a mixture of calcium and magnesium oxides, can remove sulfuric acid from water (9). Its use has to be restricted to spill situations where the heat of hydration or of neutralization does not build up dangerously and where overrun of alkali can be tolerated. Use of calcium hydroxide or milk of lime avoids heat of hydration in treatment of sulfuric acid solutions. The finer the hydroxide is ground, the better is its neutralization efficiency (10). If sulfate concentration exceeds 4 to 5 g/l, and the pH is below 7, gypsum forms on the neutralizer and slows neutralization (11). Ferrochromium slags are cheap substitutes for lime (12) in neutralization. Sodium carbonate has also been used (13). It is advantageous for spill situations in water bodies in which calcium or magnesium content is to be kept low (14).

Neutralizers for Chlorine: Use of cheap abundant basic chemicals is a viable approach to neutralization of chlorine water. A strong base is needed to effect neutralization. The result of neutralization is increase in hypochlorite ion concentration and in basicity and oxidation capacity. Calcium carbonate, hydroxide or oxide may be used. These are much more effective, available and economic in commercial forms than adventitious crustacean shells (15). Sodium hydroxide or carbonate may be more effective, but an overrun or a miss with them would be environmentally as disastrous as leaving the acidity alone. Since it is least likely to injure the environment, a most advantageous neutralizer is powdered high surface calcium carbonate in the form of limestone or dolomite.

Neutralizers for Hydrogen Fluoride: Except for calcium chloride, all of the precipitants discussed in the following section on precipitation are basic and can also be considered as neutralizers. Ammonia has been recommended in fluoride removal (16).

PRECIPITATION

Clarifying Agents

> The information in this section is based on *Manual for the Control of Hazardous Material Spills. Volume 1: Spill Assessment and Water Treatment Techniques*, EPA-600/2-77-227 prepared by K.R. Huibregtse, R.C. Scholz, R.E. Wullschleger, J.H. Moser, E.R. Bollinger and C.A. Hansen of Envirex for the Environmental Protection Agency.

The requirement for sedimentation as a pretreatment step is determined by the nature of the spill situation. A large amount of suspended solids in the influent, a strongly insoluble contaminant or sensitive downstream processes may lead to the use of a sedimentation system. Each situation must be evaluated on its own merits.

Several problems are inherent when using sedimentation processes. The first difficulty is that the batch nature of the process requires a bank of parallel tanks to produce a continuous flow system. The number of tanks is dependent on the fill and draw rate, the detention time of settling and the amount of desludging required. The desludging operation is tedious, time consuming, can be dangerous to personnel, and should be done as infrequently as possible. To reduce the amount of desludging, a presettler can be used which is set up for continuous desludging. The frequency of desludging is dependent on the nature of the sludge produced, the removal device and the type of tank used.

There is an experimental settling test which can be used to establish flow rates and number of tanks for sedimentation. It is mandatory that a mixed sample of the wastewater be used to evaluate the settling capabilities. The sample is placed in a cylinder and the position of the interface with time is recorded. This data is then translated into a settling rate and evaluated. If the settling rate is less than 0.031 mpm (0.1 fpm) then chemicals may be added to increase the settling rate. The chemicals are added, mixed and flocculated; then the clumped solids are allowed to settle. There are several commonly used chemicals including ferric chloride, alum, and polyelectrolytes. The following paragraphs describe these chemicals and their use.

Ferric Chloride: This compound is effective in clarifying both organic and inorganic suspensions. The final pH should be above 6 for best results so lime or caustic soda may be needed to control pH. Dilute suspensions require dosages of approximately 50 to 500 mg/l although larger dosages may be needed for concentrated or highly alkaline suspensions. If the wastewater is low in alkalinity, lime or caustic may be needed to raise the pH to 6 or higher. Excessive doses of ferric chloride will result in a brown colored effluent and should be avoided.

Alum: Aluminum sulfate is effective in clarifying both organic and inorganic suspensions. The pH should usually be controlled in the range of 6.5 to 7.5 and this control is generally crucial for good alum use. If a dilute suspension is to be treated, alum dosages of 100 to 1,000 mg/l should be effective. Huge dosages may be needed for concentrated or highly alkaline suspensions. As with ferric chloride, suspensions low in alkalinity may require an addition of lime or caustic to produce the final pH range of 6.5 to 7.5.

Organic Polyelectrolytes: Polyelectrolytes are available in anionic, cationic or nonionic form and may be effective alone when flocculating suspensions of inorganic materials (clay, soils, colloidals, metal salts, etc.). These polyelectrolytes are usually not effective alone when flocculating organic suspensions, but can be used with alum or ferric chloride for treating organic suspensions. Polyelectrolyte dosages vary with both the type of charge on the polymer and the type of suspension to be treated. Cationic polyelectrolytes are generally added in higher dosages, 1 to 10 mg/l in dilute situations (less than 100 mg/l suspended solids) and anionic or nonionic polymers are added at approximately 0.5 to 5 mg/l. When the solution is concentrated and the suspended solids concentration is greater than 1,000 mg/l, add 1 to 300 mg/l of a cationic polyelectrolyte or 1 to 100 mg/l of an anionic or nonionic compound.

These chemicals are also used in combinations and various types of mixtures should be evaluated prior to establishing the treatment mechanism. When the chemicals are added, mixed and flocculated, the resulting solution and solids should be examined for the following:

(1) A relatively clear internatant (i.e., the liquid between the particles)–a cloudy internatant may indicate the need for more chemicals.
(2) A medium to large but well-defined floc–this is a good sign of correct chemical dosages.
(3) Initial settling of the floc–the faster the floc drops out, the better.
(4) A relatively small sludge volume (3 to 5%)–even in a 100 ml graduate an approximate idea of sludge volume can be obtained. Excessive volumes of sludge from chemicals indicate future problems; a different chemical or smaller dosage may be desirable.

The following paragraphs describe the uses of the treatment chemicals to aid sedimentation. The total flocculation times are for full scale use. Laboratory times for mixing are 15 to 30 seconds and for flocculation are 30 seconds to 2 minutes.

Polyelectrolytes: Slowly add polyelectrolyte solutions to the waste while vigorously mixing the wastewater. Mix rapidly for 1 to 2 minutes to ensure dispersal. Then agitate the material at a speed just sufficient to keep the floc from settling and continue for 5 to 10 minutes. If more time is needed, increase the polymer dosage.

Alum Treatment: The order of addition of alum and then lime or NaOH may be critical. Generally, alum addition followed by lime or NaOH addition will give satisfactory results and allows simple pH control. Alum should be added, mixed 1 to 2 minutes and then the lime or caustic can be added to achieve the proper pH. The flocculation time should range from 5 to 15 minutes. If longer times are required, the alum dosage should be increased. If the floc is easily broken, a polyelectrolyte should be added to increase the strength.

Alum and Polyelectrolytes: The use of polyelectrolyte will allow a stronger floc and a faster settling rate. Add the alum and lime or caustic as described previously. Flocculate from 2 to 5 minutes to allow creation of the desired alum floc. Then add polyelectrolyte in concentrations from 1 to 10 mg/l. Increase the rate of agitation during polymer addition to prevent settling and mix about 1 minute. Flocculate for 5 to 10 minutes or increase dosage of polyelectrolyte.

Ferric Chloride: (Lime or caustic may be needed for pH control.) Add ferric chloride and then mix for 2 to 5 minutes vigorously. Then add lime or caustic to raise pH to the desired level, mix an additional 2 to 5 minutes and then reduce mixing speed to allow flocculation. Flocculate at a sufficient speed to keep floc from settling for 5 to 15 minutes. If additional time is necessary, increase dosages or add polyelectrolyte.

Ferric Chloride and Polyelectrolyte: Use the same procedure as ferric chloride addition. After the second 5 minutes mix and add the polyelectrolyte (1 to 10 mg/l). Mix at a higher rate to prevent settling and continue to mix approximately one minute. Then flocculate 5 to 10 minutes.

Once the chemicals and dosages have been determined, the system can be operated. To establish the efficiency of the unit, sampling should be done at both the influent and effluent of the batch. If the system is not operating properly, further bench tests may be needed.

Choice of a Precipitant–General Principles

> The information in this and the following section is based on *A Feasibility Study of Response Techniques for Discharges of Hazardous Chemicals that Disperse Through the Water Column*, CG-D-16-77 prepared by E. Drake, D. Shooter, W. Lyman and L. Davidson of Arthur D. Little, Inc., for the U.S. Coast Guard.

Chemical precipitation occurs when two reagents which are soluble in water react and form an insoluble compound. This compound then precipitates as a solid in the water column. The primary chemical reaction is generally followed by a secondary physical reaction known as coagulation where the solid particles form larger aggregates which makes them more easily separable from water by techniques such as sedimentation, centrifugation, filtration or flotation. The process of coagulation can be assisted by reagents which are different and separation is widely practiced in industrial and waste treatment processes.

In the case of a spill in a water body, it is envisioned that the addition of an appropriate amount of reagent will react with the contaminant, form an insoluble compound which precipitates, reduce the level of contaminant in the water to within acceptable limits, and not produce any additional harmful effects on aquatic life.

Precipitation has an inverse phenomenon, dissolution, which must also be considered. If the concentration of toxic ions in solution decreases, the solubility product is no longer satisfied, and the precipitate may redissolve to restore the equilibrium. This might occur in the case of a precipitate deposited in a moving water body. Solubility is also affected by temperature; generally an increase in temperature produces an increase in solubility, which may induce the dissolution of precipitated solids. The addition of an electrolyte with no ion common to the precipitate can increase the solubility of that precipitate (salt effect). This is a direct result of an increase in the ionic strength, which is a complex function of species, valence states and concentrations. An increase in ionic strength tends to increase the concentration of the sparingly soluble salt at saturation. This might occur, e.g., in a water body with a very high total dissolved solids (TDS) level. In this case, the amount of reagent necessary to attain a safe metal ion concentration would be greater than the simple calculations indicate.

Dissolved solids in a water body can also participate in the precipitation reaction, in two ways. First, natural ions in the water might be precipitated by the reagent, e.g., calcium can form an insoluble salt with phosphate reagents. And, second, coprecipitation might occur, in which a species precipitates although a solubility product has not been exceeded. This phenomenon is the result of either surface adsorption on a particle, or occlusion during crystal growth. These secondary effects cannot be predicted without a detailed knowledge of the water body in question.

Recommendations for Water-Soluble Chemical Spills

Table 3.6 gives the list of spill chemicals which are being considered for treatment by reactions; recommended allowable limits for each toxic species (from water quality criteria) are also included. The chemicals have been separated into four classes: (1) divalent metals forming insoluble products with a variety of anions; (2) metals with other valence states which might form insoluble compounds similar to the divalent metals; (3) oxyanions subject to oxidation-reduction reactions and subsequent precipitation; and (4) compounds which require individual treatment.

Table 3.6: Chemicals Subject to Precipitation

Chemical	Class*	Recommended Allowable Limit** (mg/l)	(molarity)
Cadmium chloride	1	0.0002	1.79×10^{-9} Cd^{+2}
Copper sulfate	1	0.01	1.57×10^{-7} Cu^{+2}
Ferrous sulfate	1	0.05	8.96×10^{-7} Fe^{+2}
Nickel sulfate	1	0.002	3.41×10^{-8} Ni^{+2}
Zinc chloride	1	0.02	3.06×10^{-7} Zn^{+2}
Antimony trifluoride	2	0.2	1.64×10^{-6} Sb^{+3}
Silver nitrate	2	0.001	9.27×10^{-9} Ag^{+}
Titanium tetrachloride	2	ND	–
Potassium dichromate	3	0.05 Cr	9.62×10^{-7} Cr
Potassium permanganate	3	0.02 Mn	3.64×10^{-7} Mn
Chromic anhydride	3, 4	0.05 Cr	9.62×10^{-7} Cr
Latex, liquid synthetic	4	ND	–
Sodium ferrocyanide	4	ND	–

Note: ND is no data available.

*Based on 1972 Water Quality Criteria (NAS); level for minimal risk of deleterious effects upon marine aquatic life (3).

**See text for explanation of codes.

Source: CG-D-16-77

All the Class 1 metals can generally be precipitated by the same anionic species, although the solubility product varies throughout the group. In an aerobic water body, it is likely that oxidation of the ferrous ion (in ferrous sulfate) to ferric ion will occur. However, since ferric hydroxide begins to precipitate at low pH levels (pH 4 to 5), it is unlikely that the higher valence species will pose a problem. Thus, in carrying out the analysis with respect to ferrous ions alone, a "worst case" situation is anticipated. Class 2 has three members: antimony trifluoride, silver nitrate and titanium tetrachloride. The latter decomposes rapidly in water to hydrogen chloride and insoluble titanium dioxide. Thus, treatment of this compound by precipitation is considered unnecessary, but neutralization of the hydrochloric acid formed will be required.

The Class 3 anions, permanganate and dichromate, contain manganese and chromium with valence states +7 and +6, respectively. Precipitation of these metals can be effected by reduction to a lower valence state. Chromic anhydride in solution is a mixture of chromic and dichromic acids (H_2CrO_4 and $H_2Cr_2O_7$, respectively). In Class 4, synthetic latex is discussed separately with the final recommendation that no action be taken in the event of a spill. Sodium ferrocyanide is not subject to the precipitation reactions of Classes 1 and 2, nor will it undergo the redox schemes of Class 3. However, the anionic species is subject to precipitation by substitution of another metal for the sodium; such treatment would be specific for this chemical.

Table 3.6 presents upper limits for metal concentrations to minimize the toxic action of these species on aquatic life. Typically the value will vary with the biological species under examination and is subject to some degree of interpretation. For most of the metals, meaningful data are available only for marine aquatic life. Data for freshwater aquatic life are generally presented as a function of LD_{50} values (concentration at which a material is lethal to 50% of a population sample); such values are difficult to interpret in terms of allowable limits for aquatic life.

The data presented for dichromate and permanganate are based on the toxicity of the metal ions, not the oxyanion. This may not be realistic but, unfortunately, the toxicity of the anions has not been documented. The ferrocyanide ion is similarly undocumented. It is quite stable and certainly less toxic than cyanide ion.

In the selection of a suitable treatment agent there are several criteria which must be met:

- The reagent must reduce the concentration of toxic species in solution to within acceptable limits.
- The reagent should be soluble in water to the extent necessary for application to a spill.
- Any agent remaining in solution after treatment must be at a nonharmful level, and should not produce adverse environmental effects, e.g., large pH changes.
- Misapplication of the treatment chemical should not cause toxic effects by violating the water quality criteria.
- It should be low cost, readily available, and easily handled.

Reagents which were considered as potential precipitants are discussed below.

Sulfide: Sulfide ions could be introduced into a water body in gaseous, solution or solid form. The gaseous form (hydrogen sulfide) is acidic and H_2S or

HS^- ions are quite toxic (3). Few solid sulfides are soluble in water; one exception is sodium sulfide, Na_2S. The pH of a solution of Na_2S is very close to that of sodium hydroxide, due to hydrolysis of the sulfide. Thus, a 0.1 N solution of sodium sulfide would have a pH of about 13, whereas a pH of 6 to 9 is the acceptable range. Misapplication of either of these reagents could produce adverse effects in the water body. Therefore, although a very effective precipitant, sulfide is not the first choice.

Hydroxide: For most of the metals, precipitation of hydroxides requires pH's above 9 to reduce metal concentrations to within acceptable limits. Also, because many freshwater bodies are neutral or acidic, not alkaline, there would be a distinct possibility of the dissolution of a hydroxide precipitate.

Carbonate: Sodium bicarbonate is applicable for the treatment of cadmium, iron and zinc but not for the other chemicals.

Oxalate: Sodium oxalate or oxalic acid are applicable reagents for the treatment of silver and possibly zinc. However, the oxalate ion is toxic and therefore misapplication of the reagent could produce adverse effects.

Phosphate: The phosphate ion by itself is not considered toxic, although it is a plant nutrient and under appropriate environments can lead to algae growth. This is only a minor disadvantage to its use in an emergency spill situation. The three forms of sodium phosphate deserve primary consideration:

- Sodium dihydrogen phosphate, NaH_2PO_4, is acidic in solution and precludes the formation of insoluble metal compounds.
- Dibasic sodium phosphate, Na_2HPO_4, is slightly alkaline with a 0.1 N solution pH of about 9.7. Small quantities of the monobasic phosphate can be added to reduce the pH further if desired; such mixed systems are usually effective buffers.
- Tribasic sodium phosphate, Na_3PO_4, is quite alkaline and might present some environmental hazard.

The dibasic sodium phosphate (also called disodium phosphate) is recommended as the most appropriate precipitating agent for the Class 1 metals.

Treating Agents for Precipitation of Other Spill Chemicals: For the remaining six spill chemicals, data on the solubility of potential reaction products were not available. A study of the pertinent literature produced several possible reagents for each chemical, but very little quantitative information. Therefore, test tube-scale laboratory experiments were performed to verify the efficacy of several proposed precipitating agents.

Antimony Trifluoride – Two common insoluble antimony salts were identified, carbonate and phosphate. In separate tests, the addition of sodium bicarbonate and dibasic sodium phosphate both produced precipitation from a solution of antimony trifluoride. The latter would be the preferred reagent because of the known insolubility of other phosphate salts, as well as the recommendation for use with Class 1 metals.

Silver Nitrate – The most common technique for precipitating silver is as silver chloride; sodium chloride is an ideal reagent for this purpose. It is nontoxic, inexpensive, and readily available. To reach the allowable limit for silver in Table 3.6, a chloride concentration of about 0.017 M is required. The concentration of chloride ion in seawater is about 0.5 M; therefore no treatment is necessary under these conditions.

Potassium Dichromate – There are two approaches to the precipitation of chromium. First, in alkaline solution the various chromium oxyanions (Cr^{+6}) are all converted to chromate (still Cr^{+6}). The addition of barium hydroxide will produce insoluble barium chromate. However, the chromium is still in a very toxic form. An alternative method for treatment of the dichromate ion requires reduction of the chromium(VI) in dichromate to chromium(III) with subsequent precipitation as insoluble chromic oxide, Cr_2O_3. The sulfite ion in the form of sodium sulfite or sodium bisulfite, is a suitable reducing agent. In the laboratory experiments it was found that the reaction occurred only in slightly acidic solutions (pH >5). (Sodium bisulfite is somewhat acidic.) Ferrous sulfate was also found to be a satisfactory precipitating agent.

Potassium Permanganate – The preferred treatment of permanganate ion is reduction with sodium sulfite to yield an insoluble manganese dioxide (MnO_2) precipitate.

Chromic Anhydride – Chromic anhydride can be reduced to precipitate insoluble chromium oxide (Cr_2O_3). The reduction was tested in the laboratory using sodium sulfite, ferrous chloride, and ferrous sulfate; only the latter compound gave a precipitate (also, the reaction was slow).

Sodium Ferrocyanide – The most common insoluble ferrocyanide salts are the ferrous and ferric ferrocyanide compounds (different forms of the Prussian Blue dyes). The former is precipitated as a colloidal material, which does not settle as rapidly as the ferric salt. Ferric chloride is the most effective reagent for precipitating the anion.

Coagulation and Precipitation of Synthetic Latex: Synthetic latex is, by definition, any emulsion in water of finely divided particles of synthetic rubber or plastic. The individual particles, which range from about 700 Å to over 3,000 Å in size, are not actually dissolved in the water. Most commercial latexes contain about 45 to 55 wt % solids, though some may reach 68 wt % (17).

The most common plastics used for synthetic latexes include various copolymers of styrene and butadiene, including terpolymers with vinyl pyridine; various copolymers of butadiene and acrylonitrile; and neoprene (polychloroprene) (18). Various formulations of these, and other polymers, are produced resulting in hundreds, if not thousands, of different formulations.

The formulations produced usually contain several additives including (1) emulsifiers (e.g., fatty acid or rosin acid soaps); (2) catalysts (e.g., potassium persulfate); (3) modifiers (e.g., dodecyl mercaptan); (4) antioxidants (e.g., phenyl-β-naphthalene); (5) pigments (e.g., titanium dioxide); (6) solvents; (7) fungicides or bactericides; and (8) antifoam agents (18)(19). There are hundreds of such additives currently in use. Trace levels of the original monomers may also be present.

No information on the human, aquatic or waterfowl toxicity of synthetic latexes has been found. It is anticipated that the toxicity of the polymeric material in latex will not be high since (1) it will not pass through cell membranes, and (2) it is relatively inert. The additives present may, in fact, be the determining factors in the chemical toxicity of latexes. Latex may adversely affect aquatic life by the plugging of gills, the coating of surfaces or some similar action.

Coagulation of latex is carried out in some latex manufacturing operations, and in some wastewater treatment operations. The most common method of coagulation is by acidification, to about pH 5 to 6, by the addition of alum (aluminum sulfate). Sulfuric acid, caustic or salts (e.g., calcium chloride) are also used.

There is no simple way to predict a priori whether a given latex formulation will or will not coagulate following chemical addition. Some latexes will coagulate spontaneously after a spill into water (especially salt water) while others will be quite resistant to coagulation even after chemical addition. The attainment of a good floc in any coagulation operation usually requires careful control over (1) the amounts of chemical added, (2) the mixing rate (usually slow), and (3) the mixing time (often 10 to 40 minutes in waste treatment operations). The resulting floc may float, remain as a dispersion, or sink, depending on the resulting specific gravity. Excessive turbulence in the water body will hinder phase separation.

A spill of synthetic latex, if untreated, may be expected to have a minor to moderate short-term impact (days) on aquatic life and a minor impact on aesthetic values (water color, surface fouling). The long-term impact (months to years) would probably be minor or negligible. The polymeric material in the latex would eventually be incorporated into the sediments (or collected along the shore lines) where it would undergo slow degradation via biochemical oxidation, chemical oxidation, and/or photo-oxidation.

Any attempt to treat the spill, e.g., by coagulation with alum, may create significant adverse effects. A misapplication or excess reagent would result in a local acid pH. The precipitate created could blanket the bottom of the water body (or the shore line) and have adverse impacts on benthic organisms.

Methods of Application: Two approaches for application of precipitating agents can be considered. First, the agent can be applied in situ. In this case, depending on the characteristics of the precipitant, the solids may settle to the bottom or else remain suspended. If the spilled chemical, after precipitation, is in an environmentally inert form, then in situ treatment may be feasible. A precipitate that settles out has potential for eventual removal from the bottom by dredging. However, not enough information is yet available on the character of in situ precipitates in the water column. Some experimental work will be needed before the feasibility of in situ precipitation can be evaluated.

A second possibility is to treat the contaminated water by flow through a treatment apparatus. The basic unit might include an inlet pump, a mixing tank, a separator or filter unit and a storage bin for precipitate. The weight of such a skid would be around 1 to 2 tons, and the cost should be reasonable, but weights of treating agent and precipitate will tend to be large if spills of several thousand gallons are treated.

Fate and Consequences of a Treated Spill: Optimum treatment, i.e., the addition of the proper amount of reagent in the proper location, will essentially eliminate the worst aspect of the initial spill, the excessively high concentration of toxic heavy metals (or toxic anions). In addition, some readjustment (increase) of pH may result, offsetting the initial low pH created by some spills. These beneficial effects will be slightly offset by:

- A further increase in the ionic strength of the water, to which aquatic organisms may be sensitive;
- An increase in suspended solids concentration (and turbidity); if the precipitate formed settles slowly it may have some minor adverse effect on aquatic life due to direct toxic action, fouling of gills and membranes, or decreased light penetration;
- Abnormally high concentrations of the metal in the sediments where-ever the precipitate settles (in slow moving or still waters) which may have a local effect on benthic organisms; and
- An increase in nutrient concentration (when Na_2HPO_4 is used) which might trigger increased algae growth (and, thus, an increase in the oxygen demand), and a temporary increase in the rate of eutrophication in confined waters.

It should be clear that precipitation in situ does not remove the hazardous material (e.g., heavy metal) from the water body, it only reduces the concentration of the dissolved fraction to low levels. Further, some of the material that initially precipitates may later redissolve.

The adverse effects of adding the recommended reagents in the wrong location (or in excess) should not be serious and will primarily be associated with (1) pH changes, (2) increased ionic strength, and (3) increased nutrient concentrations (with Na_2HPO_4).

Treatment of Chemicals That Sink

> The information in this section is based on *A Feasibility Study of Response Techniques for Discharges of Hazardous Chemicals that Sink*, CG-D-56-78 prepared by T.D. Hand, A.W. Ford, P.G. Malone, D.W. Thompson and R.B. Mercer of the U.S. Army Engineer Waterways Experiment Station for the U.S. Coast Guard.

An examination of the list of hazardous chemicals that sink reveals that only aluminum fluoride and barium carbonate spills are likely to be successfully treated by precipitation techniques. For the former, fluoride ions are toxic and for the latter barium ions are toxic and should be removed from solution.

Aluminum fluoride is somewhat soluble (0.559 g dissolve in 100 cc of cold water) and could be treated or covered with calcium carbonate, forming aluminum carbonate and calcium fluoride. Calcium fluoride is less soluble (0.002 g dissolve in 100 cc of cold water) and would precipitate from solution. The misapplication or overapplication of calcium carbonate could result in a pozzolanic reaction, forming a crust on the bottom. This could exert a toxic effect on pH-sensitive benthos.

The use of an active covering material that contains sulfate (e.g., calcium sulfate and ferric sulfate) is proposed for barium carbonate spills. Seawater itself may contain enough sulfate to precipitate barium ions. Barium ions could be precipitated more rapidly in a contained spill situation by the addition of a sulfate salt (e.g., sodium sulfate) that is more soluble than calcium sulfate or ferric sulfate. Misapplication or overapplication of calcium sulfate could result in a pozzolanic-type reaction on the bottom, creating a crust.

It should not cause any toxicity problems under aerobic conditions. Ferric sulfate is weakly acidic and could exert a toxic effect on pH-sensitive benthos. Under aerobic conditions, hydrated iron oxides may form and with their adsorptive properties may scavenge heavy metals from the water. They also may coat the gills of bottom-feeding fish. Sodium sulfate is soluble, but will not adversely affect the pH. It is relatively nontoxic, especially when used in a marine environment.

Technology for the delivery of precipitating agents is not well developed in a systems sense, though individual hardware components should be available. It seems likely that precipitating agents would be best applied in close proximity to the bottom, where they would intercept the toxic barium or fluoride ions prior to their entering the water column. Such an application could be achieved through the use of an active covering containing enough precipitating agents to react with the spilled chemical. This could be accomplished by pressure injection of a solution or slurry of the selected precipitating agent into the discharge pipe of a dredge being used to place covering material.

This would apply to medium or large size spills, primarily. In the event of small spills, and when containment has been achieved through the use of a hazardous material barrier or submerged dike, direct application of the slurry or solution could be accomplished with diminished chances for overapplication or misapplication of the covering material. Off-the-shelf hardware (pumps, hoses, and flow-regulating devices) is expected to be available for this application.

Treatment of Representative CHRIS Chemicals

> The information in this section is based on *Agents, Methods and Devices for Amelioration of Discharges of Hazardous Chemicals on Water*, CG-D-38-76 prepared by W.H. Bauer, D.N. Borton and J.J. Bulloff of Rensselaer Polytechnic Institute for the U.S. Coast Guard.

The following representative chemicals were considered as candidates for cleanup by precipitation methods: concentrated sulfuric acid, sodium alkylbenzenesulfonates, potassium cyanide, cadmium chloride, hydrogen fluoride, and oxalic acid. Precipitation was judged to be one of the preferred methods in only the latter two cases.

Precipitants for Hydrogen Fluoride: Many inorganic fluorides are slightly soluble or insoluble. Their precipitation is possible as long as fluoride ion concentration is high enough. If the precipitating cations are compatible with the biotic environment, the precipitate does not have to be collected by dredging and excess of precipitant is of less vital concern. If the cations are provided as bases rather than as salts, the dangers of excess increase but the undesirable acidity of the hydrofluoric acid is removed. Calcium carbonate is a cheap, effective precipitant when finely divided (20). However, when the pH has risen to 3.0 to 3.6, the precipitation rate is negligible. A second state precipitation with powdered calcium oxide is advocated which in an enclosed water body reduces fluoride concentration to 3.3 ppm (21). Calcium hydroxide and oxide have been used for the entire precipitation (16). In all cases the calcium fluoride that does not redissolve precipitates and crystallizes as the mineral fluorite. Sulfuric or other acid can be used where removal of fluoride and of excess calcium is more important than removal of acidity (22). Precipitation of magnesium fluoride by magnesium oxide (23) and of aluminum fluoride by aluminum basic chloride (24), aluminum hydroxide or sodium aluminate (25) has been advocated. No report of spill amelioration by these two precipitants has been reported.

Precipitants for Oxalic Acid: Calcium oxalate is precipitated by addition of calcium salts, or the hydroxide or oxide, to aqueous oxalic acid solutions (26). Calcium carbonate reacts slowly and is easily blocked by oxalate encrustation.

Precipitation of Heavy Metals with Sodium Sulfide

> The information in this section is based on *Methods to Treat, Control, and Monitor Spilled Hazardous Materials*, EPA-670/2-75-042 prepared by R.J. Pilie, R.E. Baier, R.C. Ziegler, R.P. Leonard, J.G. Michalovic, S.L. Peck and D.H. Bock of Calspan Corporation for the Environmental Protection Agency.

The treatment of spills of soluble heavy metal compounds, by precipitation of the generally insoluble metal sulfides, was evaluated as part of this project. The technique is potentially useful for many materials and was found to be effective in treating solutions of most of the heavy metal compounds which present a spill hazard. It was shown to be simple to apply and capable of reducing the toxic hazard presented by heavy metal ions in solution.

The treatment is applied by introducing sulfide ions, derived from a solution of sodium sulfide, into the spill. At concentrations of the heavy metal ion that present a toxic hazard, formation of the sulfide precipitate proceeds rapidly, and toxicity is reduced within seconds. The treatment chemical is stabilized by the addition of some sodium hydroxide, in order to prevent evolution of toxic hydrogen sulfide fumes.

Some method of estimating the quantity of treatment to be applied to a spill is needed when the amount or kind of material spilled is not known. The use of specific ion electrodes for the metals or for the sulfide ion itself was investigated in a spill simulation facility, using iron as a heavy metal. The flow conditions dictated a frequency of sulfide additions of at least one every ten seconds. The required operations (reading the specific ion meter, determining sulfide addition, and making the addition) proved too complex for all manual operation on this time scale, and the treatment achieved was far from ideal.

The iron concentration was nonetheless substantially reduced. Residual iron in this case was estimated to be less than 16 mg out of 520 mg added in the original spill. This is to be compared to a residue of 7 mg out of 520 mg when the exactly stoichiometric treatment was applied. The use of specific ion electrodes for the metals, the cyclic colorimeter, or of a titration and a sulfide electrode, to control localized sulfide additions should be studied further. Problems to be solved in achieving effective treatment center primarily on response rate limitations of the detector-human-treatment system and may best be alleviated by the development of a more highly automated approach.

Sulfide precipitation has been shown to be effective in significantly reducing the levels of heavy metal ions in solution and thus reducing the toxic hazard of a heavy metal spill. In the process, however, sodium sulfide, a material which is itself toxic, is introduced into the environment. It is important, therefore, to assess the hazards posed by this material and to compare them to the benefits to be derived from its use. The preparation of a completely satisfactory comparison between the hazards and benefits of the sulfide precipitation treatment of heavy metal spills would require a significantly greater body of data than that available at this time. Ideally the spreading of plumes of heavy metals and sodium sulfide over a range of stream conditions should be correlated with complete toxicity-concentration-exposure time data for these materials to compare the environmental damage caused by untreated spills, treated spills, and erroneous treatment where there was no spill.

In this study a much less rigorous approach has, by necessity, been taken. One can make a rudimentary comparison between the consequences of not treating a heavy metal spill and the consequences of applying sodium sulfide treatment where no spill had occurred. The comparison is made in terms of the volume of water contaminated at a level lethal to fish. Despite the crude nature of these comparisons, it is felt that the differences between the metal spill results and the sulfide "spill" results are sufficiently great to support the conclusion that sulfide treatment should be recommended for heavy metal spills, with some reservations, as outlined below.

The hazard involved in treating a nonexistent spill or of mistreating a spill clearly depends on the amount of material released to the environment. Experience has shown that heavy metal spills typically involve small quantities, of the order of a few kilograms. For comparison of effects, a spill of 4 kg of each heavy metal compound for which toxicity data corresponds to data available for the sulfide is taken as an example. The volume of water which could be contaminated to the reported lethal level by such a spill was calculated.

The ratios of the affected volumes vary widely, even with different data for the same heavy metal, but in no case is the sodium sulfide toxic in more than 5% of the volume calculated as lethal for the heavy metal. Further, it may be seen that the affected volume is quite small for all the sodium sulfide additions, even in the case of $ZnSO_4$ treatment where the greatest amount is added, and the sulfide toxicity corresponds only to 10% fish mortality in seven days. The hazard occasioned by sodium sulfide is further alleviated by the instability of the sulfide ions in the environment which was not accurately reproduced in the laboratory toxicity tests where contact with water of lower pH than that of the Na_2S solution did not occur.

The acute toxicity of sodium sulfide solutions in various concentrations was measured and compared with the toxicity of sodium hydroxide. Both sodium sulfide and hydroxide raise the pH of their aqueous solutions and this alone may occasion some toxicity. There is relatively little difference between the molar toxicity of sodium sulfide and hydroxide.

Many pollutants ultimately leave the environment by conversion to useful or at least acceptable chemicals. For example, both chemical and biological processes tend to oxidize most organics to water and carbon dioxide. Metals differ in this respect; those metals whose ions are toxic will continue to pose a threat indefinitely if they are not rendered insoluble and stored in an environment which inhibits redissolution. The sulfide precipitation treatment was shown to be effective in rendering many heavy metal ions insoluble. Collection of the precipitate and return of the heavy metal content to commercial channels proved to be impractical in some cases. When the sulfide precipitate is released into the environment, it may undergo slow redissolution because its solubility product is finite. Ultimately the heavy metal is converted to the least soluble species for which anions are available.

In well-aerated surface water the hydroxide ion is most available and hydroxides are commonly formed from available metal ions. These precipitate to greater depth in the oceans, eventually encountering a reducing environment where excess sulfide ion exists. In this environment, stable deposits of heavy metal sulfides are formed in sediments (27). It is believed that the processes at the interface between the anaerobic interior of sediments and the aerobic water in shallow, well-agitated waterways are similar. One may, therefore, assume that heavy metal sulfides are immobilized once they are included in a stable sediment. The organic nature of bacteria may lead to feedback, however, and this feature was investigated separately.

Several experiments were conducted to investigate the environmental effects of sulfide precipitates remaining in the environment. A series of experiments were run in which minnows were exposed to the metal sulfides individually for periods of 60 days. No significant evidence of toxicity of the sulfides of zinc, copper, lead, cadmium, or manganese was found in these tests. In the tanks containing CoS and NiS, 20% and nearly 60%, respectively, of the fish died within 60 days as compared with 5% mortality in the control tank. Maximum concentrations of dissolved ion in these tanks were 7.0 ppm Co and 8.0 ppm Ni. Significantly more fish died in the tank containing silver sulfide than in the control also, but no dissolved silver was ever detected so the reason for this result is not clear.

One further experiment was conducted as a preliminary investigation of the reentry of heavy metals into the biosphere following sulfide precipitation. A simple ecosystem was established in a tank carpeted with a sediment made up of kaolin, and two grams each of the sulfides of silver, cadmium, cobalt, copper, manganese, nickel, lead and zinc, and 5.3 grams of barium chromate. The ecosystem contained bacteria, Euglena, approximately 300 Tubifex (annelid worms), a few Hornwart plants, and 100 fathead minnows that were added after one month was allowed for the establishment of the rest of the system. The fish were observed to be quite healthy for a period of eight weeks. Samples of fish were removed from the aquarium at four weeks after their addition and at eight weeks when the experiment was concluded. Results of heavy metal analyses on these fish are shown in Table 3.7.

The analytical results strongly suggest a continuing concentration of the heavy metals in the tissue of the minnows, particularly in the case of manganese and lead. In the light of the absence of adequate controls, such as analyses of minnows aging in an environment not similarly enriched, the results of this preliminary experiment cannot be considered definitive, but further study in this area is clearly desirable.

Table 3.7: Heavy Metal Analyses of Fish from Heavy Metal Sulfide Exposure Tests (ppm of wet weight)

Element	4 Weeks' Exposure	8 Weeks' Exposure
Co	1.5	1.9
Cu	3.5	4.3
Mn	14.0	21.0
Ni	3.8	3.8
Pb	17.0	21.0
Zn	47.0	51.0

Source: EPA-670/2-75-042

A series of experiments conducted in a stream simulation test facility developed for this program provided information on the transport of spills and treatment under various flow conditions, and allowed evaluation of treatment methods. The effectiveness of sulfide precipitation with even the very simplest techniques for locating a spill and delivering the treatment was demonstrated.

Results of these channel tests indicated that under conditions typical of most natural streams, addition of the appropriate amount of sodium sulfide solution close to the plume of spilled material will still result in the removal of the metal from solution. Mixing in these highly turbulent flows rapidly eliminates any errors in the distribution of the treatment. Under these conditions, spill location using small quantities of sodium sulfide as a tracer is highly effective.

Under conditions of little or no flow encountered in ponds and lakes, sodium sulfide delivered off target will not mix with the metal for a long period of time. Accurate location of the spill, and proper distribution of the treatment will then be important. Introduction of some artificial turbulences (by outboard motor boats, e.g.) should enhance treatment effectiveness. Experiments showed that under these quiescent conditions, concentrated heavy metal solutions collected at the bottom of the water body. Measurements of ion concentration must then be made on the bottom waters of the lake or pond to be treated in order to allow proper sodium sulfide additions to be determined. Automatic spill treatment procedures may be useful in treating spills under these conditions. Some bodies of water such as canals and some rivers will have conditions between the extremes discussed above, and may or may not require precise treatment delivery and auxiliary mixing.

Field tests were conducted in a mountain brook using ferrous sulfate as a harmless simulant of hazardous heavy metal spills. Sodium sulfide was found to be highly effective for spill location and marking under these conditions. Rapid and thorough spill treatment was achieved by dumping sodium sulfide solution out of plastic bottles into the region which the sodium sulfide indicator solution showed to contain heavy metal ions.

Sulfide precipitation of spills of heavy metal compounds appears to be quite effective and ready for application in the field. Spills of heavy metal compounds are often small as shown in the spill statistics of Table 3.8 covering 1971 and early 1972 (28). They should be treated quickly, before the spill is diluted to a concentration less than the solubility of the metal sulfide. Thus, it is important that treatment material and personnel trained in its uses are available near the spill.

Table 3.8: Heavy Metal Spills

Material	Container	Cause	Quantity Spilled (lb)	Spill Site
Chromic acid	Storage tank	Rupture	140	Creek, Tennessee River
Chromium in acid	Storage tank	Overflow	*	Mill Creek
Mercury	Treatment plant	Leak	*	Detroit River
Nickel solution and ground limestone	Pile	Avalanche	*	Illinois River
Potassium permanganate	Plant	Equipment failure	11	Ohio River
Tetraethyllead	RR tank car	Derailment	*	Land
Zinc sulfate	Plant	Equipment failure	7½	*

*Not known.

Source: EPA-670/2-75-042

The recommended treatment material is a solution of sodium sulfide in water. The solution should be fairly concentrated to conserve space and reduce weight. On the other hand, crystallization of sodium sulfide at low temperature is to be avoided. A solution formulated at the freezing point, containing 0.85 lb of sodium sulfide per gallon of water, and 0.04 lb of sodium hydroxide, meets these requirements. It can be stored indefinitely at normal temperatures in plastic containers. For manual application a two-gallon bottle was found to be comfortable to carry and pour from. Two gallons of treatment are sufficient for spills of 2 to 10 lb of common heavy metal compounds. Five-gallon bottles which could treat larger spills were found to be difficult to manage in a small boat typical to a spill response situation.

Since a quick response to the spill is important, treatment should be applied by personnel near the scene. In the course of this study it was determined that firemen were interested in the spill problem and motivated to learn spill detection and treatment techniques. A police agency was interviewed and indicated that their first reaction to a hazardous spill would be to protect the public from the hazard and to ask a fire department to assist with its disposal. The agency felt that its personnel were rotated too often to benefit from spill training. These observations support the view that firemen can be trained in treating small spills.

Many volunteer and paid fire companies participate in weekly training sessions that include training films on new or difficult techniques. A training film could be prepared to illustrate detection of heavy metal spills and their treatment with sodium sulfide. There are tens of thousands of fire departments in the U.S. It is estimated that about ten thousand of these in conjunction with the U.S. Coast Guard and environmental agencies of the separate states would be sufficient to cover the waterways in which a small heavy metal spill may occur.

In the immediate future it would be desirable to conduct a pilot program involving the fire department(s) of an industrial city responding to real spills that might occur, and a few simulated (iron) spills in order to perfect training techniques and evaluate practical hazards of sulfide precipitation.

CHELATION

The information in this section is based on *A Feasibility Study of Response Techniques for Discharges of Hazardous Chemicals that Disperse Through the Water Column*, CG-D-16-77 prepared by E. Drake, D. Shooter, W. Lyman and L. Davidson of Arthur D. Little, Inc., for the U.S. Coast Guard.

Principles of Chelation

Chelating agents are compounds or ligands (generally organic) that coordinate, or bind, a metal ion in more than one position. The metal becomes the central ion in a heterocyclic ring, after reaction with two or more functional groups of the same ligand (ion or molecule of the chelating agent). This binding of the metal ion, in most cases, results in its deactivation. The metal is no longer able to react chemically, and is therefore made less toxic.

The classification of ligands is usually with respect to the number of coordinating groups–bidentate (two), tridentate (three), quadridentate (four), etc. A unidentate ligand, which coordinates with metals in only one position, forms a chelate often called a "complex" and is frequently an inorganic compound, e.g., NH_3. Chelating agents can be synthetic or natural in origin, and may react either selectively with only one or two metals or with a number of different metals.

No chelating agent is suitable for treatment of all spill chemicals listed. For the purpose of this study, reagents capable of treating at least several metals on the spill chemical list are most useful. The most important chelating agents for these metals involve coordination through either a nitrogen or oxygen atom, or both. Since several of the metals considered for chelation are transition metals, the relative order of affinities is presented here (29):

$$-NH_2 > -N= > N(\text{aromatic}) > -COO^- > -O- \gg C{=}O$$

Thus, the chelates between a metal and a compound with an amine group, NH_2, should be quite stable. The relative stabilities of metal complexes have been determined for elements in the first transition series; e.g.,

$$Mn < Fe < Co < Ni < Cu > Zn$$

This increase in stability with increasing atomic number shows a maximum at copper. The stability of zinc is usually below that of nickel. Although the interaction between a metal and a chelating agent is complex, the relationships described above do permit qualitative predictions of behavior.

Stability is affected by the metal ions in solution, the particular coordinating groups of the ligand, and by the ring size of the ligand, the number of rings, and the basic strength of the chelating agent. The latter effect is important since, in a solution with the acid form of the ligand, there is competition between hydrogen and metal ions. A stepwise equilibrium between hydrogen ions and ligands may be found to produce changes in the metal-ligand equilibria. Competition from hydrogen usually occurs at low pH levels and a decrease in pH always produces a shift towards dissociation of the complex (an increase in free metal concentration). The converse is also true; the addition of a chelating agent to a solution containing metal ions results in the displacement of hydrogen ions from the ligands and a corresponding decrease in pH. Certain chelating agents may act as buffers over a specific pH range, and thus restrict changes in pH accompanying the application of the agent.

Organic chelating agents may be divided into two classes, sequestrants and precipitants. In both cases the metal ion is firmly bound and not able to participate in the normal physical or chemical processes. Sequestrants form chelate complexes which are soluble in water; therefore, the compound still remains distributed throughout the water body although in a less toxic form.

Spills Treatable by Chelation

The chemicals containing metals subject to chelation are listed in Table 3.9, along with the recommended allowable limits for the respective metals defined by the *Water Quality Criteria* (3).

Table 3.9: Chemicals Subject to Chelation

Chemical	Recommended Allowable Limit for Metal Ion* (mg/l)	(M)
Antimony trifluoride	0.2	1.64×10^{-6}
Cadmium chloride	0.0002 (0.0004)**	1.79×10^{-9}
Copper sulfate	0.01	1.57×10^{-7}
Ferrous sulfate	0.05 (1.0)***	8.96×10^{-7}
Nickel sulfate	0.002	3.41×10^{-8}
Silver nitrate	0.001	9.27×10^{-9}
Titanium tetrachloride	no data	–
Zinc chloride	0.02 (0.07)***	3.06×10^{-7}

*Based on 1972 *Water Quality Criteria* (NAS); level for minimal risk of deleterious effects upon marine aquatic life, unless otherwise noted (3).
**Limit for freshwater aquatic life.
***EPA proposed criteria (1973): aquatic life support (4).

Source: CG-D-16-77

Selection of Sequestrants

The criteria used to screen prospective sequestrants were as follows:

(1) Any agent or complex remaining in solution after treatment must be at a nonharmful level and should not produce adverse environmental effects, e.g., large pH changes.
(2) The removal of essential trace elements, such as calcium and magnesium, from the aquatic environment must be minimized.

(3) Misapplication of the treatment chemical should not cause a violation of water quality criteria.
(4) The reagent should be soluble in water to the extent necessary for application to a spill.
(5) It should be low cost and readily available.
(6) The metal chelate should be stable in solution against degradation for a reasonable time.

Synthetic aminocarboxylic acids can form exceedingly stable compounds with a variety of metals; evaluation of various other classes of sequestering agents was carried out, but the final selection was limited to a sequestrant from this group of compounds: EDTA is an obvious candidate. Its proven applications include:

- metal cleaning–EDTA is used to complex iron, copper and water-hardness metals in removing scale;
- fruits and vegetables–EDTA is used to complex trace metals and prevent discoloration and improve flavor;
- leather tanning–EDTA complexes calcium, chrome and iron in cleaning, stripping, and stain removal; and
- medicinal uses–EDTA complexes zinc, iron, nickel, lead, and mercury, used in cases of metal poisoning.

Because of the hydrophilic anionic groups, the EDTA molecule penetrates biological cell walls poorly and is, thus, of limited toxicity. Other aminopolycarboxylic acids similar to EDTA, but not as widely used, were chosen for further investigation: nitrilotriacetic acid (NTA), 1,2-diaminocyclohexanetetraacetic acid (CDTA), and hydroxyethylethylenediaminetriacetic acid (HEDTA).

Calculations show that NTA can reduce free metal concentrations to within allowable limits at a pH of 8 or above. EDTA can reach safe metal levels at a pH of 6 or above. The efficiency of chelation increases for the metals in the following order: $Mg < Ca < Fe < Cd < Zn < Ni < Cu$.

Calculations also indicate that HEDTA is only slightly less effective than EDTA, and CDTA is the most effective. EDTA appears to be the most promising agent for the following reasons:

(1) The differences in stability from CDTA are not sufficiently great to be significant.
(2) EDTA chelates effectively in a pH range within which most natural waters fall.
(3) EDTA is a well-known, commonly used material, which has a demonstrated lack of toxic properties.
(4) EDTA functions as an effective hydrogen ion buffer, preventing undesirable pH swings.

EDTA does form complexes with silver, antimony, and bismuth (29), but there are serious limitations which must be recognized in these cases:

- Silver has a first stability constant which is less than those for calcium and magnesium. Therefore, EDTA will chelate silver ions after the more stable calcium and magnesium chelates have been formed. From a water quality standpoint, such an occurrence could be harmful to aquatic life.

- Antimony does not chelate directly with EDTA, but only through the hydrolyzed species, SbO^+. Such chelates are quite stable (K_1 = 24.8) but antimony trifluoride only hydrolyzes to a small extent. Thus, chelation would be slow under spill conditions.
- Titanium, similarly, forms stable chelates only through TiO^{+2} (K_1 = 17.5). Chelation may occur upon hydrolysis of titanium tetrachloride; however, specific data are not available.

EDTA is generally applied as a soluble sodium salt of the acid in one of the following forms:

- EDTA disodium salt–the pH of a 0.1 N solution is 5.
- EDTA tetrasodium salt–the pH of a 0.15 N solution is about 11.8.
- EDTA calcium disodium salt–the pH of a 0.1 M solution is about 7.

The order of increasing cost per unit of reagent is tetrasodium, disodium, then calcium disodium. The tetrasodium salt, while the least expensive, would produce an undesirable increase in pH in the case of misapplication to a spill. The disodium salt has a desirable solution pH and is somewhat more effective as a sequestrant than the calcium disodium salt.

The EDTA disodium salt is sold as a powder or crystal with several generic names, such as Edathamil Disodium and Disodium Edetate. Major producers in the U.S. include Dow Chemical, Eastman Kodak, and Geigy Chemical. Some of the more common trade names for the material are Questex, Sequestrene, and Versene.

The reagent can be purchased in small or large quantities and is readily available. The quantities of EDTA disodium salt required to form 1:1 ligand to metal complexes with the treatable metals of Table 3.9 are shown in Table 3.10.

Table 3.10: Quantities of Chelating Agent

Species	Na_2EDTA (kg/kg spill chemical)
$CdCl_2$	1.8
$CuSO_4 \cdot 5H_2O$	1.3
$FeSO_4 \cdot 7H_2O$	1.2
$NiSO_4$	2.2
$ZnCl_2$	2.5

Source: CG-D-16-77

Selection of Precipitants

Reagents which are potential chelating precipitants find their primary use in analytical chemistry for the separation and determination of metals. Most of the reagents described in the literature are only slightly soluble in water and precipitate by themselves if added in excess. Thus, they are frequently prepared in organic solvents or under acidic or basic conditions. The criteria for selection of a precipitating reagent are similar to those for a sequestrant, but the requirement that the agent be soluble in water may not be attainable.

The precipitants (30) selected for initial study were:

- Dimethylglyoxime–containing two oxime coordinating groups, =NOH;
- Cupferron–the ammonium salt of phenylnitrosohydroxylamine; and
- 8-Hydroxyquinoline–the phenolic form of quinoline (benzopyridine).

In addition, the recent use of natural polyelectrolytes as complexing agents has been reviewed.

Dimethylglyoxime is one of the most widely used organic precipitants. It is only slightly soluble in water, and is commonly used in a solution of alcohol or acetone. It is very effective in precipitating nickel, but not effective in complexing other metals on the chelation list. Cupferron is an organic precipitant that is freely soluble in water. Copper and titanium can be chelated by this reagent (usually from acidic solutions).

8-Hydroxyquinoline, commonly called oxine, can precipitate a wide variety of metals. It is not very soluble in water, thus solutions are usually prepared in alcohol (methyl or ethyl) or acetic acid. The precipitation itself is usually carried out in an environment well buffered with acetate ions.

Calculations indicate that oxine is capable of reducing free metal concentrations to within accepted limits, particularly if 2:1 complexes are formed. Although stability data are not available, oxine may also be capable of chelating antimony and titanium. Data on the acceptable pH range for precipitation of the metals is presented in Table 3.11. (Solutions were buffered with ammonium acetate-acetic acid.) Thus all the chelate complexes are stable within the pH range 6 to 9.

Table 3.11: Conditions for Precipitation by Oxine (31)

Metal	pH Range
Nickel	4.3-14.0
Zinc	6.0-13.4
Titanium	4.8-8.6
Copper	5.4-14.0
Cadmium	5.6-14.0

Source: CG-D-16-77

Three natural polyelectrolytes also appear promising as metal complexing agents (32)(33). They are: alginic acid–a polysaccharide found abundantly in marine algae; polygalacturonic acid–a polysaccharide similar to alginic acid, a major component of plant tissue; and starch xanthates–salt or ester of a thio acid, made from corn or wheat starch.

These materials have the advantage of being nontoxic natural substances and potentially capable of regeneration and reuse. The reagents are water-insoluble and are applied as solids, in a manner similar to ion exchange resins. Laboratory tests (32)(33) have shown that these compounds are capable of effectively precipitating the divalent heavy metals: copper, cadmium, zinc and nickel. The two acids do produce decreases in pH (unspecified) while the xanthates produce increases in pH.

Two potential treatment agents for anionic species were also identified:

(1) α-Benzoinoxime (Cupron) is a powder, slightly soluble in water but fairly soluble in alcohol. It can partially precipitate chromate.
(2) Nitron (triazonium compound) is insoluble in water, but soluble in alcohol and acetic acid. It forms insoluble compounds with chromate and ferrocyanide.

These compounds may also form complexes with dichromate and permanganate, but further data are lacking.

Based on the discussion presented in this section, the most promising chelating agent for precipitation is oxine. It is of proven value in precipitating a wide range of metals to very low levels of concentration. The toxicity of the chelating agent by itself is not great, although some questions remain concerning the environmental impact of the required buffer and of the final metal chelates.

U.S. producers of oxine (sold as a solid) include American Hoechst Corp., Ashland Chemical Co., and Merck & Co., Inc. It is sold under a variety of generic and trade names, such as: 8-Quinolinol, Oxin, Oxychinolin, Quinophenol, Tumex, 8-OQ, and Bioquin. Oxine may form chelates containing more than one ligand. However, for comparison with EDTA, 1:1 ligand to metal complexes are assumed. Table 3.12 shows the quantities of oxine required to treat the five spill chemicals.

Table 3.12: Quantities of Chelating Agent

Species	Oxine (kg/kg spill chemical)
$CdCl_2$	0.79
$CuSO_4 \cdot 5H_2O$	0.58
$FeSO_4 \cdot 7H_2O$	0.52
$NiSO_4$	0.93
$ZnCl_2$	1.07

Source: CG-D-16-77

Oxine is more economical than EDTA on a weight basis (due to its lower molecular weight), but the cost of treatment is comparable.

Solvent Extraction of Chelate Compounds: Metal chelate complexes can be extracted from an aqueous phase into an organic solvent under suitable conditions and this process is used commercially. The organic phase is frequently a common solvent, such as kerosene, while the chelating agent may be a proprietary formulation. Some examples of chelating agents are: LIX 64N–hydroxyoxime, a General Mills reagent to extract copper; Kelex 100–alkylated 8-hydroxyquinoline, Ashland Chemical Co. reagent specific for copper; and DEHPA calcium salt–di-2-ethylhexyl phosphoric acid, preferential extraction of zinc.

For the amelioration of chemical spills a nonspecific chelating agent is desirable, and for successful solvent extraction it must be preferentially soluble in the organic solvent. Data are available on the extraction of metals from aqueous solution into chloroform using oxine (34). Effective extractions of silver, titanium, nickel, copper, zinc and cadmium were obtained. The oxine concentration in the chloroform was insensitive to pH. Zinc was extracted and formed an insoluble complex in the extracting solvent. Clearly solvent extraction of metal chelate compounds shows promise but insufficient data are available to make positive recommendations.

Methods of Application: Since there are a number of specific options involved in the use of chelation as a spill response method, the application methods are varied. In a few cases, in situ chelation may be acceptable since the chelation compound may be inert enough to be acceptable or at least preferable to the untreated species.

Since most of the chelating agents that form precipitates are insoluble in water, but soluble in solvents, the solvent extraction apparatus with its initial stage for precipitate removal seems to be adaptable to use with these agents. Even with the sequestering agents which are water soluble, an extraction scheme may be desirable for recovery of the chelating agent.

Results of Treatment with a Sequestrant (EDTA)

Optimum treatment with a sequestrant should reduce the concentration of free metal cations to very low levels. All available metal cations will be chelated to some degree (i.e., those from the spill and those present naturally), though the most stable complexes will be favored. The resulting chelation complexes will tend to remain in solution since EDTA is fairly hydrophilic.

The beneficial effect of a reduction in the free toxic metal concentration will be offset by any toxic effect of the chelation complex. EDTA by itself has a NAS aquatic toxicity rating of 2 (35), indicating the toxic threshold limit is in the range of 100 to 1,000 ppm. The threshold limits for various metal ion complexes with EDTA are unknown but they are probably similar to EDTA. Thus, the treatment of spills requiring the addition of EDTA in concentrations above about 1,000 ppm should probably only be considered when human health considerations are predominant.

The extent to which EDTA will combine with heavy (toxic) metals from sediments or suspended particles is uncertain. This would result in a solubilization of additional toxic material. Transfer of metals from sediments to solution in this manner has been shown to correlate with the concentration of the chelating NTA agent present.

EDTA, and presumably its complexes with metals, have a finite lifetime in the aquatic environment. Both chemical (photo) oxidation and biochemical oxidation (a 5-day BOD of 1% has been reported) (35) will eventually break down the organic molecule and could result in the release of metal ion. The rate of release would be expected to be slow.

Chelation of metals will also tend to increase their residence time in solution by impeding the natural removal processes such as precipitation and/or cation exchange. A misapplication of EDTA (in the wrong location or in excess) will not present any additional environmental effects.

Results of Treatment with a Precipitant (Oxine)

Optimum treatment with oxine will significantly reduce the concentration of free heavy metal(s); in addition, the oxine-metal complex will precipitate. The settling characteristics of such complexes are not known; (quinoline itself is slightly denser than pure water; the specific gravity is 1.095) but it is assumed they would be poor. In that event, a colloidal suspension of the oxine-metal complex might be transported with the main flow of water in the affected water body.

The beneficial effect of a drastic lowering of the free metal cation concentration will be offset by any toxic effects of the chelation complex. No information on the aquatic toxicity of oxine or its complexes is available. Oxine, in the presence of Cu^{2+}, Fe^{2+}, or Fe^{3+}, is a powerful bactericide and fungicide–and is used as such in pharmaceutical preparations and ointments for use on human skin surfaces–which suggest that the level of toxicity to aquatic microorganisms would be quite high. Other adverse effects include: mobilization of metals from sediments; eventual release of the metal and BOD associated with the degradation of oxine; and increase in the residence time (of the metal) in the water column. A misapplication of oxine would result in similar environmental effects.

Solvent Extraction with Oxine: The use of oxine in a closed loop solvent extraction system would also be capable of chelating a significant amount of the free heavy metal cations following a spill. The extraction efficiency of such a system can only be postulated, but relatively high extraction efficiencies should be achieved in a multistage centrifugal extractor. High metal concentrations in the spill would probably require relatively high solvent-to-spill ratios.

The use of such a closed loop system would minimize the four major adverse side effects associated with in situ treatment: (1) There would be no mobilization from the sediments (since there would be no contact); (2) There would be no eventual release of the metal (which would be recovered in a concentrated solution); (3) There would be no increase in the residence time of the metals in the water; and (4) The potential hazards associated with adding large amounts of chelating agent to a water body would not be present.

The only adverse environmental effects from the use of such a system would be associated with small losses of the chelate and solvent in the discharged (treated) water. If a low solubility, low toxicity solvent such as vegetable oil is effective, solvent losses (and resulting harmful effects) would be small. Losses of oxine cannot be accurately estimated since the literature data do not give a quantitative value for its solubility.

REDOX METHODS

Oxidation of Chemicals That Sink

The information in this section is based on *A Feasibility Study of Response Techniques for Discharges of Hazardous Chemicals that*

Sink, CG-D-56-78 prepared by T.D. Hand, A.W. Ford, P.G. Malone, D.W. Thompson and R.B. Mercer of the U.S. Army Engineer Waterways Experiment Station for the U.S. Coast Guard.

Hazardous chemicals that sink and are amenable to in situ oxidation are: 2,4-dinitroaniline, 2,4-dinitrophenol, carbolic oil (phenol), trichlorophenol, cresols, phthalic anhydride and diphenylmethanediisocyanate. In each case, the oxidizing agent must be brought into molecular contact with the hazardous chemical. This would require mixing of the system with the concurrent addition of the oxidizing agent.

The most important agent for the oxidation of selected hazardous chemicals, oxygen, is commonly found in natural waters. The chemical and biological oxidation of hazardous chemicals on the bottom of a body of water will cause an increase in the oxygen demand resulting in a reduction in the concentration of available dissolved oxygen in overlying waters. To make this problem more acute, some lakes, rivers, and coastal waters already have critically low oxygen concentration levels because of high temperatures or because of industrial and municipal wastes.

Except in well-mixed or shallow water, oxygen is generally more abundant near the surface. The most important source of oxygen below the surface is photosynthesis. The rate of oxygen diffusion is extremely slow (20 ft/yr in a sharp gradient). Chemical oxidation, therefore, may be greatly hampered due to depleted oxygen levels at the bottom. In addition, the toxicity of a spilled chemical may prevent biological decomposition as well as remove the principal source of oxygen available for an oxidation reaction.

Aeration techniques plus the addition of oxygen and ozone for the oxidation of selected hazardous chemicals contained in the water column were considered because they result in oxygenic reaction products only (i.e., CO_2, H_2O and partially oxidized organics). The use of potassium permanganate was not given serious attention because of the end products of the oxidation reaction (compounds of manganese) and because of the pollution hazard of unreacted permanganate. The use of containment procedures (barriers, curtains, or submerged dikes) would allow more precise control of the oxidation reaction, thereby limiting the amount of unreacted permanganate released.

Oxygen/Aeration: There is available some technology for the introduction of oxidizing agents into the vicinity of a hazardous material spill. The addition of air and oxygen into the discharge stream of a hydraulic dredge has been accomplished in an effort to lower the Immediate Oxygen Demand (IOD) associated with the dredged material discharge process (36). The addition of oxygen into the discharge stream was accomplished through the vaporization of liquid oxygen (LOX), originally stored in banks of pressure tanks. In another experiment, an air compressor was used to inject air directly into the discharge pipe. Measurements made during the oxygenation field demonstration showed that the direct injection of oxygen significantly reduced the depletion of dissolved oxygen in the disposal area. The aeration field demonstration was not conclusive in showing that the air injection mitigated dissolved oxygen (DO) depletion in the disposal area, though characteristics of the dredged material being tested may have colored this conclusion somewhat.

Also available are a number of devices used in the aeration of sewage lagoons and similar water treatment facilities. Aeration can be an effective technique for the introduction of oxygen into water. Because molecular diffusion of oxygen occurs at a low rate, an effective aerator must continually change large surface areas through which the interchange may take place.

Ozone: Ozone is more reactive than oxygen and yields no by-products of oxidation other than carbon dioxide, water, and partially oxidized organics. Ozone has been shown in laboratory studies to reduce carbolic oil concentrations in the range of 2 to 7 ppm by more than 90%.

Commercial ozone generating equipment is available that operates on a feed stream of air or pure oxygen. At optimum efficiency and operating on pure oxygen, this equipment can deliver a stream that is from 1 to about 6% ozone. Ozone generators, however, produce a stream that is limited to pressures no greater than 30 psi. This would seriously limit compatibility with injection systems that would require pressures of 100 psi or greater and effectively limits the use of direct injection to cases involving shallow waters. Ozone is not stable at elevated pressures or temperatures and would decompose to oxygen if a booster compressor were added to the system. Since ozone has a significant inhalation toxicity, special precautions would have to be taken if it is used in a response.

Chlorination

> The information in this section is based on *Manual for the Control of Hazardous Material Spills. Volume 1: Spill Assessment and Water Treatment Techniques*, EPA-600/2-77-227 prepared by K.R. Huibregtse, R.C. Scholz, R.E. Wullschleger, J.H. Moser, E.R. Bollinger and C.A. Hansen of Envirex for the Environmental Protection Agency.

Chlorination reactions are most commonly used to oxidize cyanide to the less toxic cyanate and then to carbon dioxide and nitrogen. These reactions are most effective at alkaline pH so sodium hydroxide is often added with hypochlorite. The chlorine is most safely added in the form of liquid hypochlorite in concentrations of 5 to 6% (household bleach). On a bench scale the concentration is diluted tenfold or more. Chlorine doses are determined by doing a bench scale test until a slight residual of chlorine remains as measured by a test kit. Once operating in the field, both NaOH and hypochlorite should be added at 75% of the expected volume and then in increments to the desired end point. The system should then be mixed an additional 15 minutes and the residual chlorine tested before discharge to be certain it meets acceptable limits.

Recommendations for Representative CHRIS Chemicals

> The information in this section is based on *Agents, Methods and Devices for Amelioration of Discharges of Hazardous Chemicals on Water*, CG-D-38-76 prepared by W.H. Bauer, D.N. Borton and J.J. Bulloff of Rensselaer Polytechnic Institute for the U.S. Coast Guard.

Redox methods were investigated as potential responses to spills of the following representative chemicals: ammonia, bromine, oxalic acid, methyl alcohol, sodium alkylbenzenesulfonates, phenol, chlorine, and potassium cyanide. In the latter three cases, such methods were judged acceptable. The details are given below.

Oxidation for Phenol: Aeration can be advantageous for oxidizing phenol in appropriate spill situations (37). The process is slow and the water has to be contained. An aeration of 1 to 6 hours is apparently sufficient. Carbon acts catalytically and reduces the sparging requirements considerably (38). Ozonolysis has been extensively studied and advocated (39)(40)(41). Ozone reduces phenol levels of 1.8 to 6.5 mg/l to 0.145 to 0.5 mg/l (42). It compares favorably with other oxidizing agents in performance and does not involve some of the environmental risks use of these others might entail (43). In comparison with other oxidants (44)(45), ozone is the most advantageous oxidant. It may be used whenever the logistics permit its timely deployment. For low-concentration warm-water phenol pollution, aeration oxidation (37) enhanced by exposure to sunlight or ultraviolet light may be advantageous (46).

Reduction of Phenol: Sulfite anions and ferrous cations both reduce phenol and have been used in sewage and wastewater treatment (47).

Reducing Agents for Chlorine: Reduction changes chlorine, hypochlorous acid or hypochlorite ion to chloride ions. Reducing agents in excess are likely to injure the environment. Hypochlorite ion from chlorine spills oxidizes calcium sulfite hemihydrate to calcium sulfate hemihydrate, the mineral gypsum (48). A cheap source of sulfite ion is lignite filtrate or cellulose sulfite liquor from wood pulping.

Oxidation for Potassium Cyanide: Hydrogen cyanide can be burned to produce water, carbon dioxide and nitrogen. It can be catalytically oxidized in either the gaseous or the dissolved state, and when it is dissolved as hydrocyanic acid or as a cyanide salt or complex. In catalytic oxidation the oxidant is air, and sparging capability is required (37)(49)(50)(51)(52). Carbon is both an adsorbent and a catalyst for such oxidation (53)-(58). The process is analogous to the well-known Zimpro process of sludge oxidation of industrial cyanide sludges (59).

Carbon is also an oxidation catalyst in the Duesseldorf process for ozonolysis of reductant contaminants in water (60). Cheap, efficient ozone generation is now available on the massive scale used for water supply treatment (39)(61). Ozonolysis of cyanide-containing waters is extensively practiced and well understood (40)(62). Use of sodium polyphosphate speeds up the ozonolysis to 450% of the usual rate and even makes aeration a fast process (63)(64).

Peroxide oxidation has also been applied to detoxification of cyanide solutions (65). Hydrogen peroxide is widely available and decyanidation with it can be catalyzed (66). Perhydrol has been recommended in place of hydrogen peroxide (67)-(69). Sodium hypochlorite may be used as an oxidant (70)(71). Calcium hypochlorite, cheaper than sodium hypochlorite, has been used (72). Potassium permanganate oxidation has been used for amelioration of cyanide spills (73), as has anodic oxidation (74).

BIODEGRADATION

> The information in this section is based on *A Feasibility Study of Response Techniques for Discharges of Hazardous Chemicals that Sink*, CG-D-56-78 prepared by T.D. Hand, A.W. Ford, P.G. Malone, D.W. Thompson and R.B. Mercer of the U.S. Army Engineer Waterways Experiment Station for the U.S. Coast Guard.

Biodegradation is not at this time considered a high potential spill amelioration technique. Further, it does not have great potential as an emergency response measure. However, there are compounds, especially organics, which, over time, can be biologically degraded, either by natural or artificially induced means, and precedents for biodegradation of hazardous materials spills exist. In the petroleum industry, for example, chemical dispersants will break up a slick of spilled oil into tiny droplets which are then dispersed through the water column in a form more susceptible to biological decomposition.

Significant work on the application of biological countermeasures to hazardous material spills has been done by Armstrong et al (75)(76). The Armstrong references recognize the opinions of Dawson et al (77) which were stated as follows:

> Biological degradation, while attractive in some respects, suffers from several difficulties. In order for degradation to proceed at a rapid rate, it would be necessary to have on hand large quantities of acclimated cultures. The problems associated with stockpiling many such cultures, each of which is specific to a particular substance, are obvious. Also, many hazardous materials are apparently resistant to biological degradation.

Armstrong (75), however, points out that there is a body of literature and experience from the wastewater treatment field which tends to indicate that biological countermeasures may be effective against a wide range of hazardous chemicals. It is suggested that bacteria cultures stored in a dormant state (in either a frozen or powdered form) could be deployed easily and rapidly without highly specialized equipment. In situ deployment by spraying from a helicopter, shore, or boat was suggested. In addition to stored cultures, activated sludge wastewater treatment plants near the location of a spill might yield suitable cultures. Industrial wastewater treatment plants at the site of manufacture of certain hazardous materials (no specifics listed) could quickly provide cultures acclimated to the spilled material.

The investigation by Armstrong et al (75) centered on a list of 14 of the 20 most hazardous chemicals listed by the EPA. Included on the list were aldrin and DDT, both of which are under consideration in this report. Both were reported as biodegradable (76), but no active analysis of them was conducted, other than initial screening tests. The kinetics of biodegradable reactions on such persistent pesticides are likely to be very unfavorable.

Tests reported on phenol (76) have yielded some noteworthy results. Temperature is an important variable in biodegradation. It was noted that decreased metabolic rates at low temperatures could seriously affect biodegradation. Further, freshwater bacterial cultures might possibly not be useful in marine and coastal waters, as high salinities in water reduce the effective pH range where biodegradation might occur. A temporary dissolved oxygen impact will occur upon application of the countermeasure, though aeration might help reduce this effect.

Pond tests conducted on phenol (76) revealed the possibility of indigenous bacteria assisting in decomposition. Further, these same tests suggested in situ applications of both bacteria acclimated to the hazardous material and unacclimated bacteria (taken from a nearby wastewater treatment plant) would be effective in biodegradation. Nutrients (e.g., nitrogen and phosphorus) added to the treatment area will increase the rate of biodegradation which occurs.

Armstrong (78) notes that fairly quiescent areas will be required for any biodegradation to take place, and the spill must be contained. As to applicability to materials that sink, Armstrong notes that some bacterial cultures exist which will tend to sink to the bottom themselves, or they could be placed in sinkable bags. If cultures from sewage treatment plants were used in biodegradation, it is possible that pathogenic bacteria could be introduced into the waterway, and this could pose a problem greater than that of the hazardous material itself.

In summary, biodegradation cannot be considered a high-potential response technique at this time, though work is currently being done to refine the techniques. Limited to contained spills of organic materials in quiescent waters (drainage ditches, coves, and certain river, port, and harbor areas), the technique is plagued by problems of acclimation of bacteria cultures to specific hazardous materials, difficulties in storage and handling, and uncertainties regarding BOD levels and the introduction of unwanted pathogenic bacteria. Recognition of these difficulties is inherent in Annex X of the National Oil and Hazardous Substances Pollution Contingency Plan (40 CFR 1510) which requires EPA approval prior to the use of any biological agents.

PORTABLE TREATMENT EQUIPMENT SOURCES

> The information in this section is based on *Manual for the Control of Hazardous Material Spills. Volume 1: Spill Assessment and Water Treatment Techniques*, EPA-600/2-77-227 prepared by K.R. Huibregtse, R.C. Scholz, R.E. Wullschleger, J.H. Moser, E.R. Bollinger and C.A. Hansen of Envirex for the Environmental Protection Agency.

Once it has been established that a spill must be treated using an on-site but off-stream system, it is necessary to construct the needed process units. Prior to improvising a treatment system, it is recommended that the spill coordinator investigate the possibility of using preconstructed system components. Authorized personnel should determine the availability of equipment to be used in hazardous material spill cleanup. Table 3.13 presents a list of some of the available sources of equipment throughout the country. Prior to a spill situation, it would be desirable for the on-scene coordinator to be familiar with these sources and other local suppliers so that upon the occurrence of a spill, the availability of preconstructed equipment can be determined rapidly.

Table 3.13: Portable Treatment Equipment Sources (Used Chemical Process Equipment Suppliers)

San Francisco
Machinery and Equipment Corp.
P.O. Box 3132C
San Francisco, California 94119
Phone: 415-467-3400

Houston
Dynaquip, Inc.
1143 Brittmore
Houston, Texas 77043
Phone: 713-467-5500

Petro-Power, Inc.
6436 Rupley Circle
Houston, Texas 77087
Phone: 713-644-8271

Chicago
Aaron Equipment Co.
9301 W. Bernice Street
Schiller Park, Illinois 60176
Phone: 312-678-1500

A-1 Equipment & Chemical Co.
57 East 21st Street
Chicago, Illinois 60616
Phone: 312-842-2200

Indeck Power Equipment Co.
1075 Noel Avenue
Wheeling, Illinois 60090
Phone: 312-541-8300

Loeb Equipment Supply Co.
4131 South State Street
Chicago, Illinois 60609
Phone: 312-548-4131

Union Standard Equipment
163-167 N. May Street
Chicago, Illinois 60607
Phone: 312-421-1111

Cleveland
Arnold Equipment Co.
5055 Richmond Road
Cleveland, Ohio 44146
Phone: 216-831-8485

Cleveland (continued)
C.P.R. Machinery & Equipment Co.
5061 Richmond Road
Cleveland, Ohio 44146
Phone: 216-464-8590

Federal Equipment Co.
8200 Bessemer Avenue
Cleveland, Ohio 44127
Phone: 216-271-3500

International Power Machinery Co.
834CE Terminal Tower
Cleveland, Ohio 44113
Phone: 216-621-9514

Process Equipment Trading Co.
1250 St. George Street
East Liverpool, Ohio 43920
Phone: 216-385-2400

New York-New Jersey
Brill Equipment Co.
35-63 Iabez Street
Newark, New Jersey 07105
Phone: 201-589-7420

George Equipment Co.
27 Haynes Avenue
Newark, New Jersey 07114
Phone: 201-242-9000

H&P Equipment Co., Inc.
14 Skyline Drive, Box 368
Montville, New Jersey 07045
Phone: 201-335-9770

Keith Machinery Co.
34 Gear Avenue
Lindenhurst, New York 11757
Phone: 516-884-1200

Perry Equipment Co., Inc.
Box C
Hainesport, New Jersey 08036
Phone: 609-267-1600

Source: EPA-600/2-77-227

REFERENCES

(1) Stumm, W. and Morgan, J.E., *Aquatic Chemistry*, Wiley-Interscience, New York (1970).
(2) Park, K., Hood, D.W., and Odum, H.T., *Inst. Mar. Sci. 5*, 47 (1958).
(3) *Water Quality Criteria*, National Academy of Sciences and National Academy of Engineering, U.S. Government Printing Office, Washington, D.C. (1972).
(4) *Proposed Criteria for Water Quality, Vol. I*, U.S. Environmental Protection Agency (October 1973).
(5) Drake, E., Shooter, D., Lyman, W., and Davidson, L., *A Feasibility Study of Response Techniques for Discharges of Hazardous Chemicals that Disperse through the Water Column*, CG-D-16-77 (July 1976).
(6) Thomas, P., "Neutralization of Alkaline Wastewater with Carbon Dioxide," *Ind.-Anz., 96* (45), 1022 (1974).
(7) Kruz, G. and Trenkle, V., "Neutralization of Alkaline Wastewater by Flue Gas or Carbon Dioxide," *CZ-Chem.-Tech., 2* (10), 393-395 (1973).
(8) Ramishvili, I.M., Gryaznova, Z.V., Antipina, T.V. and Tsitsishvili, G.V., *Acidity of Some A, X and Y Type Zeolites*, Deposit No. 7099-73, VINITI, Moscow, USSR (1973).
(9) Colas, R., "Deacidification of Corrosive Waters with Superfritted Dolomite," *Ind. Chim. Belge, 32* (Spec. No., Pt. 1), 538-539 (1967).
(10) Ranskii, B.N. and Popov, S.I., "Effects of the Quality of Milk of Lime on Its Consumption During Flotation and Neutralization of Acidic Waters," *Tsvet. Metal., 42* (9), 39-40 (1969).
(11) Micheli, K. and Fritzsche, E., "Incrustations Taking Place when Neutralizing Acid Sewage by Lime Hydrate and Their Suppression," *Wasserwirt.-Wassertech., 19* (1), 8-10 (1969).
(12) Kristenko, B.N., "Ferrochromium Slags as Substitute for Lime in the Neutralization of Acid Wash Waters and the Technology of Its Use," *Stroct. Mater. Izd. Met. Sklakov, Ural Nauch.-Issled. Proekt. Inst. Stroct. Mater.*, 282-286 (1965).
(13) Ozerov, A.I., Yashanov, G.G. and Porubaev, V.P., "Automation of Wastewater Purification in the Balkash Mining-Metallurgical Combine," *Tr. Nauch.-Issled. Proekt. Inst. Obogashch. Rud Tsvet. Metal.*, (6), 231-235 (1971).
(14) Tyco Labs., *Electrochemical Treatment of Acid Mine Waters,* W72-07799, EPA-14010-FNQ-02/72, Contr. No. EPA-14-12-859, Waltham, Mass., (microfiche) NTIS Doc. No. PB 208820, Springfield, Va., Feb. 1972 (hard copy) Supt. of Docs. No. EP2.10: 14010FNQ, USGPO, Wash., D.C. (1972).
(15) Bennett, G.F., ed., *Proceedings of the 1974 Conference on Control of Hazardous Material Spills*, San Francisco, Calif., AICHE and EPA (Aug. 25-28, 1974).
(16) Gunarsson, K.E., *Treating Industrial Wastewater Contaminated with Metal Ions, Nitrate Ions and Fluoride Ions*, U.S. Patent 3,647,686 assigned to Nyby Bruks Aktiebolag, Sweden (March 7, 1972).
(17) Kirk, R.E. and Othmar, D.F., eds., *Encyclopedia of Chemical Technology*, Vol. 6, p. 1-24, The Interscience Encyclopedia, Inc., New York (1971).
(18) *Chemical Economics Handbook*, Stanford Research Institute, Menlo Park, Calif. (July 1974).
(19) Parks, K., Hood, D.W., and Odum, H.T., *Inst. Mar. Sci. 5*, 47 (1958).
(20) Bauman, A.N. and Bird, R.E., *Purification of Fluorine-containing Industrial Wastewaters*, U.S. Patent 3,551,332 assigned to Int. Minerals and Chemical Corp. (Dec. 29, 1970).
(21) *Idem.*, Ger. Offen. 2,029,056 (Jan. 7, 1971).
(22) Nagano, T., Endo, T. and Muramichi, S., *Removing Fluoride or Fluoride-containing Compounds from Wastewaters by Precipitation*, Japanese Kokai 74-48,151, assigned to Nissan Engrg. Co., Ltd. (May 10, 1974).
(23) Thergaonkov, V.P. and Nawalakhe, W.G., "Activated Magnesia for Fluoride Removal," *Indian J. Environ. Health, 13* (3), 241-243 (1971).
(24) Levitskii, E.A. and Lazovskii, Ya.B., *Defluorination of Water,* USSR Patent 261,996, assigned to Pikalevevo Alumina Combine (Jan. 13, 1970).
(25) Watanabe, M., Okamoto, T. and Ida, K., *Removing Fluorides from Wastewaters by Treating with Sodium Aluminate and Acids*, Japanese Kokai 74-32,472, assigned to Sumitomo Chemical Co., Ltd. (March 25, 1974).

(26) Pakhrutdinova, M., Mansurov, P.Kh., and Rustamov, Kh.R., "Kinetics of the Extraction of Oxalic Acid from Anabasinine Sulfate Production Wastes," *Uzb. Khim. Zh. 18* (3), 26-27 (1974).

(27) Sears, M., ed., *Oceanography* AAAS Publ. No. 67, p. 555, Washington, D.C. (1961).

(28) Bernard, H., EPA, Private communication.

(29) *Stability Constants of Metal Ion Complexes* (supplement #), The Chemical Society, London (1971).

(30) Willard, H.H., and Diehl, H., *Advanced Quantitative Analysis*, Van Nostrand, New York (1956).

(31) Albert, A., *Selective Toxicity and Related Topics*, Methuen & Co., Ltd., London (1968).

(32) Wheatland, A.B., Glidhill, C., and O'Gorman, J.V., "Developments in the Treatment of Metal-Bearing Effluents," *J. Chem. E. Symposium Series #41*.

(33) Fletcher, A.W., "Metal Recovery from Effluents–Some Recent Developments," *J. Chem. E. Symposium Series #41.*

(34) Stary, J., "Systematic Study of the Solvent Extraction of Metal Oxinates," *Analytica Chimica Acta* 28, pp. 132-149 (1963).

(35) *CHRIS Hazardous Chemical Data*, U.S. Coast Guard, CG-446-2 (Jan. 1974).

(36) Neal, R.W., Pojasek, R.B., and Johnson, J.C., *Oxygenation of Dredged Material by Direct Injection of Oxygen and Air During Open Water Pipeline Disposal*, D-77-15, U.S. Army Engineer Waterways Experiment Station, CE, Vicksburg, Miss. (Oct. 1977).

(37) Popov, K., "Purification of Different Kinds of Wastewaters Containing Cyanides, Phenols and Other Compounds and Investigation of the Dependence between the Purification Effectiveness and the Redox Potential of the Water," *Tr. Nauchnoizsled. Inst. Vodonabdyavane, Kanaliz. Sanit. Tekh. 9* (2), 91-108 (1973).

(38) Murkami, K., and Yasui, T., *Oxidation of Phenols,* Japanese, Kokai 74-36,642 assigned to Kuraray Co., Ltd., Japan (April 5, 1974).

(39) Karelina, O. Ya., "Modern Current Status of Ozonation in Water Treatment," *Izv. Vyssh. Ucheb. Zaved. Stroit. Arkhitekt, 14* (12), 131-134 (1971).

(40) Bischoff, C., "Purification of Wastewater by Ozone," *Fortschr. Wasserchem. Ihrer Grenzgeb.* (9), 121-130 (1968).

(41) Korolev, A.A., Abinder, A.A., Bogdanov, M.V., Zakharova, T.A., and Khitrov, N.K., "Hygienic and Toxicological Features of Products of Phenol Degradation During Ozone Treatment of Waters," *Gig. Sanit.*, (8), 6-10 (1973).

(42) Pasynkiewicz, J., Grassman, A., and Nowara, S., Application of Ozone in Purification of Drinking Water Containing Phenols, *Przem. Chem., 47* (4), 227-231 (1968).

(43) Atkinson, J.W. and Palin, A.T., "Chemical Oxidation in Water Treatment," *Intl. Water Supply Assn. Congr., 9th, 1972,* E1-E9 (1973).

(44) Tada, K., *Purification of Wastewater Containing Phenol Derivatives*, Japan. Kokai 73-104,352, assigned to Inuiu Yakuhin Kogyo Co., Ltd. (Dec. 27, 1973).

(45) Apostolov, S.A., Kozhevnikov, A.V., and Statslaua, N.M., Removal of Phenol Dissolved in Small Amounts in Industrial Wastewaters, *Tr. Ses.-Zapad. Zaoch. Politekh. Inst.* (21), 32-34 (1972).

(46) Leygue, G., Ollivier, J., and Seris, J.L., *Elimination of Phenols and Other Organic Micropollutants from Water*, French Demande 2,158,714, assigned to ELF, Enterprise de Recherches et Activities Petrolieres (July 20, 1973).

(47) Asonov, A.M., Kagasov, V.M., and Derbysheva, E.K., "Purification of Phenol-Containing Waters from the Coke and Chemical Industry Using Hydrated Ferrous Sulfate," *Gidrokhim Urola*, (3), 79-81 (1973).

(48) Murayama, N., and Sato, K., *Porous Polyolefin Containing Sulfite for Reducing Chlorine in Water*, Japanese Kokai 73-91,853 assigned to Kureka Chemical Industry Co., Ltd., Japan (Nov. 29, 1973).

(49) Schindewolf, U., "New Process for Disposal of Cyanide Wastes," *Chem.-Ing.-Tech., 44* (10), 682-683 (1972).

(50) Ohtsubo, K., *Purification of Wastewater Containing a Complex Cyanide*, Jap. Kokai 74-39,264, assigned to Hitachi Plant Construction Engineering Co., Ltd. (April 12, 1974).

(51) Uss, D., *Detoxification of Wastewater*, Ger. Offen. 2,261,133 assigned to Deutsche Gold- und Silver-Scheideanstalt vorm. Roessler (June 20, 1944).

(52) Jola, M., Catalytic Hydrogen Cyanide Oxidation, *Galvannotecknik, 61* (12), 1003-1008 (1970).

(53) Bernardin, F.E., Detoxification of Cyanide by Adsorption and Catalytic Oxidation on Granular Activated Carbon, *Proc. Mid-Atlantic Ind. Waste Conf., 4th, 1970*, 203-228 (1971).

(54) Hoeke, B., and Wittbold, H.A., Catalytic Oxidation of Nitrite and Cyanide Ions in Neutral and Alkaline Aqueous Solutions, *Galvannotecknik 61* (6), 468-474 (1970) and German Offen. 1,912,473, assigned to F. Krupp GmbH (Sept. 24, 1970).

(55) Laube, K., and Oehme, F., *Catalytic Decontamination of Aqueous Cyanide Solutions,* Ger. Offen. 2,009,120, assigned to Zellweger A.-G. Apparate-und Maschinenfabricken Uster (March 18, 1971) and *Reactivation of Active Charcoal for Removal of Cyanide from Wastes,* Ger. Offen. 2,046,005 (June 24, 1971).

(56) Avias Fabrik fuer Kunstoffverarbeitung GmbH, *Lowering the Toxicity of Wastewaters Containing Cyanide*, Fr. Patent 1,494,619 (Sept. 8, 1967).

(57) Buksteeg, W., and Thiele, H., *Detoxification of Wastewaters Containing Cyanide*, British Patent 1,140,963 (Jan. 22, 1969).

(58) Bloem, G., Laube, K. and Oehme, F., *Depoisoning of Cyanide-containing Wastes*, Ger. Offen. 2,046,006 (June 24, 1971).

(59) Huesler, H., "Zimpro Process, Moist-Air Oxidation of Industrial Sludges," *Abwasserteknik, 22* (1), III-IV (1971).

(60) Thiele, H., "Detoxification of Cyanide-containing Water by Catalytic Oxidation and Adsorption Process," *Fortschr. Wasserchem. Ihrer. Grenzgeb.* (9), 109-120 (1968).

(61) Diaper, E.W.J., "Ozone, Practical Aspects of Its Generation and Use, *Chem. Techol., 2* (8), 498-504 (1972).

(62) Khandelwal, K.K., "Ozonation of Cyanides," Order No. 56-18024, *Univ. Microfilms*, Ann Arbor, Mich. (1956).

(63) Komendova, V., *Neutralizing Cyanides in Wastewaters,* Czeck. Patent No. 130,792 (Jan. 15, 1969).

(64) Valtr, Z., "Catalyzed Ozonization of Cyanides," *Zb. Pr. Chemikotechnol. Fak. SVST (Solv. Vys. Sk. Tech.), 1969-1970*, 31-41 (1971).

(65) Henry, C., and Boeglin, J.C., "Decyaniding with Peroxidized Products," *Trib. CEBEDEAU (Centr. Belge Etude Doc. Eaux), 24* (331-332), 282-294 (1971).

(66) Lawes, B.C., and Mathre, O.B., *Cyanide Removal from Wastewaters*, Ger. Offen. 2,109,939, assigned to E.I. du Pont de Nemours and Co., Wilmington, Del (Sept. 16, 1971).

(67) Zumbrunn, J.P., "Purification of Cyanide-Containing Wastewaters," *Chem. Ind., Genie Chim., 104* (20), 2573-2584 (1971).

(68) Conrad, J., and Jola, M., "Cyanide Detoxification Today," *Oberflaeche-Surface, 13* (7), 143-148 (1972). Sakota, K., and Chomei, K., *Treating Wastewaters Containing Cyanides by Reacting with Chlorine*, Japan. Kokai 74-27,060, assigned to Teihoku Piston Ring. Co., Ltd. (March 11, 1974).

(69) Fischer, G., "Liquefied Chlorine for Cyanide Oxidation During the Lancy Process," *Wasser, Luft. Betr., 13* (10), 389-392 (1969).

(70) Kieszkowshi, M., "Treating of Cyanide Effluents with Sodium Hypochlorite," *Koroze Ochr. Mater., 11* (4), 77-80 (1967) and *Pr. Inst. Mech. Precyz., 15* (1), 62-71 (1967).

(71) Koniecka, K., "Treatment of Chromium and Cyanide Wastewaters," *Gaz, Woda, Tech. Sanit., 42* (8), 280-281 (1968) and Schulze, G., East German Patent 58070 (Sept. 20, 1967).

(72) Moore, S.L., and Kin, S.R., "Cyanide Pollution and Emergency Duty Train Wreck, Dunereith, Indiana," *Engrg. Bull. Purdue Univ. Engrg. Ext. Series*, (132, Pt. 1), 583-600 (1968).

(73) Kiezkowski, M., and Krajewski, S., "Purification of Cyanide-Containing Water with Permanganate," *Tech. Eau (Brussels)* (259-260), 21-27 (1968) and *Pr. Inst. Mech. Precyz., 14* (3), 68-73 (1966).

(74) Han, Q.-S., "Treatment of Cyanide-Containing Wastewater with Electrolytic Method," *Hsa Hsuch Tung Pao*, (3), 44-45 (1974).

(75) Armstrong, N.E., Wyss, O., Gloyna, E.F., and Behn, V.C., "Biological Countermeasures for the Mitigation of Hazardous Materials Spills," *Proceedings of the 1974 National Conference on Control of Hazardous Material Spills,* San Francisco, Calif. (Aug. 25-28, 1974).

(76) Armstrong, N.E., Wyss, O., Gloyna, E.F., and Behn, V.C., "Biological Countermeasures for the Mitigation of Hazardous Materials Spills," *Proceedings of the 1976 National Conference on Control of Hazardsous Material Spills*, New Orleans, La. (April 25-28, 1976).

(77) Dawson, G.W., Shuckrow, A.J., and Mercer, B.W., "Strategy for Treatment of Water Contaminated by Hazardous Materials," *Proceedings of the 1972 National Conference on the Control of Hazardous Material Spills*, Houston, Texas (March 21-22, 1972).

(78) Armstrong, N.E., Private communication.

SORBENTS, GELS AND FOAMS

SORBENTS FOR REPRESENTATIVE HAZARDOUS CHEMICALS

Chemicals That Mix with Water

The information in this section is based on *Agents, Methods and Devices for Amelioration of Discharges of Hazardous Chemicals on Water,* CG-D-38-76 prepared by W.H. Bauer, D.N. Borton and J.J. Bulloff of Rensselaer Polytechnic Institute for the U.S. Coast Guard.

Identification of High Potential Sorbents for Chemicals that Mix: Specific high potential sorbents are listed below for each of 13 miscible/soluble representative hazardous chemicals and for three other representative chemicals which mix, but are representative of other behavior categories. The behavior categories were defined in Table 1.7.

Concentrated sulfuric acid (IV A) – Activated carbon, Dowex 1 x 10, Amberlite IRA 402-Amberlite IRA 93 (bilayer), De Sal Process mixture (2 weak-base anion and a strong-base cation exchanged).

Sodium hydroxide (IV B) – Activated carbon, Amberlite IRA 93, Amberlite IRA 400.

Potassium cyanide (IV C) – Activated carbon, Amberlite IRA 93, Amberlite IRA 400.

Cadmium chloride (IV D) – Activated carbon, sheet poly(styrenesulfonate), Dowex 50W x 8, Dowex 50W x 4, Amberlite IR 252, Amberlite IRA 900.

Bromine (IV E, IIIB) – Activated carbon, polyurethane, polyolefin, 5A type molecular sieve.

Oxalic acid (IV F) – Activated carbon, Dowex 5W x 4.

Methyl alcohol (IV G) – Activated carbon.

Ethyl acetate (IV H) – Activated carbon, Amberlite XAD or other macroreticular resins.

Ethylene diamine (IV J) – Activated carbon, Amberlite XAD or other macroreticular resins, Dowex 50W x 8, Amberlite IRA 900.

Dimethyl sulfoxide (IV K) – Activated carbon, polyurethane foam.

Epichlorohydrin (IV L) – Activated carbon, polyurethane foam.

Sodium alkylbenzenesulfonates (IV M) – Activated carbon, Amberlite XAD or other macroreticular resins, Amberlite IRA 93, Amberlite IRA 400.

Phenol (IV N) – Activated carbon, Floridin XXF, Tonsil AC, polyurethane foam, Amberlite XAD-1-8 and other macroreticular resins, poly(styrenesulfonate), Amberlite IR 45, Dowex 1.

Chlorine (I B) – Activated carbon, polyurethane, polyolefin, Dowex 1.

Hydrogen fluoride (I C) – Activated carbon, carbonized sulfonated sawdust, Dowex 1.

Ammonia (I E) – Activated carbon, Zeolite F (potassium form), Dowex 50W x 8.

The sorbents identified as most advantageous were activated carbon, polyurethane foam, macroreticular resins, propylene fibers, zeolite molecular sieves, sorbent clays, polyolefins, polyethylenes, polyisobutylenes, poly(methyl methacrylates) and poly(styrenesulfonates).

Activated Carbon – Activated carbon sorbs every one of the representative hazardous chemicals, but different activated carbons are selective for different hazardous chemicals. A carbon surface can be acidic or basic, hydrophilic or hydrophobic or oleophilic or lipophobic. It can vary in porosity. The surface area per unit weight is a function of the size of the carbon particles and of the area generated by the process of activation. Further, activated carbon is sold with its particles in various states of agglomeration and aggregation. It can come in bead, pellet, rod, sheet and other forms and shapes. Thus, the classification of carbons is a task much like that of classification of hazardous chemicals (1) to (3). Similar complications beset the evaluations of ion exchange sorbents (4) to (7).

Apart from this variability and its particular sorbent capacity, activated carbon can be viewed as a stable chemical element of excellent resistance to chemical attack. Its particle porosity and interparticle packing sets its bulk density. This density can vary from 0.09 to over 2.0 relative to water. Wet density is a different matter. A hydrophilic carbon of low bulk density might sink by displacement of all absorbed and adsorbed air by water. A higher density hydrophobic carbon might not suffer the displacement. The latter carbon might sorb a floating spill without sinking.

If a sorbent floats, it can be collected by booms and recovered by skimmers whether it is in piece, fiber or powder form (8). With activated carbons, the highest surface forms, disaggregated submicronic powders, are most difficult to keep undispersed, floating and skimmable. Flocculation methods and other methods (9) to (13) developed to apply, float and recover sorbents and ion exchangers have not been studied for activated carbons. It can be presumed that in moving or turbulent water, existing methods of application would be less effective for amelioration of spills of hazardous chemicals than would be indicated by equilibrium sorption performance data derived from literature based on laboratory evaluation.

The variance of field conditions of spill and amelioration and the lack of a standard classification of spills and receiving waters or of the sorbencies and forms of commercial activated carbons make estimation of the degree of improvement to be expected in spill amelioration difficult. Manufacturer's data, literature publication data and the data obtained from laboratory testing necessarily involve equilibrium sorption which may be regarded as a measure of maximum effectiveness.

Activated carbon has the advantage that it can be obtained from many manufacturers and distributors, since it is industry's prime and most universal sorbent. These firms sell many complex forms of activated carbon for specialized use or for use in fixed-bed or fluidized-bed application. A typical variety of versatile activated carbons can be purchased from the following firms: American Norit Co., Inc.; Calgon Corporation, Pittsburgh Active Carbon Division and Water Management Division; ICI America, Ltd.; Westvaco Chemical Division, Carbon Department; Union Carbide Corp., Carbon Products Division; and Barneby-Cheney Co.

Price of activated carbon sorbent depends on the size and nature of the purchase. Without regeneration it can be quite costly for any but small spills. On the other hand, stocking two or three varieties developed for versatile sorption of many hazardous chemicals could be far cheaper and of far higher short term potential than investment in a number of ameliorants each optimized to fit a narrow amelioration response category of spill. The Environmental Protection Agency has taken the versatile agent approach (14).

Regeneration of activated carbon after use and recovery of the sorbed hazardous chemical is very difficult. Regeneration might not lower costs significantly or at all for hazardous chemicals that are likely to be hard to desorb. For some sorbates continued regeneration of the sorbent by heat or by elution produces a carbon of diminished or even continually declining effectiveness. Thermal desorption is risky for flammable sorbates.

All methods of using activated carbon for amelioration of spills of hazardous chemicals on water must include a means of removal or separation of sorbent and sorbate from water. Three basic types are in use. In the simplest, the activated carbon is placed in a porous bag, and this is dragged through the polluted water or the water is allowed to flow through the bag (14). In a second method of application, polluted water is passed through activated carbon in beds or columns from which the cleaned water emerges and the hazardous sorbate is retained on the carbon sorbent.

Such systems may be set up in fixed locations, or they may be mobile. These techniques will be discussed later in this chapter. Probably the most efficient system for applying activated carbon in a mobile apparatus with very high flow through treatment rate is the Dynactor, now in use with proven application to spills of hazardous chemicals on water. The Dynactor will be described in the following chapter, which deals with mobile units.

Polyurethane Foam – Polyurethane polymer can be produced in open pore, closed pore and nonporous particulate form. All of the forms adsorb a variety of chemicals from aqueous systems, but the open pore form also absorbs liquids like a sponge. Polyurethane is not as versatile as activated carbon, but it is a most advantageous sorbent for benzene (15), (16), chlorine (17), kerosene, naphtha:solvent, hexane, n-butyraldehyde, dimethylsulfoxide, epichlorohydrin and phenol. Polyurethane sorbents use in amelioration of petroleum oil spills has led to the development of many existing application methods. Polyurethane belts sorb some floating chemicals, flocculate dispersions and adsorb many dissolved solutes. The sorption action is mostly by absorption, depending on the openness of the pore structure and its connectivity, and also upon the viscosity and the wetting power for polyurethane of the sorbate. Polyurethane foam belts laden with recovered sorbate chemical and some water can be squeezed, causing removal of the absorbed chemical which is led off to storage. Return of the wrung belt permits continued autoregenerative skimming or extraction (18), (19).

Continuous operation by continuous belt pickup is also possible for polyurethane that is broadcast by blower over the spill. The broadcast polymer may act to immobilize a spill (8), (20). The polyurethane may be shredded at the spill site. To avoid bulky transport of a foam that can be mostly air, the polyurethane can be formed on site, or even in situ (20). The foam can immediately immobilize the entire spill, an advantage over the fixed belt system. Once the polyurethane foam is loaded with sorbate, it is picked up and transferred to a mechanical belt that transports it to a squeezer. Once the squeezer removes the sorbate, the foam can be rebroadcast. Thus, continuous recovery is possible.

Repeated application may effect the sorption of practically all of a spill of hazardous chemical on water. Since polyurethane is essentially oleophilic and hydrophobic, as long as it is in a contained or recovered mode of operation, water is not contaminated by it. The more efficient the method of application, the greater the reduction of the only hazard present, that of the fire hazard or the health hazard of the spilled material itself. The potential hazard of remnant polyurethane can be reduced in two ways. Its flotability can be assured by providing closed as well as open pores. Laminating a thin sheet of closed pore polyurethane to a thick web of open pore polyurethane with a polyurethane adhesive is an effective and economical way of doing this.

The sorption capacity of polyurethane foam is a function of its open-volume porosity, foam densities ranging from 1.0 to 30 or 40 pounds per cubic foot, and of its pore size that can range from micronic size up to the centimeter range. Post-formation etching processes can enlarge the pore size and increase the porosity. Sorption rate and capacity also depend upon the viscosity of the spilled hazardous chemical. In situations that set up gravity flow, retention is enhanced but rate of pickup decreased by smaller pore structure. Ideally, a different balance of pore size and connectivity should be supplied for every different hazardous chemical to give optimum sorption rate and capacity, but this is not necessary for practical use. Depending on this complex of sorbent, sorbate and ambient interaction, polyurethane foams can pick up from 0.1 to 80 times their weight of spilled chemical in a single operation. Polyurethane may be used in continuous cyclic operation for those hazardous chemical spills that do not involve chemical solvency for polyurethane. Scrap polyurethane foam, generally available, which is cheaper than new open market or custom produced material, may be of some use in ameliorations.

Polyurethane can be purchased from manufacturers, bulk and custom fabricators and distributors. Two general suppliers are BASF Wyandotte Corp., Industrial Chemicals Group, and the B.F. Goodrich Chemical Co.

For hazardous chemicals that float or float on water and vaporize, and to which existing devices for oil recovery apply, polyurethane foam can be applied in the same modes that are used against oil. For hazardous chemicals that mix or float and mix, devices for recovery of sorbate-loaded floating polyurethane can be readily applied, if fire hazards are controlled. Though polyurethane foam is not as universally applicable as carbon black, the art of using it for spill amelioration of floating hazardous chemicals, especially regeneratively and continually, is more advanced. Thus, its use is feasible for most spill situations of liquids such as benzene, kerosene, naphtha:solvent, hexane, n-butyraldehyde, dimethyl sulfoxide, epichlorohydrin, phenol or chlorine and by inference, of most of the 167 floaters on the CHRIS 400 list.

Macroreticular Resins – Crosslinking of essentially linear polymers by bifunctional monomers added in the polymarization process can produce polymers with a spongy or reticulated structure that is a small-scale replication of the macroporous structure of polyurethane foam or a large-scale replication of the molecular porosity of activated carbon. Two common macroreticular polymers of this description are poly(styrene-co-divinylbenzene) and poly(methyl methacrylate-co-ethylene dimethacrylate). Further, the monomers can be made into polar sorbents or ion exchangers by substitution of polar groups. For example, sulfonation produces a resin that behaves chemically like poly(styrenesulfonate) ion-exchange resin of a similar degree of sulfonation.

Such resins sorb liquids or solutes like ethyl acetate, hexane and ethylene diamine and dissolved solids like aldrin, sodium alkylbenzenesulfonates and phenol. If sulfonated they sorb ions like cadmium more readily the higher the sulfonation. Thus the macroreticular resins are more versatile than polyurethane foam, but not as universally so as activated carbons. Little art has been developed for their use and practically none on a large scale. The degree of improvement they effect in one-pass operation should compare favorably with that effected by activated carbon or polyurethane foam, but regeneration capabilities remain unascertained.

The commercial source for macroreticular resins is the Rohm and Haas Co. Presumably their XAD line of resins are poly(styrene sulfonate-co-divinylbenzenes) of varying degree of crosslinking and sulfonation, with the XAD-4 resin acting as sorbent for all the representative hazardous chemicals except phenol which is sorbed by XAD-1-8.

Application, recovery and regeneration equipment and techniques would resemble those described for carbon and for ion-exchange sorbents. Desorption by heating is not possible for organic resin sorbents because of their thermal sensitivity to bond cleavage and oxidation. Thermal desorption is too risky for use with flammable sorbates. Air stream desorption is a possibility. Eluent desorption seems to have some potential, as does vacuum desorption with recovery of sorbate for disposal. Use of XAD-type and similar macroreticular resins has a good overall feasibility for present day use, and most likely for future use in amelioration of hazardous chemical spills on water.

Polypropylene Fibers – Polypropylene is essentially a linear hydrocarbon polymer that is inherently oleophilic and thus an adsorbent for covalent liquids or solutes. But it can be fabricated in forms that make it an absorbent as well. Thus, it is a sorbent for benzene, kerosene, naphtha:solvent, aldrin, dimethylsulfoxide and epichlorohydrin, of somewhat less versatility in range than polyurethane foam. It cannot be exposed to high solvency liquids because of lack of crosslinking desolubilization (and increase of chemical resistance). It is also less versatile in application than polyurethanes. The means employed for collecting and reusing polyurethane should apply to polypropylene.

Polypropylene may be procured from many firms in a variety of forms and fibers. It can be procured in a variety of molecular weights and distributions. Polypropylenes of higher weight-average molecular weight, isotacticity and crystallinity have better solvent and chemical resistance. Such improved resistance widens their applicability, but it may make formulation or fabrication more difficult or most costly for certain methods of application. Two general suppliers of polypropylene are the Dow Chemical Co. and Phillips Petroleum Co., Plastics Division.

As is the case with other polymers, by-product polypropylene and recycled polypropylene are commonly cheaper than new polymer of the same type and grade. Correspondingly, polypropylene tailored to specifications or delivered in unusual form is often much more expensive. Waste polypropylene is sometimes usable with little loss of control performance capability. Overall, especially for application in existing devices, polypropylene is more feasible for present day use than macroreticular resins of higher versatility. It is a tenable alternative for polyurethane or activated carbon for sorption of chlorine, kerosene, naphtha:solvent, aldrin, dimethyl sulfoxide and epichlorohydrin.

Zeolite Molecular Sieves – Zeolites sorb either ions or molecules, and they can sorb them selectively by size as well as by polarity or electrical charge. This wide versatility can fail for zeolites that sorb water strongly, or that can sorb ions or molecules present in natural water better than molecules of spilled hazardous chemicals. Zeolites can act as sorbents, ion exchangers and molecular sieves (sieving occurs when the pore sizes permit ions or molecules smaller than the pores to sorb, but exclude larger ions or molecules) and they can simultaneously act in more than one of these modes. Zeolites are hydrous silicates that may contain various cations. They have a crystalline structure that contains empty channels of atomic, ionic or molecular dimensions. The sizes and admittances of the channels can be adjusted by substitutions of cations of different diameters. Thus, each of a great variety of zeolitic materials can yield sieves of a range of sizes. Some readily available minerals and synthetic silicates have been adapted to large scale processes as diverse as softening of water, refining of petroleum, production of pure heavy chemicals, treatment of polluted plant effluents and of wastewater and sewage. Others have remained expensive small scale chemical reagents. Their uses and capabilities have attracted much research and development effort. They have not been applied to hazardous chemical spill amelioration.

Molecular sieves of Type 5A (pore size of about 5×10^{-8} cm diameter) have been applied to sorb bromine and they might apply to chlorine. These chemicals are sinkers as well as vaporizers and mixers. Their immobilization, where current dredging techniques such as water-bottom excavation or pipeline slurry suction apply, would permit their improved removal, recovery and disposal. Use of regenerable sieves for these ameliorations is expected to provide data and experience for other applications of zeolites and for ameliorations of other sinking chemicals. Zeolites can be used in available impactors and other closed mobile hazardous spill amelioration systems. Where dredging could be applied, methods would have to be developed to float loaded sieves for collection. These methods would be similar to some developed for floating activated carbon, ion-exchange resins and other normally sinking sorbents.

All available synthetic zeolite types can be procured from Union Carbide Corp. Information on Permutit-type and other ion exchanging materials can be procured from Calgon Corp., Water Management Division.

Water sorption competition and desorption of sorbates by water and natural solutes in natural water reduce the effectiveness of molecular sieve zeolites. Their use is feasible only for spill situations in which they are immersed in a separate phase of an immiscible chemical, a floating or a sunken layer. Application methods must not involve prewetting by water, or postsorption desorption by water. So used, molecular sieve zeolites might apply to many floaters and sinkers for immobilization and recovery.

Sorbent Clays – Clays sorb like zeolites except that the sorption channels are interstitial between polysilicate sheets that are forced apart by the sorbate in a swelling action. Sorbent clays have been used for sorption in a variety of large scale separations. The only clay application to amelioration of water spills of hazardous chemicals encountered is that of removal of phenol from effluent by Floridin XXF and Tonsil AC clays.

Clays are sold by Ashland Chemical Co., Industrial Chemicals and Solvents Division, and Georgia Kaolin Co. Costs vary with the order size and the kind and type of clay. Except where Floridin and Tonsil supplies are close to a phenol spill site there seems to be little potential for using sorbent clays of this type for spill amelioration.

Polyolefins, Polyethylenes, Polyisobutylenes, Poly(Methyl Methacrylates) and Poly(Styrene Sulfonates) – No currently available polymeric sorbents can be viewed as more readily available, better tested or better in performance or potential for use against spills of hazardous chemicals on water than polyurethane. Thus, use of polyolefins against chlorine, polyethylenes and polyisobutylenes against aldrin, poly(methyl methacrylates) against sodium alkylbenzenesulfonates and poly(styrene sulfonates) against cadmium chloride, though possible, is not advantageous for amelioration except in conditions where local availability may compensate.

Ion-Exchangers – Ion exchangers, resin or zeolite, have been judged most advantageous for use against ammonia, concentrated sulfuric acid, sodium hydroxide, potassium cyanide, cadmium chloride, oxalic acid, ethylene diamine, sodium alkylbenzenesulfonates and phenol. They are sorbents for the ionic, most polar and soluble solutes, and complement polymeric sorbents for covalent nonmixers. Polyurethane and polypropylene sorbents, macroreticular resins, zeolites and ion-exchange resins cover the sorption spectrum of activated carbon, but in each of the sorption ranges, a particular one or another of these can be more advantageous or more feasible for use in amelioration than carbon.

There are strong and weak acidic and basic anion and cation exchangers and bonding and exchanging neutralizer resins, and a variety of deionization, deacidification and dealkalization layers or mixtures sold in the trade in many ion-exchange resins bead sizes and other forms. Thus, prices are difficult to compare. Regenerative use seems necessary, so regeneration costs, as well as regenerability, have to be considered.

Ion-exchange resins are sold by many manufacturers and distributors. The Dowex resins are sold by Dow Chemical Co. Amberlite resins, which include macroreticular resins as well, are sold by Rohm and Haas, Co. Between them, these two companies cover the range of types of commercially available ion-exchange resins. Water-softening ion-exchangers, some of zeolitic type, like Permutit, are cheaper by the pound and still cheaper in larger quantities. Ion-exchange sorption is not cheap though it is feasible for present use in amelioration. Because of costs, more versatile ion-exchange resins like Amberlite IRA 93 for sulfuric acid, sodium hydroxide and sodium alkylbenzenesulfonates and IRA 900 for cadmium chloride and ethylene diamine are more advantageous and more feasible for ameliorations.

Ranking of High Potential Sorbents: Because of the lack of information concerning the comparative efficiency of the sorbents identified as high potential, a requirement for general feasibility ranking, it is not possible at the present time to make valid comparative rankings. One specific property which is highly important may be evaluated. This property is the breadth of application measured by the number of hazardous chemicals to which the individual sorbent would apply. Some sorbents are highly efficient for only one hazardous chemical, while others although at lesser efficiency will act to ameliorate spills of a very large number of hazardous chemicals. Using a rationale based upon applicability and on availability, a ranking may be made. Such a ranking is given in the following list in which the sorbents are listed in rank of diminishing feasibility.

(1) Activated carbon

(2) Polyurethane foam

(3) Polypropylene fiber, polyolefin, cellulose fiber, Amberlite XAD

(4) Zeolite 5A

(5) Poly(methyl-co-dimethacrylate), Amberlite IRA 93, Amberlite IRA 900, Dowex 50W x 8

(6) Carbonized sawdust, Zeolite F (K-form), clinoptilolite, polyisobutylene fiber, Dowex 1 x 10, Amberlite IRA 402, De Sal process resin, Amberlite IRA 400, poly(styrene-sulfonate), Dowex 50W x 4, Amberlite IR 252, poly(methacrylate) Floridin XXF, Tonsil AC

Chemicals That Float

The information in this section is based on *Feasibility Study of Response Techniques for Discharges of Hazardous Chemicals That Float on Water*, CG-D-56-77 prepared by J.S. Greer of MSA Research Corporation for the U.S. Coast Guard.

Choice of Sorbent: Sorption has been a basic tool in water treatment processes for many years. The objectives of these processes, to reduce contaminant concentrations and remove nuisance vapors, are similar to those of this investigation.

Sorption is a surface process and is controlled by the physical and chemical properties of the sorbing surface. It is usually expressed as a capacity, mass of sorbate per unit mass of sorbent, which is determined for a dilute solution under equilibrium conditions. Since there is no way to completely separate the phases for this investigation, this term indicates both liquid and vapor sorption for a hazardous chemical spill. The value may be used to indicate the relative effectiveness of a sorbent to ameliorate chemical spills.

Normal sorbent capacities are in the range of 10% by weight, which places a severe limit on the efficiency of the sorption technique. This lack of capacity is balanced by their flexibility. With the proper selection of compatibilities, sorbent materials may be used without modifications or serious corrosion with all the floating, hazardous chemicals.

Activated carbons, which would provide the greatest flexibility of response, normally have densities greater than water and present difficulties for treating those chemicals which float. Floating, activated carbons are in the process of development, and will be described in the sections which follow.

The sorption technique, to ameliorate spills of floating, hazardous chemicals, is influenced by five factors:

(1) Physical and chemical attraction

(2) Surface geometry or morphology

(3) Surface area

(4) Contact time

(5) Density ratio of sorbent/sorbate

Physical and chemical attraction are the binding powers of adsorption and chemisorption, with varying degrees of interplay between the two. These binding forces influence both the initial pickup of the hazardous material and its subsequent removal for recycle or disposal.

Surface geometry is a general term used to cover pore conformation, surface activity and surface cleanliness. Correlations between molecular and surface pore geometry indicate the degree of sorption possible. Surface activity and surface cleanliness affect sorption processes insofar as they relate to surface contact. The viscosity of the hazardous chemical also influences this contact as to whether it "wets" the sorbent.

Surface area is a controlling factor in sorption processes since sorption is directly affected by the total area available. The surface area of the sorbent depends upon the surface geometry, but is primarily controlled by the type and amount of material.

Contact time is an indication of the time available for the physical and chemical interactions of the sorption process. Greater contact times permit more flexible response towards those chemicals where interactions are the slowest.

The ratio of sorbent and sorbate densities is a factor affecting sorption as used in spill response techniques. These two densities must be similar to obtain an adequate contact time for the sorption process and must be less than the density of water, if recovery of hazardous floating chemical spills is to be practical.

Several environmental factors also affect the sorption process. Ambient temperature affects both the rate of reaction and sorbent capacity. The pH and salinity of the water affect the chemical and physical interactions and may also influence the sorbent capacity.

Combining normal sorption and sponge-like entrapment offers a means of increasing the efficiency of sorption processes. This combination increases the total capacity of the sorbent to a point where it may hold nearly its own weight of spilled chemical. This has been developed into the "imbiber" beads and other macroreticular resins used as sorbents.

Direct cost and other economic factors have been the criteria used for selecting sorbent materials to ameliorate oil spills, creating favorable circumstances for using straw, seaweed, leaves, corncobs and similar materials. None of these have the capacity or efficiency required to reduce the health and flammability hazards associated with the floating hazardous chemicals.

Activated carbons provide the most versatile approach to the sorption of hazardous chemicals, but the density of carbon is a most serious drawback. Carbons must be treated to retain hydrophobic character throughout the exposure to water while retaining their capacity. Floating activated carbons are being produced in experimental quantities at the present time. Alternative application methods to circumvent the problem of density are the "tea bag container" which will be discussed in detail later in this chapter, or the "Dynactor" manufactured by R.P. Industries, which combines vapor and liquid removal in venturi-scrubber-type equipment. The Dynactor will be discussed in the following chapter, which deals with mobile units.

Molecular sieves offer good capacities, but must be treated the same as carbons to maintain a low density and prohibit sinking. These materials are also more selective sorbents and would require maintaining a larger stockpile to permit responses to the wide variety of hazardous chemicals.

Polyurethane foams provide a combination of physical and chemical properties which are a definite advantage for hazardous chemical spills. Owens-Corning Fiberglas Corporation has developed the "Glasorb" blanket and several manufacturers market systems to utilize small pieces of urethane for continuous recycling sorption processes. Another approach would be to foam the urethane in situ as a container and cover for the spill.

The macroreticular resins, ion-exchangers and various foamed plastics would provide alternative sorbent characteristics which would benefit spill response techniques. Most of these do not have the versatility of activated carbons, but could have greater capacities for individual chemicals and all have lower densities.

The best sorbents would interact with both the liquid and vapor phases of the spilled chemical to reduce the vapor hazards resulting from the spill. The volume of sorbent used would range from one to ten times the volume of the spilled chemical. Calculated from the typical spill characteristics used for this investigation, the maximum amount of sorbent material could be as much as 380 m^3 (1.34 x 10^4 ft^3) spread over an area as large as 3,800 m^2 (40,700 ft^2). The average distribution rate would be approximately 0.1 m^3 sorbent per m^2 of spill (0.33 ft^3/ft^2).

The sorbent material must have a good capacity, buoyancy and compatibility with spill removal equipment. Polyurethane foams, macroreticular resins and ion-exchange resins have the desired characteristics. A combination of buoyancy (to maintain the excess sorbent in the gas phase) and particle geometry (to aid close packing) influence the ability of the sorbents to reduce evaporation. If they become completely saturated or sink beneath the layer of floating, hazardous chemicals, their effectiveness is lost.

Dispersing Methods: Dispersing sorbent materials over the surface of a spill is a difficult procedure interrelated with several complex variables. The amount of sorbent material required to ameliorate a given volume of spilled chemical may be easily determined. The rate of coverage required in real spill situations becomes quite complex, however, when wind, waves and spreading actions combine to change the spill dimensions. This interrelationship of complex variables makes modeling and calculating design specifications almost impossible.

Pneumatic equipment appears to be the best means for dispersing sorbent materials. Hydraulic systems are similar in all respects, but the presence of water disrupts the spill layer (as well as the sorbent cover) permitting vapor losses.

Snow blowers, forage blowers and mulch spreading equipment could be used to deploy the sorbent materials. Capacities are in the range of 1.4 x 10^4 to 4.5 x 10^5 kg/hr (15 to 500 tons/hr). Sizing the equipment or number of units to be used would be predicated upon the characteristics of the sorbent and amount of cover required.

American Snowblast Corporation manufactures a typical snow thrower unit which would be capable of handling sorbent materials. Their model RHS-300-76 is capable of dispersing 2.7 x 10^5 kg/hr (300 tons/hr) of snow. This unit weighs approximately 730 kg (1,600 lb) and occupies approximately 2.1 m^2 (23 ft^2) of deck space. A 37 hp gasoline motor is used to give a range of 3 to 18.3 m (10 to 60 ft).

Kool Brothers Corporation manufactures a typical forage blower unit. Their model KB-1000 is capable of dispersing approximately 1.4 x 10^5 kg/hr (150 tons/hr) by using a 90 hp power takeoff. Electric motors have also been used as a power source to make an independent system. This unit weighs slightly less than 450 kg (1,000 lb) and occupies approximately 3.8 m^2 (41 ft^2) of deck space. Sorbents could be deployed a distance of 15 to 30 m (50 to 100 ft) from the blower.

Finn Equipment Company manufactures an "Eagle" model, typical of mulch spreaders, which could be used to deploy sorbent materials. This unit is capable of handling 1.4 to 10^4 kg/hr (15 tons/hr) with either 110 hp gasoline or 104 hp diesel power. It weighs 2,360 kg (5,200 lb) and would occupy approximately 10 m^2 (109 ft^2) of deck space. Sorbents could be blown up to 26 m (85 ft) from this system in standard form, and over 46 m(150 ft) with optional extensions.

Storage and transfer systems would be additional accessories required for any of the pneumatic systems used for deploying sorbents. Storing up to 380 m^3 (13,400 ft^3) of sorbent materials would require barges or separate cargo holds and either mechanical or pneumatic systems to transfer the sorbent to the disperser unit.

Recommendations: The evaluation of sorbents has shown that effective materials exist, but are quite costly. Reticulated resins are available and work is already in process in developing buoyancy in other effective sorbents such as activated carbon. The cost and logistics of this type of material may legislate against their use. Their applicability may depend on the availability of equipment capable of efficient application with small losses outside of the spill area.

Other sorbents, particularly urethane foams, offer a less efficient medium but also a less costly medium. Urethane chips have been used as a sorbent and technology exists to produce these over a wide range of pore sizes. These materials also would benefit from efficient dispensing equipment. One difficulty exists, however. As has been described as part of the evaluation, as the urethane absorbs the spilled chemical and possibly water, it submerges into the spill and can exaggerate the vaporization rate by essentially increasing the surface area available for evaporative losses.

Proposals have been made to circumvent the difficulty. They would generate the urethane on site, forming a continuous urethane blanket rather than a mat of chipped pieces. These have failed to take into consideration that the urethane must be rigidized before it contacts liquid. Before being fully reacted it can degrade releasing isocyanates which are toxic chemicals of themselves and soluble and reactive in water. To build a blanket there must be a suitable substrate.

The potential of continuous urethane foam blankets warrants some consideration of mechanisms to effect them. One approach is to develop an initial mat of urethane chips with the subsequent overlay of a continuous blanket generated in place. The technology for generating urethane foam continuously in the field is well developed and urethane chips are available. Thus, this program would be minimal. Other approaches may be possible, and a program is recommended that would evaluate mechanisms for developing urethane foam blanket over spill surfaces.

Application of any of these sorbents is dependent on the availability of suitable dispensing equipment. Those which are now available have been derived from other systems. They tend to be inefficient and there are large losses of materials outside of the spill area. Equipment is needed which can dispense the sorbent directly into the spill surface from close proximity to it with a narrow range of discharge.

It appears that booms or bridge units will be the type of units to be evolved, systems which can stretch out over the spill or completely span it to discharge a granular or powdered material from multiple ports directly down onto the spill surface. They will need propulsion to move them parallel to the spill flow direction. The propulsion and the sizes involved will pose the basic problem in development. Technology for the pneumatic, mechanical, or hydraulic transport of bulk materials is available.

Based upon costs and development times for similar equipment for land spills and other equipment for water use the evolution of prototype equipment including field tests could take up to three years and cost in excess of $500,000. This, plus the cost of sorbents, provides a costly spill control technique. A decision in this area needs to weigh the cost of clean-up against the potential damage to the environment.

The type of dispensing equipment needed for the sorbents would also be needed with gels, solid CO_2 and packed particles. If those procedures are found to have practical application to spill control, the cost of developing dispensing equipment will be further justified.

Chemicals That Sink

> The information in this section is based on *A Feasibility Study of Response Techniques for Discharges of Hazardous Chemicals That Sink,* CG-D-56-78 prepared by T.D. Hand, A.W. Ford, P.G. Malone, D.W. Thompson and R.B. Mercer of the U.S. Army Engineer Waterways Experiment Station for the U.S. Coast Guard.

Application of sorbents to spills of sinking chemicals poses some problems. Since most of the sorbents available were originally intended for use on oil spills, many of them float, making their use on the bottom impossible unless special delivery systems are designed. With regard to oil spills, sorbents are most useful in cleaning up small spills and spills in obstructed and difficult-to-reach areas. In the context of spills of sinking hazardous materials, the use of sorbents will likely be most efficient in nonnavigable waters, difficult-to-reach port and harbor areas, and along or in close proximity to riverbanks. Application to confined spills would prevent problems with sorbents degrading nearby clean water and land areas. Use in open waters will be expensive and, perhaps, unwieldy.

Delivery systems for sorbents used on oil spills will provide some insight as to their application for sinking hazardous spills. Sorbents are available in boom form, being designed to contain the spill as well as to sorb it. Sorbents also come in floating pads of varying sizes easily reuseable by wringing out. Strips and "chips" of synthetic sorbents have been arranged into "pillow" form or can be applied directly to the spill. Organic sorbents (e.g., straw, ground corncobs, peat moss, and sawdust) are particulate in nature. These, plus chipped foam, may be blown into the surface of a spill using a device similar to a hayblower used in highway landscaping work.

The use of sorbents in pads, pillows, and booms in control of sinking spills appears to have some promise. Such devices could be weighted, if necessary, to permit their deployment on the bottom. Buoys could be attached to lines to mark them and to permit easy retrieval. Best use would occur in relatively shallow waters, where deployment and retrieval would be less of a problem, and accurate deployment to the location of the spill would be simplified. Use of a boom would have the added advantage of acting to confine the spill. Some difficulty might be expected in getting other than the outer layer of confined sorbents to perform effectively using the above arrangment, especially if powdered or fine granular forms are used. High surface area, low volume containment devices would prove advantageous.

Shuckrow, et al (21) have suggested injection of slurries of sorbents (e.g., activated carbon) below the surface of the water. It was noted that air-deployable slurry and injection units would be most valuable, as these could be transported to the site of a spill quickly enough to be of some use. Such units could be mounted on workboats used in combatting the spill. Activated carbon is available with densities sufficiently greater than water that a slurry made with it would remain on the bottom. Use of such mechanical injection methods would allow the activated carbon (or other sinking sorbent) to be used in medium as well as small spills and also in deeper and more open waters of ports and harbors. Due to dispersion problems, such a technique might not be successful in high current areas, where the sorbent could spread into virgin areas. Certain confinement techniques (e.g., submerged dikes, trenches or booms, or the hazardous material spill barrier) could be used in conjunction with such a method to aid in controlling dispersion. Removal of the sorbent slurry could be accomplished using a handheld vacuum pump, or in the case of larger spills, a dredge.

Suggs, et al (22) suggested the use of coated cotton meshwork to remove inorganic mercury from contaminated water and sediments. Such a technique would appear to have direct application to the present study especially since many of the sinking chemicals (e.g., heavy metals and pesticides) will bind themselves strongly to sediments. A high surface area, sulfur-coated cotton mesh was successfully devised as a "getter" system for mercury. Suggs, et al (22) reported that, of the possible mesh materials tested (including plastic and metallic materials as well), cotton accepted coatings the best. Coating was accomplished by dipping the mesh into a solution of sulfur in volatile solvents. Such a process appears to have potential in sorbent application to sinking chemicals in the case of small and medium spills. The meshwork net could have the added advantage of erosion control, preventing or lessening resuspension of spilled materials. The cotton net could be incinerated after retrieval from the spill site. The potential of the various sorbents as coating materials needs to be evaluated.

Another potential technique for sorbent deployment and recovery discussed in the literature is magnetized activated carbon, a technique presently being developed with EPA support for use on Kepone amelioration in the James River, Virginia. Details of this method and results of research are unavailable at the time of this writing. However, it has been reported that the method, which includes tilling the special carbon into sediments and subsequent removal, "shows promise" (23).

The following subsections present brief descriptions of selected high potential sorbents. Much of this discussion has been summarized from information found in Bauer, et al (24). This has been supplemented through discussions with selected manufacturers of the sorbents. Sorbents to be discussed include activated carbon, polymeric sorbents, macroreticular resins and molecular sieves.

Activated Carbon: Bauer, et al (24) suggests that activated carbon may be useful in absorbing virtually all sinking chemicals, the exceptions being the sinking solids barium carbonate, lead arsenate, phosphorus, sulfur, and liquid mercury. Activated carbon has also been reported effective on a large number of other chemicals, including o-cresol, chloroform, DDT, and toxaphene (25), (26).

Activated carbon is available in a number of forms, including powdered, granular, rod, sheet, and others. Techniques under consideration here (slurry or bagged application) use either the powdered or granular forms. Carbons of various densities (including heavier than water) and various sorption properties are also available, making the task of optimizing the use of activated carbon to the spill of a particular hazardous material difficult. Due to its widespread use throughout the country, common forms of activated carbon are usually available on one-day notice or less.

In summary, activated carbon can best be applied to chemical spills in the form of booms, pillows, or other bags, or within confined or curtained-off spill areas. Recovery will be enhanced if such procedures are followed. To be effective, application should surround, and/or blanket, the spill areas. Negative impacts of carbon on biota surviving a spill would stem from burial, if a slurry or blanket technique was used. Activated carbon also has the disadvantage of turning water black due to carbon fines. Overall, activated carbon must be rated as having a high potential as a sorbent for use in this type of response technique.

Polymer Foams and Fibers: A wide range of polymeric sorbents has been developed in conjunction with oil spill recovery work. Among these have been polyurethane foams and polyethylene and polypropylene fibers. These have been manufactured in a number of forms, including belts, pillows, pads, and roll blankets. It is expected that, because of the oleophilic properties of these materials, they will likely find at least some applicability to spills of liquid organics as well as spills of selected organic solids. Bauer, et al (24) report that "no currently available polymeric sorbents can be viewed as more readily available, better tested, or better in performance or potential for use against spills of hazardous chemicals on water than polyurethane." Other types of foams or fibers may work, though local availability will play a large part in their potential for use in a response technique.

Polyurethane foams have reportedly been formed on site which could prove advantageous. Their use is not considered to contaminate water, though it may pose a fire hazard to handlers. Many of these polymeric materials are available from oil spill contractors in ready-to-use form, and their availability through this network makes the transfer of material and technology to the control of hazardous materials a simpler task than it otherwise would be. As a result, polyurethane foams must be considered as having a moderate to high potential as an adsorbent in a spill response.

Macroreticular Resins: Bauer, et al (24) characterized macroreticular resins as "polymers with a spongy or reticulated structure that is a small-scale replication of the macroporous structure of polyurethane foam or a large-scale replication of the molecular porosity of activated carbon." It is interesting to note that such resins may easily be given ion-exchange characteristics, so that the distinction between sorbent and ion-exchange resin becomes somewhat unclear.

It was noted (24) that these resins are capable of sorbing dissolved aldrin and phenol and are considered "more versatile" than polyurethane foam, but less so than activated carbon. The discussion indicated that they have not been used on a large scale, and thus there is little information on deployment techniques or impacts on biota. It is expected, however, that deployment techniques and likely use scenarios would be similar to those noted for activated carbon. Desorption would likely pose many of the same problems as it would with activated carbon, again due to the nature of the sorbed material.

Due to the present low level use of such resins, availability on short notice is expected to be a problem. These resins should be considered for use on an "as available" basis only. The commercial source for such resins is the Rohm and Haas Company (24).

Zeolite Sieves: Zeolites are traditionally used in water treatment, chiefly for softening. They sorb ions or molecules and can be made size specific. Once again, the distinction between sorbent and ion-exchange becomes difficult. Bauer, et al (24) recognized the value of a type 5A molecular sieve (5×10^{-8} cm pore size) for the sorption of bromine. No other application to sinking hazardous chemicals was noted, and no application to actual hazardous material spills was noted in any case. Technology related to these zeolite sieves as they apply to sinking chemicals must therefore be considered undeveloped.

It was noted that zeolites have a tendency to sorb water as well as the intended chemicals, and desorption by water was also considered a drawback. Such sieves would have the highest potential in cases where they could be completely immersed in the sinking chemical basically limiting their use to immiscible liquids. Special deployment systems, not requiring or allowing prewetting, would have to be designed. None were suggested.

Such agents as zeolite sieves are not expected to be generally available, and will probably be expensive. Lack of technology available for deployment further reduces the practicality of their use. Use would have to be on an "as available" basis. In any event, their use would likely be confined to small spills in confined or other hard-to-reach areas.

Ion-Exchange: Several different ion-exchange resins (24) have been used to remove carbolic oil (phenol) from wastewater and drinking water. There are ion-exchange materials available for ionic, most polar, and soluble chemicals. Hazardous chemicals that sink, with their inherent low solubilities, could preclude the use of ion-exchange in many cases. Pilie, et al (27) have investigated the use of ion-exchange resins as a substitute for activated carbon in the amelioration of heavy metal spills in water.

Given the wide range of ion-exchange resins available and because each of them may be applicable to only a small range of hazardous chemicals, it might be worthwhile to develop a "universal exchange resin," patterned after the universal gelling agent. Several resins, anionic and cationic, could be combined to produce this agent. Delivery could be accomplished in a manner similar to that for activated carbon sorbent. Confinement of the exchanger in readily recoverable booms, pillows, mats, or similar devices would promote easy retrieval at some loss of efficiency.

Ion-exchange resins are uniformly available on a nationwide basis, due in large part to the wide range of use in water purification and treatment. On a cost and flexibility basis, ion-exchange resins must be considered inferior to activated carbon. Development of a "universal exchange agent" might increase the flexibility of exchange resins, but would do nothing to improve the cost differential. Impaired efficiency in salt water would correspondingly reduce its usefulness in that environment.

Universal Sorbent for Land Spills

The material in this section if taken from a paper presented at Conference 1978 by R.E. Temple, W.T. Gooding and P.F. Woerner of Diamond Shamrock Corporation and G.F. Bennett of the University of Toledo. The title of the paper was "A New Universal Sorbent for Hazardous Spills."

In the manufacture, handling, transportation and use of hazardous materials, it is inevitable that there will be spills. Once a spill occurs there are four basic steps that must be taken to keep the environmental damage to a minimum. These are:

(1) stop the discharge;
(2) contain the spill;
(3) collect the spilled material for safe removal; and
(4) safe disposal.

In the *Proceedings of the National Conference on Control of Hazardous Material Spills* 1972, 1974 and 1976, there are many papers dealing with methods for stopping discharges and for containing spilled liquids. There are accounts of collecting spills mechanically, of removing insoluble spilled material floating on water or that has sunk, and removal by activated carbon adsorption of toxic materials dissolved in water.

As for amelioration of on-land spills, it has long been a popular practice (particularly by fire departments) to wash the spill into drainage ditches or sewers to remove an immediate public danger. This will no longer be an acceptable practice, but the alternative technologies for collecting and removing on-land spilled liquids are limited.

In the above paper the authors present information on a new, broad spectrum sorbent. This material has the potential to render spilled hazardous liquids easily and safely collected and removed. Because of the inert nature of this material, and its low density, it can be readily stored in plants or carried on fire trucks and other emergency response vehicles. It will also make safe disposal less of a problem for many hazardous liquids.

Description of the Product: This universal sorbent material (USM) consists of all off-white free-flowing granules with a loose bulk density of about 2 lb/ft^3. It is an amorphous silicate glass foam consisting of spheroid shaped particles with numerous cells. The particles range from 8- to 200-mesh. This product is very safe to handle. Studies show the material to have very low toxicity. Results of toxicity tests (performed by Industrial Bio-Test Laboratories), are shown in Table 4.1.

Table 4.1: Toxicity Properties of USM

Acute inhalation toxicity	$LC_{50} > 2.32$ mg/l air
Acute oral toxicity	$LD_{50} > 15,380$ mg/kg
Eye irritation	Mildly irritating (19.7/110.0)
Primary skin irritation	Slightly irritating (0.7/8.0)
Skin Sensitization	Not a skin sensitizer

Source: Conference 1978

Adsorption Characteristics: Because there are presently no accepted standards for measuring capacity for sorbent materials, it was necessary to invent a meaningful test procedure. Because of the wide variety in the properties of the liquids to be tested, it is nearly impossible to derive a universal procedure that would give an absolute measure of its performance. Therefore, using the principle of parsimony, a simple comparative test was used.

The procedure consisted of immersing a weighted quantity of sorbent material in the test liquid for a given time, then removing the saturated sorbent from the liquid allowing it to drain for a given time and determining the increase in weight. The increase in weight was converted to volume and the results reported as gallons of liquid adsorbed per pound of sorbent.

Through trial and error it was determined that five minutes of soaking was sufficient for saturation except when viscous liquids such as No. 6 fuel oil were used. Again, for other than viscous liquids, a drain time of 20 minutes was found to be sufficient.

Liquids representing several classes of regularly transported hazardous chemicals were tested using various sorbent materials. The results of these tests are shown in Table 4.2. As the data in the table shows, USM is an excellent adsorbent for many classes of compounds including acids, alkalis, alcohols, aldehydes, arsenates, ketones, petroleum products, chlorinated solvents, etc. One pound of USM sorbed amounts ranging from 0.77 gallon of perchlorethylene to 1.82 gallons of phosphoric acid. With nearly all liquids tested, USM sorbed more than 10 times the amount of liquid absorbed by expanded clays. Although the list of chemicals tested is far from complete, this work is continuing and will be reported in the future.

Table 4.2: Sorbent Properties of USM

Liquid	Gallon Sorbed Per Pound Sorbent USM	Expanded Clay
Sulfuric acid 96%	1.25	0.11
Nitric acid 70%	1.69	0.19
Phosphoric acid 85%	1.82	0.13
Methanol	0.99	0.08
Phenol 83%	1.74	0.13
Ferric chloride 40% solution	1.66	0.12
Daconate 6*	1.64	0.11
Perchlorethylene	0.77	0.10
Sodium hydroxide 50%	1.70	0.15
Formaldehyde 37%	1.11	0.09
Gasoline	0.95	0.08
No. 2 fuel oil	1.10	0.09
Chlorosulfuric acid	1.28	0.10
Acrylonitrile	1.51	0.10
Benzene	1.44	0.10
BTX (Benzene/Toluene/Xylene)	1.54	0.11

*Diamond Shamrock TM monosodium acid methane arsenate

Source: Conference 1978

Advantages and Disadvantages: In working with USM in the laboratory, it became obvious that it had many advantages and some disadvantages. One very important feature is its ability to wick liquids. When poured on top of a liquid, the liquid is wicked through the USM and adsorbed. Once it is adsorbed, the material becomes dry and easily handled.

The very light weight of the USM is both an advantage and disadvantage. It gives the material its ability to adsorb large quantities of liquid, but it also makes it very difficult to handle under windy conditions. For this reason it became necessary to develop unique packaging so that the material could be handled in the field no matter what the weather conditions. Developmental work is underway to solve this problem.

Another observed disadvantage was that once the USM has contacted water, it no longer has good adsorbent properties for petroleum products. For this reason the potential for adsorbing oil spills on water is very limited. One area of great interest is the ability of USM to promote combustion of crude oil on water. For this purpose it would be necessary to apply the material directly on to the oil spill. Once this is done the wicking ability of the material allows it to promote combustion of crude oil.

Applications: The two areas of most interest are those spills which occur in plants and those that occur with overland transportation. In-plant spills occur around manufacturing equipment and where loading and unloading of liquids takes place. In these areas pillows of USM can be spread to soak up any spilled material. Once they are saturated, these pillows can easily be removed to disposal areas.

Once spills occur on streets, highways or railroad right-of-ways and the material has been contained, it is very important to remove as much of the material by mechanical means as possible. The remaining isolated puddles of material too shallow to pump can easily be soaked up with USM, leaving it in a dry, easily handled form which can be safely removed to disposal areas.

Disposal: There are three areas of disposal where USM can be very advantageous. With products that can be landfilled, USM makes them easily handled, safely removed to the landfill area and there is no hazard involved with the sorbent material.

With products that can be incinerated, USM because of its very low density, adds little ash to the incinerator and can be easily handled by most combustion equipment. Products that cannot be safely landfilled without being fixed can be adsorbed on USM and then solidified in a system of Portland cement and sodium silicate to yield a nonleaching, safely disposed material.

Summary: Information about the properties of a sorbent material for hazardous liquids has been presented. Although the data are limited, they do indicate that this product will adsorb a wide variety of hazardous liquids. The material is safe to handle and does not present any environmental hazard in itself. Although there are some mechanical problems with handling the material, they do not appear to be insoluble. Work is continuing in both areas of packaging and handling and testing sorbency on many more hazardous liquids.

ACTIVATED CARBON AND ION-EXCHANGE RESINS

Activated Carbon Columns and Tanks

> The information in this and the following section is based on *Manual for the Control of Hazardous Material Spills, Volume 1: Spill Assessment and Water Treatment Techniques*, EPA-600/2-77-227 prepared by K.R. Huibregtse, R.C. Scholz, R.E. Wullschleger, J.H. Moser, E.R. Bollinger and C.A. Hansen of Envirex for the Environmental Protection Agency.

Activated carbon adsorption is a physical phenomenon which removes organic and some inorganic chemicals from water. These chemicals are physically adsorbed on the large surface area of the carbon (typically 500 to 1,000 m^2/g). The activated carbon can be produced from various cellulosic materials including wood, coal, peat, lignin, etc. These are prepared using dehydration and carbonization, followed by activation to enlarge the pore openings, which increases the surface area and therefore increases the adsorptive capacity.

The adsorption process is dependent on the nature of the material being adsorbed, the solution and the carbon used for adsorption. Critical factors include molecular size and polarity, type of carbon, pH of the solution, carbon contact time and solubility of the contaminant. The adsorption rate increases with increasing temperature and decreasing concentrations. In general, concentrations greater than 1,000 mg/l of soluble contaminant require excessive detention times and produce large amounts of spent carbon.

Once the capacity of the carbon has been reached, the carbon must be replaced and the spent carbon disposed of or regenerated for reuse. Regeneration can be done using various physical and chemical techniques. However, thermal regeneration is the most common method. This process requires high temperatures and a controlled atmosphere and is therefore unsuited for field implementation unless a preconstructed mobile system is available. Instead, carbon should be removed and hauled to an established site for regeneration or incineration.

Offstream treatment is typically done using either powdered or granular carbon. Usually offstream treatment is performed in column tanks which provide efficient use of the carbon in the system. Carbon columns are similar to filters in many ways:

(1) efficiency of the bed is dependent on good flow distribution which will provide uniform contact time for the entire fluid stream;

(2) an underdrain system is necessary to prevent the carbon from exiting with the effluent water and to distribute backwash water; and

(3) initial backwash is required to remove fines and air pockets, as well as to stratify the bed.

In other ways, carbon columns are distinct from filter operations:

(1) Termination of the cycle is established by "breakthrough" which indicates that the adsorptive capacity of the bed has

been reached. Once spent, the carbon must be transported out of the bed and replaced with fresh media.

(2) For efficient adsorption, the carbon must be "wetted" prior to use. This process may require up to 24 hours at room temperature with the carbon submerged in clean water (or less time at higher temperatures). Therefore, a source of clean water must be available on site for use in wetting the carbon prior to startup of the system.

(3) Use of the carbon column as a filter causes inefficient use of the adsorption capabilities. Therefore, clarification processes including dual media filtration are necessary pretreatment steps prior to carbon adsorption.

(4) Carbon columns possess more versatility than filters and can be operated in either downflow or upflow modes. Suspended solids are not removed during upflow operation due to bed expansion and extra contact time is generally necessary for this operation because of the expanded bed condition.

When a carbon adsorption process is constructed in the field, the first priority is the ordering of carbon, which may require a 24 to 48 hour lead time and an additional 24 hours to wet prior to use.

Once the carbon is placed in the column, then the actual carbon requirement of the system must be tested. Since it is recommended that the carbon columns be run in series with an equalization tank between, samples can be taken periodically from the effluent lines of the columns, composited, and sent to the laboratory for analysis. These analyses will indicate when the first carbon column has broken through and future carbon changes can be based on that time period or additional sampling. The second column will allow the operation to safely continue in the interim until the samples can be analyzed.

When using powdered carbon during offstream operation, carbon can be injected into a tank, mixed via hydraulic, air or mechanical means and then collected prior to discharge of the wastewater. The carbon is not used as efficiently with this method, but the same weight of carbon should be initially ordered. As the carbon is spent, the data can be extrapolated and the additional amount ordered.

Activated Carbon Column Troubleshooting: Excessive Clean Bed Head – If backwashing does not permit return to the clean bed head, surface blinding may be occurring which can be alleviated by removing and replacing the top 5 to 12 cm (2 to 5 in) of carbon from the bed. If suspended material is clogging the column, steps should be implemented to improve the feed water quality.

Poor Effluent Quality Suggests –

(a) too high a flow rate;

(b) channeling of the bed (uneven distribution of flow);

(c) excessive mixing of media; and

(d) exhausted activated carbon.

Ion-Exchange Columns and Tanks

Ion exchange is a process in which ions held by electrostatic forces to functional groups on the surface of a solid are exchanged for ions of a different species in solution. This process takes place on a resin which is usually made of a synthetic material. The resin contains a variable number of functional groups which establish both the capacity of the resin and the type of group removed. Various kinds of resins are available including weakly and strongly acidic cationic exchangers and weakly and strongly basic anion exchangers. The ions are exchanged until the resin is exhausted and then the resin is regenerated with a concentrated solution of ions flowing in a reverse direction. Various specific reactions occur, but generally the reaction is as follows:

$$RI_x + I_c \rightleftarrows RI_c + I_x$$

where R = resin, I_x = exchangeable ion, and I_c = contaminating ion.

The ion exchange process is dependent on the type of resin involved, the specificity of the resin and the general ion content of the wastewater. Capacities of resins also vary with the manufacturer of the resin, the distribution of flow and concentration of contaminant.

The amount of resin required must be established by chemical tests done on wastewater for ion content. The best type of resin to use is established mainly by the specific contaminant to be removed, the amount of wastewater involved and the other ionic demand on the resin. A resin manufacturer must be contacted to allow the correct resin to be chosen. The following information must be given to the manufacturer:

(1) name of compound to be removed,
(2) concentration of contaminant,
(3) amount of wastewater to be treated, and
(4) chemical analysis of ions.

The resin manufacturer can then specify the amount and type of resin required to remove the entire contaminant from the waterway. Unless absolutely necessary, the resin will not be regenerated on site; once the capacity is depleted, the resin will be replaced, hauled away for regeneration and either returned for reuse on site or sent to storage.

Two types of off-stream treatment are available: (1) column exchange; and (2) distribution of uncontained media into a tank. Column treatment is more common and more efficient. There are many similarities between ion-exchange and carbon columns and some similarities to filters. The three systems have the following features in common:

(1) efficiency of the bed is dependent on good flow distribution which will provide uniform contact time for the entire fluid stream;
(2) an underdrain system is necessary to prevent the media from exiting with the effluent water and to distribute backwash water; and
(3) initial backwash is required to remove fines and air pockets, as well as to stratify the bed.

The carbon and ion exchange systems are similar in the following ways:

(1) Termination of the cycle is established by "breakthrough" which indicates that the exchange capacity of the bed has been spent. This procedure is indicated by an increase in the concentration of the contaminant to be removed or by a change in pH (when strongly anionic or cationic resins are involved).

(2) Use of the column as a filter causes inefficient use of the exchange capabilities. Therefore clarification processes including dual media filtration are necessary pretreatment steps.

(3) Backwashing of these systems can be done; however, it is not recommended and the necessity of frequent backwashing indicates the malfunction of upstream processes.

Ion-exchange does have a high potential for fouling since the size of the resin particles is approximately the same as that of filter sand.

Ion-Exchange Column Troubleshooting: Excessive Clean Bed Head – If backwashing does not permit return to clean bed head, 5 to 12 cm (2 to 5 in) media may be removed and replaced. This procedure will only be necessary if feed water quality is high in suspended material which is causing surface blinding. If suspended material is clogging the column, steps should be implemented to improve the feed water quality.

Poor Effluent Quality Suggests –

(a) too high a flow rate (insufficient detention time);

(b) channeling of the bed;

(c) exhausted ion-exchange resin; and

(d) improper choice of resin—wrong type of anion/cation removal.

Carbon "Teabags" and Other Packaging Concepts

The information in this and the following section is based on *Methods to Treat, Control, and Monitor Spilled Hazardous Materials,* EPA-670/2-75-042 prepared by R.J. Pilie, R.E. Baier, R.C. Ziegler, R.P. Leonard, J.G. Michalovic, S.L. Peck and D.H. Bock of Calspan Corporation for the Environmental Protection Agency.

Early experiments demonstrated that the total capacity of activated carbon for many classes of pollutants and the rate of removal of pollutants from water were both consistent with anticipated needs for cleaning spill-contaminated water. Attention was, therefore, directed toward the development of practical methods for the dispersal of activated carbon in water and the removal of the carbon from the stream or lake after adsorption was complete.

Powdered Carbon: In the laboratory studies, it was shown that most efficient adsorption is accomplished by powdered carbon in polluted water that is sufficiently turbulent to keep the powder in suspension. However, 1,000 ppm of powdered carbon reduced the transparent depth of water to less than 2 mm. One percent of that concentration still produced a totally unacceptable turbidity. Addition of flocculents created a thick, unacceptable bottom sludge, and was never more than 99% efficient, leaving turbid water. Bioassay experiments with sludge formed of carbon previously used to remove phenol from water, showed some toxicity to fish (50% fatal to fathead minnows in 30 days).

Froth flotation techniques in which compressed air is bubbled through the water in order to float carbon particles to the surface were found to be less efficient than flocculation. 50% recovery of carbon was achieved only with the addition of surface active agents at concentrations which would pose a secondary pollution problem, sometimes equal to the first.

It was concluded, therefore, that free powdered carbon was not suitable for use in natural water, except as a last resort or where the pollutant laden powder would settle to the bottom where it could be located and removed.

The Carbon Teabag: While granular carbon adsorbs pollutants less rapidly than powdered carbon, its coarse grain size permits greater flexibility in the design of dispersal and retrieval techniques. Among the first concepts considered was a carbon filled porous cloth "teabag". The initial design involved a vertically suspended bag with flotation at its top and well ventilated pockets of activated carbon extending to the desired depth in the water.

Small-scale experiments were followed by a series of tests of the teabag concept under conditions that more nearly simulate natural environments expected in the field. One set of experiments, channel tests, was performed in a 1,000 liter, racetrack shaped channel 28' long, 8' wide and having a 1' x 1' cross-section. Stream velocity through the channel could be controlled between zero and 1.5 ft/sec and bottom roughness could be adjusted by adjusting the distribution of stones placed on the channel bottom. Several teabag configurations were tested with three different types of carbon. In all cases, the carbon-to-pollutant ratio was 10:1. The bags were permitted to move freely with the channel water. Ventilation of water through the bags was produced only by the shear and turbulence in the channel.

Comparison of channel tests 1 and 2 illustrates the importance of bag design and packing. In test 1, the bags used were constructed with 1-inch wide vertical pockets fully packed with carbon. Without changing other experimental conditions, test 2 was performed using bags with 1-inch wide horizontal pockets that were half filled with the same carbon. The loose packing permitted less restricted flow of polluted water through the charcoal, which was free to move inside the pockets. The horizontal configuration prevented packing of the charcoal by gravity. Obviously, the test 2 configuration is superior.

With the importance of teabag ventilation thoroughly established, some concern developed as to the potential value of the concept for treatment of spills that occur in still water such as ponds, lakes, or slow moving streams. A final series of experiments was therefore performed in an outdoor swimming pool.

It is apparent from the "still water" tests that even a small amount of agitation is effective in increasing pollution removal rate. While totally inadequate for effective treatment of spills, the adsorption rate during the first eight hours of the experiment, a period of moderate breeze, was significantly greater than that of the next eight hours during the nocturnal calm. A truly significant increase in adsorption rate resulted from the artificial generation of 2 cm high waves, 40 cm long in the second pool experiment. Similar results were obtained with small artificial waves in the channel under conditions of zero flow. These tests also demonstrated the need for mooring the bags to prevent aggregation and drift due to wave action. A nylon fishing line with fish hooks attached at 1-foot intervals was sufficient to maintain proper separation and position.

This suggests that natural wave action in a lake should be adequate for ventilation of carbon filled teabags. Under very calm water conditions, the waves produced in the wake of a few small outboard motor boats could be used to provide sufficient agitation to permit removal of a significant fraction of spilled material from solution in a few hours.

A teabag design of general applicability would consist of an arbitrarily long bag (20 to 100 feet) consisting only of carbon-containing compartments similar to those illustrated in Figure 4.1. Separate flotation units which could be readily attached to the bag at any desired location could be provided to permit adjustment of the depth to which the bag would hang to accommodate the depth of the water body and the nature of the spill.

For example, in a 5-foot deep stream, flotation units could be attached every 8 feet to permit the bag to hang to the 4-foot level, and just clear the bottom. If the spilled material were more dense than water and likely to be dissolving from a pool on the stream bottom, the flotation units could be attached at, for example, 15-foot intervals, so that most of the carbon would remain at lower levels where concentration is greatest. Attachment of the units at 2-foot intervals could accommodate spills in shallow water or of materials that are less dense than water and going into solution from floating pools of the concentrated pollutant.

While it was impossible to test the effectiveness of this concept in treating a spill or simulated spilled hazardous material, two 10-foot long bags were constructed to evaluate the mechanics of the idea. Tests in the Buffalo harbor indicated that the desired flexibility can be achieved and that the amount of agitation desired from wave action, either natural or artificially produced in the boat wake, would produce the required ventilation.

Carbon Fibers: Recently, several different types of activated carbon filters have become available in small experimental quantities from the Carborundum Company. Beaker tests were performed with a variety of these materials to determine their potential for removal of pollutants from water. One variety which resembles loosely packed, fine grain steel wool in appearance shows excellent potential. Test data for this material were compared with channel tests data obtained with freely dispersed powdered carbon. Within experimental accuracy the carbon fibers are as effective as powdered carbon, both in removal capacity for the test pollutant and in the rate of removal of the pollutant from water.

The wool-like fibers in the tested samples have a density very nearly equal to the density of water. The structural strength of the strongest samples was adequate to permit compression of the matrix for storage and shipment. When placed in water, the matrix expanded almost to its initial configuration and floated with the uppermost fibers at the water surface. Ventilation of the loosely packed fibers, therefore, was excellent. Agitation of this sample did not produce significant fiber fracture. Besides normal stirring, the wool was repeatedly lifted from the beaker with a spatula and replaced without excessive fragmentation. In the field, the material could be readily removed from water using a coarse-mesh net or perhaps a grappling hook.

This steel-wool-like material, therefore, appears to have many of the properties of activated carbon desirable for field use. Other types of carbon fibers tested were either too frangible for consideration or less effective chemically. Since the samples were produced by manual processing on an experimental basis, the variability is quite understandable. Further experimentation and development should be encouraged.

Figure 4.1: Teabag Configuration

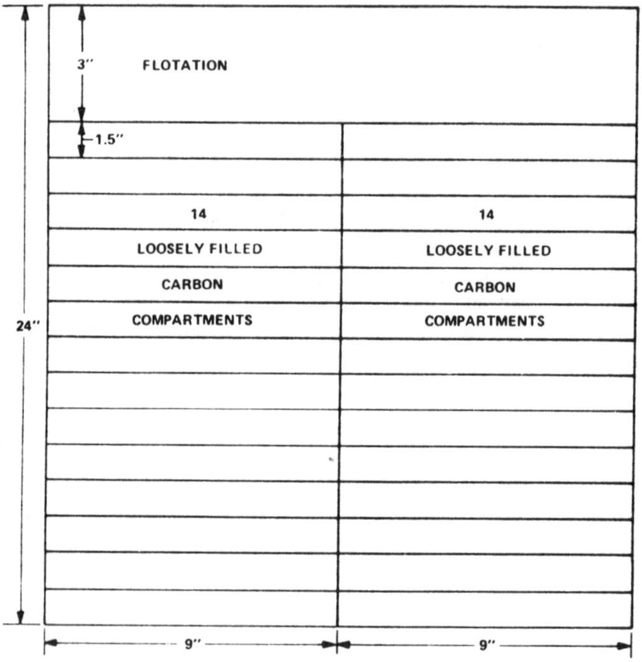

Source: EPA-670/2-75-042

Other Packaging Concepts for Activated Carbon: Two additional concepts were investigated for packaging activated carbon in such a way that it might be retrieved from natural water after treatment was completed. With the aid of the Carborundum Company, activated carbon was bonded to waterproof paper using techniques designed for sandpaper manufacture. Laboratory tests showed that the product was in general less efficient than the carbon-filled teabag, primarily because the bonding agent occupied some of the available adsorption sites and restricted access of the pollutant to approximately half of the exterior surface of each carbon grain. Similar results were achieved in in-house attempts to coat reticulated foam with granular carbon.

Logistics: This analysis was intended only to determine whether the use of the concept of properly packaged activated carbon showed enough promise of practicability to warrant further development and testing, keeping in mind that the potential advantages of using the newly developed activated carbon fibers suggest that a vastly superior treatment may become available in the near future.

The analysis was based on the assumption that 20,000 lb of properly packaged carbon would be stored at each of 200 locations throughout the United States, and that 100,000 lb of additional carbon would be stored at each of five Air Force Bases or commercial airports, ready for shipment to any spot in the nation. The purpose of this distribution would be to provide adequate supplies of carbon at "local" warehouses to permit rapid treatment of small spills; i.e., spills of up to 2,000 lb of material, in a time frame which is consistent with the rate at which the concentration of spilled material is reduced to levels that are too low for effective treatment.

The local supplies would also be available for initiation of the treatment of large spills at about the maximum rate which could be delivered in a very rapid response situation. Even if the 200 caches of carbon were uniformly distributed throughout the contiguous states, there would be 100,000 lb of material within 120 miles of any spill. The time required to make the decision to treat with carbon, design a specific treatment procedure, and implement the treatment procedure is estimated to be comparable to, or greater than, the time required to transport the carbon 120 miles over land to the spill site (assuming 1 hour for loading, 1 hour for transport of the closest carbon, and 3 hours for transport of the remaining 80,000 lb from nearby caches).

It was assumed that the first 50 tons of carbon to arrive at a large spill would be dispersed from ten small boats supplied by local agencies (or by local sportsmen to protect their waters). An 18-foot runabout could carry 1,000 lb of material safely with a crew of three for dispensing the carbon. Assuming 20 minutes for each trip, the 10 boats could disperse the carbon available through ground transportation within four hours after shipment is initiated. The first air shipment of carbon could be available at most sites within that period.

Ion-Exchange "Teabags"

Heavy Metals: It was shown previously that activated carbon is not effective for the treatment of most heavy metal spills in water. Therefore, laboratory investigations were conducted to assess the relative effectiveness of ion-exchange resins for the treatment of water spills as compared to activated carbon. Ion-exchange resins were investigated for removal of As^{+5}, Cr^{+3}, and Cr^{+6}. Tests were conducted in which ion-exchange resin was freely dispersed in water or contained in "teabags" to facilitate recovery of resin with exchanged pollutants. Parallel tests were run using powdered carbon as the treatment agent to compare effectiveness of carbon and ion-exchange resin.

The exchange resins used for the experiments were a mixture of cationic and anionic resins normally used for demineralizing of water (Corning No. 3508A).

"Teabags" for holding the resins were made from dacron polyester fabric having a mesh size of 0.4 mm x 0.5 mm.

By comparing removals of arsenic, trivalent and hexavalent chromium by ion-exchange resins to removal by activated carbons, it is seen that ion-exchange resins, as expected, are much more effective. This is so even when powdered carbon is used and applied loosely, rather than in bags. For a contact period of one hour using ion-exchange resin in teabags, As^{+5}, Cr^{+3}, and Cr^{+6} removals were 85%, 91.5%, and 97% respectively. By comparison, removal of As^{+5} by 8 x 30-mesh granular carbon in a teabag was 5.6% in one hour and removals of Cr^{+3} and Cr^{+6} in one hour using loose powdered carbons were 47.5% and 36% respectively. Dosages were the same for all treatments (i.e., 10 g/l).

Studies were made to determine the effect placement of the resins in teabags has on the rate of pollutant removal as compared to free dispersion of the resin in water. Freely dispersed resin resulted in 90% removal of Cr^{+6} in 10 minutes, whereas resin contained in the bag required 40 minutes for an approximately comparable removal of 88%. Containment in bags is desirable, of course, to facilitate removal of the pollutant from the water body after it has been exchanged onto the resin.

Because carbon treatment was the principal method of spill treatment studied in this project, available resources did not allow for further evaluation of ion-exchange "teabags" in channel or pool tests. In view of experimentation which demonstrated greater removal of some metals with ion-exchange resins, further evaluation and development of spill treatment with exchange resins is recommended. An important advantage of resins is that they are easily regenerated for reuse.

Removal of Phenol: Phenol removal by both cationic and anionic exchange resins was evaluated in laboratory beaker tests and compared to removal with activated carbon. Exchange resins used were Amberlite IR-120-H, a cation exchanger, and Amberlite IRA 400-OH, an anionic exchanger. These resins are manufactured by the Rohm and Haas Company.

All treatment agents were applied in loose form at a dosage of 10 g/l. Initial phenol concentrations for all tests were 1,000 mg/l. Agitation was provided by magnetic stirring. From the results of these tests, it is seen that the cation exchange resin does not remove phenol. However, anionic exchange resin, applied loosely, removed a greater amount of phenol than powdered carbon, but does not remove it quite as fast. After five minutes contact, removals of phenol by anion exchanger and powdered carbon were 76 and 97%, respectively. After ten minutes contact, the resin removed almost as much as the powdered carbon. After ten minutes contact, the powdered carbon approached its ultimate capacity for phenol whereas the anion exchange resin continued to remove phenol to the conclusion of the experiment (85 minutes). Granular carbon attained nearly the same removal of phenol as powdered carbon but required nearly an hour more contact time.

The dosage of anion exchange resin as a ratio to initial pollutant concentration in this test was 10:1. On a performance basis then, it would be beneficial to further evaluate anion exchange resins for removal of phenol and substituted phenols which behave similarly in water. Phenol in water behaves as a weak acid and dissociates:

The negatively charged ring is exchanged on the anion exchange resin. In a similar manner, a large number of organic amines which are weak bases could be exchanged on cation exchange resins. The dissociation of two widely used organic amine dyes in water, aniline and pyridine, are:

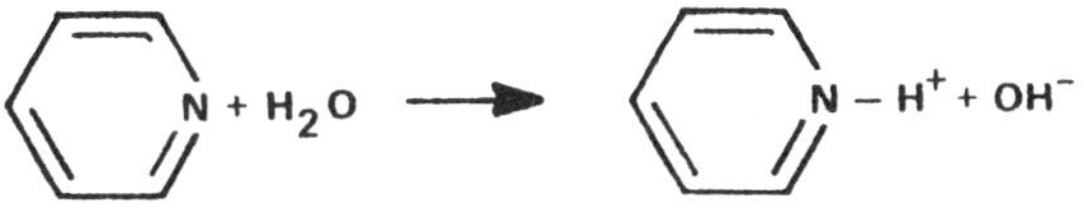

The positively charged rings might be exchanged on cation exchange resins. Available resources did not permit evaluation of ion-exchange resins in the "tea-bags" for organic pollutant removal either in laboratory tests or channel tests. Continued investigation of exchange resins for the treatment of hazardous pollutant spills in water is recommended.

Flotation Methods for Activated Carbon

> The information in this section was based on a paper given at Conference 1972 by R.C. Ziegler of Cornell Aeronautical Laboratory, Inc., and J.P. Lafornara of the Environmental Protection Agency. The tital of the paper was "In Situ Treatment Methods for Hazardous Material Spills."

In the laboratory experiments conducted as a part of this project, the activated carbon was mixed directly into the pollutant-water solution. In a spill situation in the real world, dumping activated carbon into streams and lakes as a treatment method at some of the dosages considered (e.g., 5,000 to 10,000 ppm) causes a severe blackening of the water. Thus, from a purely esthetic point of view, this treatment method is in itself undesirable. However, carbon treatment might be acceptable if suitable means can be found to remove most of the carbon following spill treatment, or, if it can be shown that the pollutant-laden carbon, after settling to the bottom of the watercourse, will not have an adverse effect on the aquatic environment. Desorption of the pollutant from the settled carbon may occur eventually, but if the rate of desorption is low enough, pollutant concentrations in the bulk water might be reduced to acceptable levels. However, until the long-term effects on the aquatic environment can be evaluated fully, it is desirable that means be found for preventing the carbon from remaining in the water for long periods of time. Therefore, methods for floating carbon particles were investigated. In these investigations, emphasis was placed on air flotation techniques whereby flotation is achieved by attachment of the carbon particles to air bubbles.

The initial laboratory studies of flotation of carbon were conducted in a cylindrical column 6 cm in diameter and 53 cm high fitted to a fritted glass Buchner funnel. The fritted glass funnel was of medium porosity to insure small air bubble size and uniform dispersion of bubbles. Compressed air was introduced through the end of the funnel at flow rates ranging from 2 to 8 cfh. Collecting agents tested were trimethyldodecylammonium chloride (Arquad 12-50), dimethyldi(hydrogenated-tallow)ammonium chloride (Arquad 2HT-75), and Cetol. The collector causes the small air bubbles produced by the fritted disc to be attached to the carbon particles, thus causing the carbon particles to float. The air-particle mixture, upon reaching the surface, forms a foam which can be removed by skimming or suction techniques. The amount of carbon collected with the foam is a function of collector dosage, air flow rates, and initial carbon concentration.

Following the flotation experiments using carbon alone, additional flotation experiments were conducted using phenol solutions which had been treated with activated carbon. For an initial phenol concentration of 1,000 ppm and a carbon dosage of 10,000 ppm, the effects of various increments of Arquad 12-50 collector on carbon removal are shown in Table 4.3. After the addition of each increment of collector and establishment of equilibrium, samples were withdrawn for analysis of residual phenol, residual carbon, and residual collector. It is evident from the table that the addition of collector causes the residual phenol concentration to increase slightly, indicating that the adsorbed phenol is displaced slightly by the collector.

The collector dosage required for carbon flotation in the carbon-phenol-water system was significantly less than in the carbon-water system. It appears that the phenol itself is acting as a collecting and flotation agent. It was also observed that a gas flow rate of only 2 cfh was sufficient for foaming in the carbon-phenol-water system, whereas a gas flow rate of 6 cfh was generally required in the carbon-water system.

On the basis of the above flotation data, it appears that the removal of a high percentage of carbon from a water system after treatment requires large amounts of collecting agent and large gas volumes. Considering a spill of 1,000 lb of phenol and assuming that treatment will occur when the concentration is reduced to 1,000 ppm of phenol in the water, it can be determined that 10,000 lb of carbon would be required to adsorb 95% of the phenol. From the data in Table 4.3, which correspond to a 10:1 carbon-to-phenol weight ratio, 390 lb of collecting agent would be required to remove all but 16 lb of the carbon. Since the effect of a collecting agent is to displace some of the phenol, the phenol removal would be slightly less than 95%. It must be realized that the data presented in Table 4.3 correspond to ideal laboratory conditions. In a practical field situation, it is expected that this efficiency would be reduced considerably. From a logistics point of view, it also must be realized that to obtain the required air supply in real world conditions would be extremely difficult. Let us consider an example. Assume that phenol spilled into a stream has been treated with activated carbon and that, downstream from the point of treatment, air flotation is to be attempted. Assume further that the stream is 100 ft wide and has a velocity of 0.5 ft/sec. On the basis of laboratory data, it is estimated that the carbon and the collector agent must be subjected to air bubbles continuously for at least 3 minutes to achieve adequate flotation. Then, the air bubble system must provide a continuous curtain of air for a distance of 90 ft along the direction of stream flow. That is, it must supply air bubbles continuously over an area 90' x 100' or 9,000 square feet.

Laboratory experiments indicate that the minimum flow rate required to achieve flotation in the 6 cm diameter air flotation column is 2 cfh. This corresponds to a flow rate of 1 cubic foot of air per minute per square foot of area. Therefore, for the case just considered, an air supply of 9,000 cfm would be required to achieve flotation of the phenol laden carbon. This value represents an extremely large compressor system, and the logistics of installing a continuous air system of such magnitude into a stream on short notice would be extremely difficult to handle. Consequently, alternative methods for treating spills with carbon have been considered. These include the attachment of carbon particles to floating beads, the use of floating granular carbon, and the use of water-permeable recoverable bags containing the activated carbon. This latter method is discussed in further detail in the following section.

Table 4.3: Flotation of Activated Carbon and Adsorbed Phenol*

Collector** Concentration	Residual Phenol Concentration	Residual Carbon Concentration	Residual Collector Concentration
ppm.			
0	40	10,000	0
260	52	2,000	3
390	66	16	380
460	--	***	–

*Initial phenol concentration 1,000 mg/l; carbon dosage 10,000 mg/l.
**Collector, Arquad 12-50.
***Overdose, no liquid.

Source: Conference 1972

Carbon Bags: An alternative to dumping powdered or granular carbon directly into the stream with the intent of later recovery is to introduce carbon into the stream in water-permeable bags which would allow the pollutant laden water to pass through the bag material and interact with the contained carbon. Such bags could be attached to floats to permit them to be suspended at desired depths in the water. After treatment of the spill is complete, the bags could be easily retrieved from boats, since the floats would allow their locations to be readily determined.

Tests are currently underway on bag design; however, a few desirable features are obvious. First, the bag material should be as permeable as possible to allow free exchange of water in and out of the bags. However, there is a limitation to mesh size, since if the mesh is too large, carbon will be lost from the bags. Using large carbon particles is generally undesirable because, for a given weight of carbon, greater active surface area is available with smaller particles. Thus, the porosity of the bag must be consistent with free interchange of water and ability to retain carbon. Large bags which would lump the carbon inside are undesirable because the carbon particles at the center of the "lumps" could not interact as freely with the pollutant laden water. The bags should be thin and at the same time large in surface area. A large bag having many small pockets might be a desirable compromise.

It does not appear practical to use the bags as a "dam" which would require the water to pass through the individual bags since contact times would be too short to achieve sufficient adsorption of the pollutant. Under real world conditions, contact times on the order of tens of minutes would be required. Furthermore, the resistance of flow through the bags would create large tension on the supporting cables, and on the bags themselves. A more practical approach to the use of carbon bags would be to allow them to float along with the spill. This would greatly increase the contact time. As the spill, and therefore the bags, disperse, additional bags could be added to fill in the voids and to promote the effective pickup of the pollutant. In this intended use, however, since the bag is moving with the spill, there would be less interchange of water in and out of the bag than if the bags were held fixed. A factor of importance is the bag size relative to the turbulent cell structure of the stream.

Initial investigations of the carbon bag approach consisted of fabricating bags of different cloth types containing different kinds of carbons and studying the adsorption efficiency for phenol as a function of time of interaction. The bag sizes were 2" x 3", and the mesh size varied from 0.8 mm x 0.5 mm to 0.25 mm x 0.3 mm. These mesh sizes did not permit the retention of powdered activated carbon, so granular carbons of various mesh sizes were used. The carbons used were Darco 8X35-mesh, Filtrasorb 300 (effective particle size 0.8 to 0.9 mm), Filtrasorb 400 (effective particle size 0.9 to 1.1 mm), Nuchar 8X30-mesh (effective particle size 0.85 to 1.05 mm), and Nuchar 20X50-mesh (effective particle size 0.35 to 0.45 mm). Data on the effective particle size of Darco 8X35-mesh was not available.

For each test, 2 g of carbon were placed in a bag which was then immersed in 200 ml of a 1,000 ppm phenol solution. This carbon dosage corresponds to a concentration of 10,000 ppm. All samples were attached to a tumbler to allow uniform mixing at contact times ranging from 10 minutes to 3 hours. The results indicate that Nuchar 20X50 in a coarse bag is the most effective of the configurations tested. After about 60 minutes of contact time, the maximum adsorptive capacity of the carbon is approached, leaving only about 5% of the phenol remaining in the bulk water. One can expect rather good mixing conditions in the laboratory experiments because of the tumble-type mixer that was used. It is expected that in a real stream a longer contact time would be required to achieve the same adsorptive effectiveness. Plans are now underway to test various types of carbons and bags in a stream flow facility which will permit the turbulence conditions of real streams to be simulated.

In summary, activated carbon appears to be a useful means for treating spills in streams, provided that the carbon can be removed or that it can be shown that allowing the carbon to remain in the waters will not adversely affect the environment. While generally large doses of carbon are required to effect a high removal percentage, carbon has the advantage that it can be used to treat many hazardous substances. Therefore, stockpiling large quantities of activated carbon for spill treatment would not be unreasonable. The water-permeable bag approach would have a number of advantages, but it must be shown that appropriate interaction can be obtained between the pollutant and the carbon.

Floating Sorbents

> The material in this section is based on a paper presented at Conference 1972 by A.J. Shuckrow, B.W. Mercer and G.W. Dawson of Battelle Pacific Northwest Laboratories. The title of the paper was "The Application of Sorption Process for In Situ Treatment of Hazardous Material Spills."

Sorption processes are highly effective for removing many different substances from water, and application of sorption techniques has been identified as one of the most potentially promising approaches to in situ treatment of hazardous material spills (28). The principal advantages of using sorption processes for this purpose are: treatment can be accomplished without the addition of chemical reagents to the water; one sorbent can be used to remove a large number of different substances from water; and the sorbent collects and holds a hazardous material in a relatively innocuous form for long periods of time. The term "sorption" as used in this paper applies to both physical sorption (e.g., surface attraction as in activated carbon) and ion-exchange, although the latter may be interpreted differently as a matter of personal preference.

Sorption may be accomplished either by batch treatment or by column treatment. Batch treatment is carried out by mixing the sorbent with the solution to be treated for a specified period of time, after which the sorbent is separated from the solution for disposal or regeneration and reuse. This method is simple and may be carried out with relatively low cost equipment.

Column treatment is accomplished by percolating the solution through a bed of the sorbent until some level of the sorption capacity is utilized. The sorbent in the column is usually regenerated at this point and the loading cycle repeated. Column treatment is generally more efficient than batch treatment, and lower effluent concentration values may be attained. The major disadvantage of column treatment with respect to field use in treating spills of hazardous materials in water is the relatively long time period required to set up and operate the equipment needed. Because response and treatment time must be rapid to minimize dispersal of the hazardous material from the site of the spill, batch treatment appears most feasible for field use.

Recovery of the sorbent after treatment is necessary to avoid problems related to slow desorption of highly toxic material and for aesthetic reasons. Most sorbents will sink when applied to water, which complicates retrieval of the sorbent. The use of floating sorbents, however, would facilitate recovery, because oil skimming or similar equipment may be used to retrieve the sorbent following in situ use. This paper deals with the development and evaluation of floating sorbents for use in treatment of hazardous materials spills in water.

It is anticipated that floating sorbents will be used in situ by injection at the bottom of a watercourse and allowed to float to the surface through the zone of contamination. A major concern with a floating sorbent relates to the kinetics of removing various hazardous polluting substances as the sorbent moves from the bottom of a watercourse to the surface of the water. In the case of columns, kinetic problems are solved by adjustment of the throughput rate to achieve the necessary residence time to effect the desired degree of removal. But kinetic problems are not so simply solved in a buoyant sorbent system. The sorbent must perform its function with only one pass to the surface.

The controlling parameters involved in a free water mass transfer system include: temperature, concentration of solute, particle velocity, and particle surface area. Obviously, the first two parameters are functions of the location and nature of the spill and cannot be practically controlled in a free water environment. Indeed, the purpose of applying a countermeasure is to reduce the concentration of the solute (hazardous polluting substance) to some acceptable level. However, particle velocity and particle surface area can be controlled through adjustment of the particle size.

Since mass transfer rates are proportional to surface area, the division of a large particle into several smaller ones increases the surface area and thus accelerates the mass transfer rates. Reduction in particle size also reduces the rate of rise of a flotable medium, since buoyancy is a function of r^3 while drag is a function of r^2. Reduction of the rate of rise increases the residence time as the medium rises to the surface. Thus, reduction of particle radius causes an increase in the rate of mass transfer to occur. Adjustment of medium density can also be utilized to increase the contact time.

Floating Ion-Exchange Resins: Boyd, et al (29) note that for organic zeolites in the 60 to 70-mesh range, up to 90% of the theoretical capacity can be achieved within 30 seconds. Kressman and Kitchner (30) found phenolsulfonic resins reach 90% of capacity in 2 to 9 minutes with simple cations. Helfferich (31) makes similar observations for strong acid and base exchanges. It should be possible to adjust the rise rate of the selected medium to provide the several minutes required for utilizing the major portion of the medium's capacity.

Two types of resin were selected for study in preparation and evaluation of floating ion-exchangers. These were: (1) a weakly acidic carboxylic resin prepared from acrylic acid and a suitable crosslinking agent; and (2) an intermediate epoxy-polyamine resin prepared from epichlorohydrin and polyethyleneimine. The exchange resins were rendered buoyant by incorporating hollow glass microspheres (Eccosphere, manufactured by Emerson and Cuming Inc.) in the resin granules or beads. A bulk polymerization technique was employed to produce the floating acrylic cation exchange resin whereas a suspension polymerization technique was employed to prepare the floating epoxy-polyamine anion exchange resin. The latter produces spherical beads suspended in hot mineral oil during the polymerization step.

The ultimate exchange capacity of a floating acrylic cation exchange resin, prepared from acrylic acid and ethylene glycol dimethacrylate was found to be 8 meq/g of dry resin (includes weight of hollow glass microspheres). Two different mesh sizes of this resin in the H^+ form were floated through a 5 ft column of water (2" diameter) containing 200 mg/l of NaOH. The application consisted of 2,000 mg/l of 25 x 70-mesh resin and the same quantity of 20 x 50-mesh resin both released dry from a bottle at the bottom of the column. The finer 25 x 70-mesh resin removed 82% of the alkali compared to 66%

removal by the coarser 20 x 50-mesh resin. The time for the bulk of the resin to float to the surface was approximately ten minutes for both mesh sizes.

Another floating acrylic cation exchange resin was prepared using acrylic acid crosslinked with divinylbenzene with hollow glass microspheres incorporated in the resin granules. An alkali removal of 89% was obtained with a 30 x 70-mesh size with this resin.

A floating anion exchange resin was prepared by suspension polymerization in mineral oil using epichlorohydrin and tetraethylene pentamine in a 3:1 mol ratio plus hollow glass microspheres. This resin was found to have an ion-exchange capacity of 4 meq/g of resin. The resin beads used for evaluation were largely in the 30 to 50-mesh size with 12% + 30-mesh and 9% - 50-mesh. When released to float up through a 5 ft column of water containing 183 mg/l HCl, the resin removed 61% of the acid at an application ratio of 2,000 mg of resin per liter of water. This removal is somewhat less than the removal of NaOH by floating acrylic cation exchange resin under comparable conditions which is believed due to the smaller ion-exchange capacity of the epoxy-polyamine resin (4 meq/g vs 8 meq/g). In both cases, however, only a small fraction of the available ion-exchange capacity was utilized.

A smaller mesh size is considered to be more important in increasing the removal than adjusting the resin bead density to a level just under that of water. Resin bead density cannot be adjusted too near to that of water since the density of the bead will increase as it sorbs ions from the water to replace either H^+ or OH^--ions. The resin bead will rise more rapdily at first when its density is the least.

Floating Physical Sorbents: Physical sorption results from the attractive forces that are present at the interface of two phases. The surface of a solid will attract and hold molecules present in either gases or liquids. The amount held per unit area of the surface is relatively small, but the amount becomes a significant fraction of the total mass of the solid phase when the surface area per unit of mass becomes very large. The surface area of a good activated carbon, for example, will be on the order of 1,000 m^2 or more per gram of carbon.

A number of physical sorbents are available for removing nonpolar materials such as organic substances from water, but the best known and most widely used physical sorbent is activated carbon. An investigation of available active carbons with the desired floatability disclosed two commercial granular carbons which resist wetting and will remain floating for long periods of time. These are Nuchar C-190 and Nuchar WA (Westvaco) + 30-mesh. Various mesh sizes of these carbons both wet and dry were evaluated for their uptake of phenol by floating the carbons through a 5 ft column of phenol solution (185 mg/l) buffered at pH 6.5. The carbon was released from a bottle at the bottom of the column at an application ratio of 2,000 mg carbon per liter of solution. Nuchar C-190 was selected for further study on the basis of its superior sorbtive ability and floatability. In addition, the Nuchar C-190 was found to contain a smaller percentage of sinking granules than the Nuchar WA. A mesh size of 100 x 325 was selected for application.

Larger scale experiments were conducted in a 6" diameter, 6' deep column of tap water containing 200 mg/l of: phenol, oil emulsion of Malathion (trade name for the insecticide O,O-dimethyl dithiophosphate of diethyl mercaptosuccinate), and an oil emulsion of Diazinon (trademark for the insecticide O,O-diethyl O-(2-isopropyl-4-methyl-6-pyrimidinyl) phosphorothioate, manufactured by the Geigy Chemical Company). Removals of 86%, 82% and 86%, respectively, were obtained by floating 1,000 mg/l of 100 x 325-mesh carbon through the water. The concentration of phenol and pesticide did not vary significantly at various depths through the column after flotation of the carbon. Carbon containing sorbed phenol was observed to remain floating for at least two days following application. The carbon was found to be effective for removing phenol in the 0° to 25°C temperature range.

Delivery Systems: Mechanical subsurface injection by means of slurry pumps installed on ships is probably the most desirable method of delivery in areas such as ports and harbors where such equipment is readily available. However, in remote areas, rapid transport of subsurface injection equipment may not be feasible. Therefore, techniques are under development which lend themselves to air transport and injection of the floating solvents.

Systems currently under active consideration for delivering the floating sorbent to the bottom of body of water where a spill of hazardous material has occurred include the following:

ice cakes containing floating sorbent and sufficient ballast such as sand or gravel to sink the ice cake;

unfired clay-sand containers which disintegrate under water to release the floating sorbent; and

plastic bottles which separate from the ballast and float to the surface for recovery following release of the sorbent. Water-soluble seals could be employed to contain the sorbent and ballast during delivery.

These systems were selected to avoid littering the area concerned with empty containers or polluting the water with foreign matter such as gelatin (capsules) and cardboard. The use of ice cakes represents the most efficient system from the viewpoint of the least amount of ballast required. In addition, the sand or gravel ballast is the least objectionable material that can be deposited in most cases. The major disadvantage of the ice is the refrigeration requirement during storage. Preliminary results with unfired clay-sand containers are promising and this system also involves natural materials for ballast. Plastic bottles attached to concrete ballast with water-soluble material have been successfully tested in the laboratory. All systems are undergoing further investigation to assure that the containers can withstand an air drop since air transport is considered to be the most probable means of delivery in remote areas.

It is to be noted that a considerable amount of ballast is required to sink the floating sorbents. Dry Nuchar C-190, for example, has a bulk density of 7 to 10 lb/ft^3. Therefore, approximately 100 lb of concrete ballast (SG = 2.3) will be required to sink a cubic foot of dry Nuchar C-190. Wetting carbon to fill the interstices with water reduces the amount of ballast required but eliminates the use of soluble seals for releasing the carbon under water from bottles and also the use of unfired clay-sand containers. The use of ice, whereby water fills the interstices, circumvents this problem to a large degree.

Summary and Conclusions: The results of laboratory studies show that floating sorbents are highly effective for removing hazardous materials from water when applied as small particles beneath the surface. The sorbents remove the hazardous materials while rising through the contaminated water. A light weight commercial activated carbon was found to be particularly effective for removing organic substances from water. The active carbon can be pulverized to a small mesh size (100 x 325) which floats slowly to the surface of water. The small mesh size both enhances the speed of sorption and increases the contact time with the water by flow floatation.

Floating ion-exchange resins were prepared which are effective for removing acid or alkali from water. Hollow glass microspheres are incorporated in the resin granules for buoyancy. Due to the use of larger particle sizes (minimum 70-mesh), only a fraction of the resin's total exchange capacity is utilized in floating to the surface.

Delivery systems are being investigated which employ air drops of the floating sorbents into the water where the spill occurred. This method allows the shortest response time to a reported spill. Delivery packages under study for the floating sorbents include: ice cakes containing the floating sorbent and ballast; unfired clay-sand containers; and recoverable plastic containers. Field demonstrations are planned on the use of floating sorbents for removing 10 to 500 lb of hazardous materials spilled in water contained in large concrete lined basins.

Comparison of Buoyant and Sinking Carbon Methods

> The information in this section is based on *In Situ Treatment of Hazardous Material Spills in Flowing Streams*, EPA-600/2-77-164 prepared by G.W. Dawson, B.W. Mercer and R.G. Parkhurst of Battelle-Northwest for the Environmental Protection Agency.

Early studies of both the buoyant and sinking carbon systems were focused largely on concept development and limited testing in ponded waters. Historical data, however, indicate that a preponderance of spills occur in flowing waters. Dawson and Stradley have estimated that 82% of all freshwater spills occur in rivers and streams (32). Of the remaining 18%, some spills will actually involve reservoirs on navigable waterways. It was the purpose of these studies to evaluate the two most promising in situ treatment techniques operable in flowing waters. Of particular interest with regard to the use of buoyant activated carbon were several technical issues: the need (or lack thereof) for ballast and packaging; the efficiency of contact; probability of unsightly carbon build-up along stream banks; and the efficiency of spent carbon collection. If the need for ballast were eliminated as a result of media suspension sponsored by natural turbulence, delivery systems could be greatly simplified. The major question to be resolved with respect to sinking carbon was the ability of natural currents to supply sufficient turbulence to enhance the kinetics of adsorption.

The work reported herein was performed under two separate contracts with the U.S. Environmental Protection Agency and was directed to the evaluation of both buoyant and sinking activated carbon systems in flowing streams. All work was conducted in a simulated flowing stream maintained on the Hanford Atomic Energy Reservation. Early studies to verify the potential for application of floating carbon without ballast were followed by parallel evaluations of the two systems under varying conditions.

Practical Aspects of In Situ Treatment in Flowing Streams: Field trials have shown that buoyant sorbents and porous fiber bags, if given sufficient contact time, can be effective in removing organic materials spilled into flowing streams. Removal, however, is highly dependent upon the prompt location of the contaminant plume, even dispersion of media or bags over the surface, and favorable environmental conditions.

Wind will prove to be one of the major obstacles to efficient use of floating media or bags. Not only will air movement result in significant effects on the contact period, it can severely hamper collection efforts by herding the buoyant carbon or bags away from the boom. If aerial application is attempted, wind complications will be further amplified. It is also important to note that collection booms have a limited operation range and cannot deal with excessive currents (~5 knots). Therefore, quiescent or slower moving reaches of the stream must be sought for media retrieval.

Use of buoyant media also carries the potential for leaving unsightly carbon residuals along shorelines and beaches. Similarly, fiber bags may be left along the shore or hung up on shoals. While these effects were minimal during field trials, they must be considered prior to application in any public waterway.

The studies made to date suggest that removal efficiency will be greatly affected by scaling. Small spills such as those employed in the test program amplify the sensitivity to dosing and environmental considerations. Larger spills are marked by much larger spill plumes and higher concentrations. The dilute edges of the plume represent a much smaller percentage of the total spill. Therefore, removal in the center of the plume where the carbon is most efficient and where movement of the carbon does not separate it from the contaminated water is a greater part of the total removal. This indicates that average removal is likely to be much better on a larger scale than that noted in static water testing. A small acid spill in a semiconfined basin revealed very poor removal when compared to laboratory work. A much larger pesticide spill, however, resulted in removals comparable to those obtained in confined column work in the laboratory (21). The edge and dilution effects become very important as the size of the spill is reduced, and subsequently removal is less efficient in small scale application.

While some of these factors suggest that removal may in fact be better than is suggested by field trials, one important feature of actual spills will complicate response greatly. That is the location and tracing of the contaminant plume. In the field trials, dye was used to facilitate accurate application. In the field, response personnel may find it very difficult to locate the spill and to define the boundaries of the plume. Thus, all the media could be applied in the wrong area and hence be totally ineffectual. This underscores the need for systems to monitor spills. Recent work with remote sensing devices and detection kits could satisfy this need (33) to (35).

It is obvious that under no conditions will removal ever be complete. Therefore, one cannot assume that a response effort will eliminate a spill. It will only reduce its impact. Certain circumstances can be expected to maximize that reduction, and those are the cases where response should be promoted. From the observations made in this study, buoyant carbon is preferable to porous bags in the flowing stream system when contact times are very short (30 minutes) just as it was in static waters. Because the major difference appears to be one of kinetics, the longer contact times achievable in natural waters may change this somewhat.

Application of either buoyant carbon or porous bags may pose a logistics problem. Dr. Allen Jennings of the U.S. EPA Hazardous Substances Branch estimates that spills average 3,500 pounds. At a ratio of 10:1 carbon to contaminant, this would require 35,000 pounds of media or 17.5 tons. If air transport is desired, this will require specialized equipment. Larger spills may exceed the present transport capacity and thus suggest the potential use of containerized media supplied by shuttle to the response component.

All of the considerations offered here have bearing on the use of in situ response techniques on real spill events. They are pointed out to facilitate informed decisions about spill mitigation. Despite the restrictive nature of some of these observations, both the floating carbon approach and the porous bag approach are deemed promising for given spill scenarios. Present information needs center on the evaluation of these techniques under actual spill conditions.

Application and Retrieval: Buoyant Carbon — Major problems are associated with the application of packaged buoyant media for flowing streams. As a result of movement of the contaminated plume with the current, timing of media release becomes critical. If media were delivered in packages, the release of the media from the packages would have to be timed exactly to coincide with the passage of the plume in order to achieve effective treatment. In order to avoid this problem, two alternative methods of application were explored: (1) surface application of the media with contact dependent upon the natural turnover of the stream water; and (2) subsurface injection of slurried media.

Surface application relies on two mechanisms to provide intimate contact between the contaminated water and the buoyant sorption media. Both of these are related to the natural turnover of the water as it flows downstream. The first mechanism involves the vertical velocity components of the flow itself which disperse fine media particles downward where they contact contaminated waters and sorb the contaminant. In the case of the second mechanism, the larger particles float on the surface and sorb contaminant from the deeper waters as the latter comes to the surface and rolls back to the bottom.

With subsurface application, the media is slurried and pumped into the deeper portions of the contaminant plume. Intimate contact is achieved initially through the dispersion of the slurry itself in the receiving waters and its subsequent ascent to the surface.

Optimal particle size will depend in part on the mode of application anticipated. Since surface collection of spent media in quiescent reaches is the mode of retrieval, particles must be sufficiently buoyant (a function of particle diameter when density is held constant) to rise to the surface during residence in the selected quiescent zone. On the other hand, if particles are too large they will not be carried down into the water column or will rise too quickly after subsurface injection to achieve the required contact. These system requirements, therefore, determine physical limitations on the acceptable particle size range for buoyant sorption media. They, in turn, are influenced by the velocity components of the receiving waters.

Sinking Carbon – The use of sinking activated carbon for in situ treatment of waters requires removal of the spent carbon from the bottom of the watercourse, abandonment of the spent media at the bottom, or incorporation of the carbon in a package which allows retrieval at some later time. The latter approach (packaging) is the most practical and was selected for development here. Each package must be designed to allow contact with contaminated water without releasing media. These requirements are similar to those which have led to the development of teabags. Indeed, the teabag concept is very appropriate for application of activated carbon.

Activated carbon is placed in a porous fiber bag with a preselected thread count just sufficient to hold the smallest granule sizes to be used. Contaminated water can thereby flow into the bag, contact the carbon, and flow out. The bags do not perform well in static water because of a lack of flow through the bag and media. In flowing streams, however, the natural turbulence should be sufficient to constantly exchange waters in immediate contact with the bag. Contact is further stimulated by filling the bags only partially full and thereby leaving ample room for the media to fluidize. In this expanded state, adsorption kinetics are enhanced.

Adsorption will be greatly affected by carbon particle size as well as the aforementioned factors related to water exchange. The finer the particles employed, the greater the ease of fluidization and surface contact, and the more rapid the adsorption. Fine particles, on the other hand, require fine materials to contain them, which in turn, discourages water flow. In balancing these interests, researchers at Calspan Corporation determined that commercially available granular carbons were adequate (27). These are marketed in two size ranges: 8 x 30-mesh and 12 x 40-mesh. Early investigations were conducted with the larger carbon. The smaller 12 x 40-mesh material was selected for use in flowing streams to enhance contact with the water.

Retrieval of the "bags" can be provided for by attaching them to floats or other bouyant devices. Visual observation of the floats allows constant knowledge of where and how the bags are moving. At the end of the contact period, bags can be retrieved by placing a boom or other collection device across the channel such that the floats are snared and held against the current. Since the floats allow the bags to move with the current, the carbon is kept in constant contact with the contaminant plume. Thus, contact times can be made as long or as short as desired depending upon where collection booms are deployed. A final degree of treatment is also provided at the boom where the snared bags are analogous to a fixed carbon bed through which the contaminated water must flow.

Conclusions: The removal of a soluble organic hazardous material from a flowing stream environment has been demonstrated by two different approaches–one utilizing buoyant activated carbon, and one employing sinking carbon in porous fiber bags suspended from floats. The following conclusions are based on results of these studies.

Buoyant Activated Carbon –

- Buoyant carbon can be effectively employed on flowing streams without the use of ballasted packages. Carbon can be applied directly to the surface or slurried and injected beneath the surface.
- Natural turnover is sufficient to provide intimate contact of carbon with contaminated water in shallow streams.
- Floating carbon was capable of achieving 50% removal at a carbon to contaminant ratio of 10:1 under the range of flow rates and spill conditions studied.
- Some floating carbon was captured in eddies and debris along the side of the water body, but carbon recovery is generally in excess of 90%. Containment booms must be placed in a quiescent stretch where velocity components will not exceed the capabilities of the device. Simple oil booms are sufficient if weather and current conditions allow operation.

Porous Fiber Bags –

- Porous fiber bags can be effectively applied to spills in flowing streams, but removal is directly related to the turbulence and current structure in the receiving water. Removal rose from insignificant levels at low flow rates to 20% at a flow rate of 15 cubic feet (425 liters) per second.
- Fiber bag efficiency appears to be limited by kinetic considerations. Removal may improve with longer contact times than those which can be achieved in the test facility employed here.
- Up to 25% of the fiber bags were lost during the tests as a result of shore capture and snagging in shallow areas. This effect diminished with higher stream velocities.
- Fiber bags can be collected after use with simple booms or a wire strung perpendicular to the flow just below the surface of the stream. Positioning in a quiescent stretch is not required.
- Fabric for the fiber bags must be carefully selected to avoid decomposition by the material being removed. At the same time, use of a heat and pressure resistant material allows regeneration in the bag, thus avoiding the necessity for empty and refill sequences.

Comparative Evaluations:

- Buoyant carbon was superior to porous fiber bags in terms of removal efficiency under the spill and flow conditions tested.
- Little difference between methods with respect to media loss along shorelines was noted, but fiber bags can be retrieved more simply than buoyant carbon and under more extreme flow conditions.

- Fiber bags can be loaded, unloaded, and handled with greater ease than buoyant carbon.
- Both approaches rely heavily on the ability of the response team to locate and trace the movement of the contaminant plume.
- The ability to inject slurried buoyant carbon at depth renders this approach more attractive than fiber bags for use in deeper channels.
- Buoyant carbon will be more greatly affected by adverse weather conditions than will fiber bags.
- Both approaches may be difficult to apply to very large spills simply as a result of the logistics of ferrying large quantities of carbon to the site.

Recommendations:

- A survey should be conducted to assess the availability of buoyant carbons for use in spill response.
- An air deliverable slurry injection system should be developed and tested for routine application of buoyant carbon to spills.
- A study of the treatment effectiveness of porous bag packaged carbon should be performed at a site where longer contact times than were possible for these tests can be achieved.
- Further work should be conducted to develop remote sensing and/or other techniques for the identification, location, and monitoring of spills. Some consideration should also be given to the development of methods for making spill plumes.
- A decision framework is needed to determine when spill response is warranted and what the most effective means of response is for specific spills on a real time basis.

Deployment Techniques

The information in this section is based on *Treatment of Hazardous Material Spills with Floating Mass Transfer Media*, EPA-670/2-73-078 by B.W. Mercer, A.J. Shuckrow and G.W. Dawson of Battelle-Northwest for the Environmental Protection Agency.

Sorption may be accomplished either by batch treatment or by column treatment. Batch treatment is carried out by mixing the sorbent with the solution to be treated for a specified period of time after which the sorbent is separated from the solution for disposal or regeneration and reuse. This method is simple and may be carried out with relatively low cost equipment. Column treatment is accomplished by percolating the solution through a bed of the sorbent until some level of the sorption capacity is utilized. The sorbent in the column is usually regenerated at this point and the cycle repeated.

Column treatment is generally more efficient than batch treatment and lower effluent concentration values may be attained. However, in the case of a spill situation, column treatment would involve the transport of large volumes of water to a treatment facility and then return of the treated water to the lake or stream. Another disadvantage of column treatment in the field is the substantial time period required to set up and operate the necessary equipment. Because response and treatment time must be rapid to minimize dispersal of the hazardous material from the site of the spill, batch treatment appears most feasible for field use.

Recovery of the sorbent after treatment is necessary to prevent slow desorption of highly toxic material and for aesthetic reasons. Most sorbents will sink when applied to water, which complicates their retrieval. Moreover, dredging may result in the removal of large quantities of benthic life and bottom materials with the media. Floating sorbents, however, can be recovered by oil skimming or similar well established surface collection techniques.

Effective use of floating media requires that: buoyant sorption media be deposited at the bottom of the water column so that it removes selected dissolved contaminants (i.e., toxic ions or organics) as it rises to the surface; means be provided to uniformly distribute the media during the deposition step or as it rises to the surface; and the application rate be properly proportioned to the amount of pollutant to be removed or neutralized per unit area of water surface.

Application Methods: Two general methods are envisioned for applying the mass transfer media to achieve such a distribution: (1) pumping bulk media as a slurry through a pipe which terminates near the bottom of the watercourse and which is installed on a boat or vessel propelled at a speed proportional to the application rate; and (2) dropping on the water surface packages or media capsules weighted with ballast, the packaging material of which disintegrates or opens upon exposure to water after sufficient time to allow the packages or capsules to reach the bottom.

Mechanical subsurface injection by means of slurry pumps installed on ships is probably the most desirable method of delivery in areas such as ports and harbors where equipment of this nature is readily available. However, in remote areas rapid transport of subsurface injection equipment may not be feasible. Therefore, techniques which lend themselves to air transport and injection of the floating sorbents were selected for development in this program.

Air delivery would involve "bombing" the affected area with packages containing the sorption media. In addition to the active media each package unit must contain an appropriate amount of ballast so that it will sink to the bottom of a waterway prior to release of the media. The optimal size of an individual package is dependent upon two separate considerations: (1) achieving uniform distribution of the packages (and thus the media) over a spill area; and (2) overall economics of the concept. While the former consideration tends to favor smaller packages, the latter biases the decision toward larger units.

Recovery: Once the media has been applied and allowed to perform its function, relatively long periods can elapse before recovery of the spent media, because the potential pollutant will be sorbed or fixed on a material which is essentially insoluble. The physical recovery of the spent media can be achieved either manually or with mechanized equipment.

When treating a flowing stream, a simple but probably effective approach would call for positioning of floating baffles or booms at an angle across the stream and downstream of the spill treatment area. This would serve to direct the floating media to an accessible shore area. Pickup of the media could then be achieved either manually or by using simple agricultural or earth-moving equipment (e.g., drag lines, back hoes).

In the case of static waters such as lakes, ponds, reservoirs, bays, and harbors, a means would be needed for collecting or concentrating the media and then removing it from the water surface or from adjacent shore areas. This could be partially accomplished by techniques similar to those mentioned for a flowing stream. However, there is the added requirement for concentration and possible removal of the spent media from the static water surface. The techniques required for this are parallel to those necessary for recovering floating oil slicks. In fact, application of much of the technology developed for oil spill countermeasures to the recovery of spent sorption media appears feasible.

Example: The simplicity of application and ultimate effectiveness of floating sorption media can best be illustrated with a hypothetical spill incident. Consider a barge which founders while carrying 50,000 gallons of benzene.

Assuming the nature and location of the spill is reported promptly and accurately, countermeasures could be taken to minimize the damage to marine fisheries in the area. Upon identification of the spilled material as an aromatic compound, transport aircraft would be loaded with packaged floating sorbents and dispatched to the release site.

Personnel at the spill site would maintain radio contact with the aircraft and direct their drops to cover the contaminant plume. If the aircraft arrived approximately six hours after the spill occurred and the stream had a one knot current and high diffusivities, the spill would have moved approximately 11,000 meters downstream and would be 580 meters long and the width of the stream. While making low passes over the stream, the aircraft would release a pattern of media packages over the affected area.

Individual packages would sink to the bottom, disintegrate, and release the floatable carbon. Particles of carbon would disperse with the eddy currents and slowly rise to the surface, sorbing benzene as it passed.

A screen skimmer or boom could be deployed downstream to guide the spent carbon to a surface pumping device. The collected media could then be stored for drying and subsequent disposal or regeneration and reuse. While retrieval is best accomplished as soon after deployment as possible, the carbon could be allowed to drift with the current for several days while skimming devices were located and put in place.

Packaging and Ballast: It is evident that potential packaging techniques for the floating media and ballast for aerial delivery fall into two categories: (1) methods employing retrievable containers, and (2) methods employing nonretrievable containers or containers which will decompose in water over a period of time. Of prime importance in the program has been the identification of a packaging technique which would not result in significant degradation of water quality if the packages were used in spill response activities. Packaging techniques which were considered include plastic containers, soluble synthetic films, unfired clay containers, and ice cakes.

Plastic Bottles – Plastic bottles are an inexpensive, standardized container form potentially useful for packaging floating media.

Plastic containers were employed in column tests to evaluate the effects of various bottle mouth widths on media release rate and release patterns. It was observed that media release from wide mouthed bottles was accompanied by simultaneous release of quantities of large air bubbles. These air bubbles tended to convey clusters of media rapidly to the surface, thus preventing dispersion of the media in the water column. Further tests indicated that this problem could be virtually eliminated by using containers with walls which sloped upward to a mouth of about one inch diameter. Using this type of container, there was a short period of media release followed by a longer period in which small air bubbles escaped but little media was emitted. This venting period preceded a sustained release of clouds of media from the bottle which distributed the media throughout the water column.

Either cement or sand and gravel proved satisfactory as ballast in partially filled plastic bottles. One gallon polyethylene bottles half-filled with cement were found to withstand the impact of a fall from a 50' height to a gravel bed. Similarly, these containers withstood drops from a 100' height into water.

It was determined that the mouths of the plastic bottles could be sealed with a water-soluble material, Quicksol A, to prevent the media from falling from the containers during the air drop. However, in field trials with carbon filled containers, it was observed that no significant quantity of media was spilled during a drop from bottles with open mouths.

The major drawback associated with plastic containers is the potential aesthetic and/or ecological damage which might result if weighted plastic bottles were allowed to remain permanently in a waterway after treatment of a spill.

Retrieval of plastic containers could be accomplished either manually (e.g., by divers or with grappling hooks) or by use of containers which will surface and float some time after release of the media, thus permitting surface collection.

Another approach which was considered but which was not examined in great detail in this program is the use of biodegradable containers. If a container could be identified which would eventually degrade without creating a significant oxygen demand or other pollution problems, it might offer an attractive alternative. Since such materials are still in the experimental development stage, an adequate technical and economic assessment of the feasibility of this approach cannot be made at this time.

Soluble Films – Contacts with several major pharmaceutical firms indicated that both the technology and equipment are presently available for large scale encapsulation of a mixture of ballast and floating media in water-soluble capsules.

Several samples of water-soluble films and bottles were obtained from vendors and subjected to a cursory examination to determine their suitability for containing floating media. Individual packets with volumes of approximately four cubic inches were prepared from each material and loaded with sand ballast and polystyrene chips to simulate media "bombs". Each packet was then immersed in 4 gallons of tap water (@25°C) to determine the time required for release of the polystyrene chips.

Most of the films dissolved uniformly and released the polystyrene chips cleanly with little or no partially dissolved film adhering to them. The immersion time required for release could be adjusted by using thicker films or multiple layers of film.

One pound of polymer will yield about 7,500 square inches of film with a thickness of 3 mils. If individual packets with volumes of 4 cubic inches each were fabricated, a pound of polymer would provide containment for a volume of approximately 1,200 cubic inches. The weight of dry carbon thus packaged would be approximately 5 lb or a 5:1 ratio of carbon to soluble film. This quantity of soluble film would be equal to or perhaps greater than the spilled organic material being removed from the water.

Experiments were conducted to study the effects resulting from dissolution of one of the films, Quicksol A. It was determined that 160 mg of Quicksol A dissolved in water yielded a total organic carbon concentration of 88 mg/l. Activated carbon at 2,000 mg/l removed 47% of this TOC. The same dose of activated carbon removed 91% of the 171 mg/l phenol solution, but could adsorb only 75% of a phenol-Quicksol A mixture. The dissolved film showed a definite tendency to use up available carbon sorption capacity. Biochemical oxygen demand (BOD) tests indicate that a 1,000 mg/l solution of Quicksol P has a five day BOD of only 8.2 mg/l. Although the use of these films should not cause excessive oxygen depletion at a spill site, the addition of relatively large quantities of soluble organic matter to the water is a questionable approach. In addition, the tendency of the dissolved films to use up a significant portion of the available sorption capacity of activated carbon is not a desirable characteristic.

Clay Containers – In an attempt to find a packaging agent which could also serve as the ballast material, an effort was initiated to study the feasibility of unfired clay containers. A local pottery shop was engaged to produce various clay container shapes and sizes through a slip molding technique. The resulting products were dried but were not fired so that they displayed an affinity for water which soon caused them to collapse when immersed.

It was found that sand could be added to the clay in varying amounts to increase the speed with which the package lost integrity. Inclusion of approximately 30% sand resulted in a decomposition time of 2 to 5 minutes. Pyramidal and cylindrical shapes displayed the cleanest release characteristics, trapping little of the floating media under the collapsed clay.

An impact evaluation was made to determine whether clay containers could withstand the forces encountered when dropped into water. Vessels shaped somewhat like artillery shells with a one quart capacity (approximately 1 ft high x 3 inches o.d.) were dropped from a height of 100 ft into the Columbia River. Each clay container was attached to 185 ft of light cord so that it could be retrieved and inspected. Vessels with a wall thickness of 0.25 inch or greater withstood the impact. Those having wall thicknesses of 0.125 inch or less collapsed. Several alignment modifications could potentially increase the effectiveness of the clay cylinders. Walls could be tapered so that heavy sections would absorb the impact while thinner walls at the top would rapidly disintegrate when immersed in water. Alternatively, lids could be loosely molded into place so that soon after coming to rest on the bottom of the water body they would fall off, allowing the media to escape. It might also be desirable to add fins to the design since the models tested continued to tumble when initially given angular momentum.

Clay containers would require use of dry media and would thus have a low payload to ballast ratio. A further disadvantage of unfired clay is the turbidity caused by suspension of the fine clay particles in water. The mixing action caused by the release of air and media suspends a significant amount of clay in the water. Use of a fired clay container with an unfired clay seal is a possible alternative to minimize the amount of suspended clay formed.

Ice Cakes – The search for a completely innocuous encapsulation agent led to an attractive possibility—the use of ice as a binding agent. Ballast and media could either be layered and frozen into discrete packages for aerial distribution or intimately mixed and frozen into blocks which could be shattered into any size desired at the time of release.

Advance production and stockpiling of media ice cakes would have associated high storage costs since refrigerated storage would be required. In addition, sublimation might be a problem associated with long term storage of ice cakes. One possible solution to these drawbacks would be dry storage of components (media and ballast) and subsequent quick freezing when use of the media is required. Transport of the ice cakes might not involve too many difficulties since low temperatures prevail in the higher altitude flying lanes.

Column evaluations showed media release from ice packages to have several excellent characteristics. Media release begins almost immediately after the ice cake reaches the bottom of the water column and the media soon distributes itself evenly throughout the water column. Release of the media is spread out over the period of time required for the ice to melt and thus the release is smooth. Problems with air bubbles are avoided and ballast requirements are much less than when dry media is utilized.

Freezing the media in an ice matrix was found to have no debilitating effects on the performance of floating ion-exchange resins. Packaging in ice did, however, appear to result in some loss of capacity in the case of the floating carbon. Parallel column removal tests run with a standard phenol solution showed that unfrozen carbon removed 65% of the phenol while an equal quantity of carbon frozen in an ice cake achieved only 51% phenol removal. Removal differences fluctuated in subsequent tests, but the carbon ice cakes were always found to be somewhat less effective than unfrozen carbon.

Parallel column runs with unfrozen carbon at solution temperatures of 0°C and 17°C showed little difference in phenol removal efficiencies (75% at the lower temperature and 77% at the higher). This observation led to the conclusion that the reduced efficiency of the carbon frozen in ice cakes was not a function of temperature so much as the result of interactions between the ice and the carbon. It was noted that finer carbon particles appeared to floc when released from the ice cake. Flocculation of carbon particles was not observed in release from plastic bottles. This flocculation effect could reduce the actual coverage of the media and hence reduce removal efficiency. In addition, it is possible that residual ice crystals on the carbon may interfere with surface forces and hence reduce sorption.

In deference to the major objectives of the program, these apparent interactions were not fully explored. Because of the lower ballast requirements and superior dispersion of the carbon near the bottom of the stream or lake, the carbon ice cake approach appears preferably in spite of the somewhat lower sorption capacity or rate. However, further work is needed to establish the feasibility of the carbon ice cake packaging concept. Clearly, there is great promise in the use of ice with floating ion-exchange resins.

GELS

Multipurpose Gelling Agent

> The information in this section is based on *Multipurpose Gelling Agent and Its Application to Spilled Hazardous Materials,* EPA-600/2-77-151 prepared by J.G. Michalovic, C.K. Akers, R.E. Baier and R.J. Pilie of Calspan Corporation for the Environmental Protection Agency.

The purpose of this investigation was to develop materials and equipment which could be used to gel and, thereby, immobilize a wide variety of liquid chemical compounds that may be spilled to the detriment of the environment.

The ideal countermeasure to any spill of a hazardous material is to terminate the spill as soon as possible by sealing the material in its original container. Once hazardous chemicals are on the ground, the most attractive secondary countermeasure consists of immobilization of the chemical to minimize the affected land area, to prevent flow of hazardous liquids to surface water, and to minimize percolation of liquids into the surface and subsoils. Immobilization of a spilled liquid with a gelling agent ideally leaves the material in a form that can be easily removed by mechanical means.

Under an earlier EPA sponsored program on "Methods to Treat, Control and Monitor Spilled Hazardous Materials" (27) a preliminary blend of materials was formulated that was useful for gelling both organic and inorganic liquids. The mixture was tested on 35 hazardous chemical compounds that were successfully gelled. This original blend, because of its ability to thicken and gel all hazardous liquids tested, was referred to as the "Multipurpose Gelling Agent" (MGA).

The first objective of this study was to determine the gelling agent formulation which would provide an optimum balance of gelling speed, ease of application and product cost. Generic components and suitable generic substitutes were identified as far as possible. The requirements for storage were identified and are reported herein. Consideration was also given to the physical form of the agent to make it more amenable to effective delivery to an interaction with a spilled liquid. Various physical forms of the MGA were evaluated with respect to gelling efficiency and ease of dispersion.

The second objective was to evaluate various types of dry dissemination devices that would be suitable for dispensing the gelling agent in a spill situation. Consideration was given only to "off-the-shelf" equipment. The types of devices showing the most promising principles of operation were field tested on simulated spill targets to define dispersion, delivery capacity, rate of delivery, portability, and unit cost.

The MGA developed under EPA Contract Number 68-01-0110 (27) was optimized to obtain a balanced formulation that would effectively gel a large variety of spilled hazardous liquids.

A formulation, Blend D, consisting of a polyacrylamide, a poly-tert-butylstyrene, a polyacrylonitrile rubber, a polycarboxymethyl cellulose, and a fumed silica was determined to be the optimum gelling agent based on its ability to immobilize the greatest variety of hazardous spill liquids with the least amount of material.

Also evaluated were alternate materials that could possibly be substituted for components of the MGA. Commercial sources for each of the ingredients were identified and preliminary generic descriptions obtained.

It was determined through early field testing that the powder form of the MGA could be deployed using air conveyance modes of dispersal but only with high losses due to effects of wind. Three compressed and granulated forms of the MGA were developed which are clearly superior to the original powdery blend for field dispersal onto liquid spill targets.

Various off-the-shelf dry solid dispersal devices were evaluated with the most promising dispersal modes field tested. Auger-fed, pneumatic conveyor systems (such as "Rockduster") originally developed for the coal mining industry were shown to be best suited for distribution of MGA at high rates and acceptable accuracy. The portable "sandblaster" was also found to be an acceptable delivery device.

Delivery nozzles bent at 30° to the horizontal plane of the dispersing tube provide better accuracy and agent distribution than straight pipe nozzles.

The MGA must be stored in moisture-proof containers at temperatures less than 50°C (122°F).

The MGA is now sufficiently well defined to be manufactured and stored in reasonably large batches. It is recommended that approximately 1,350 kg (3,000 lb) of this material be procured and utilized in large scale field experiments to more completely demonstrate its efficacy.

Auger-fed, pneumatic conveyor systems should be obtained and field tested, followed by redesign, if necessary, for MGA distribution to large spill targets.

Ultimate clean up measures to be applied following successful use of the MGA should be experimentally determined, and the prospects for economical recovery of the spilled liquid from the containing gel mass should be investigated.

Background: Under a completed EPA sponsored research program on "Methods to Treat, Control, and Monitor Spilled Hazardous Materials" (27), Calspan developed and evaluated a mixture of several dry chemical materials for immobilizing spills of hazardous liquid chemical substances. The hazardous liquid chemicals considered in that program are listed in Table 4.4. They were selected for study on the basis of toxicity, volumes shipped, and previous histories of spills.

There were a number of criteria used in the development of the MGA. One was that the gelling agent should be able to immobilize a broad spectrum of organic and inorganic liquids. The development of many specific agents would create a perhaps insurmountable logistics problem. A second criterion was that the presence of, or interaction of, the treatment agent should not create a serious secondary hazard, such as increased danger of fire or explosion.

The MGA developed represented a new composition of matter, created by the mechanical blending of equal weights of five specific components having the following properties.

The first component was a material of the highly water-soluble polyelectrolyte-type, typified by polyacrylamide. This material was selected for its ability to thicken water and aqueous solutions.

The second component of the blend was a loosely crosslinked copolymer of the class typified by poly-tertiary-butyl-styrene copolymerized with divinylbenzene. This material was selected to interact most strongly with those liquids having almost no polarity and only poor solvent power such as cyclohexane, gasoline fractions, and a variety of other inert spirits.

A third component was a material of the polyacrylonitrile butadiene copolymer class chosen to be especially effective against the very strongly polar organic chemicals such as acrylonitrile, ethylene dichloride, and other chlorinated or polar liquids.

The fourth required component of the blend was a material to cope with the most difficult of all hazardous liquids to thicken, solidify, and immobilize in place, as typified by methyl alcohol and other chemicals of the alcoholic class. Materials suitable for this use included the polycarboxymethylcellulose polymers.

This four-component blend uses the commercial products with the following trade names:

(1) Gelgard (Dow Chemical), to combat spills of aqueous liquids;

(2) Imbiber Beads (Dow Chemical), to combat spills of the inert spirits-type liquids;

(3) Hycar 1422 (BF Goodrich), to combat the polar organic chemical spills including the chlorinated hydrocarbons; and

(4) Carbopol 934,(BF Goodrich), to combat spills of alcoholic liquids.

For ease of delivery of this four-polymer blend, it required fluidization to ensure rapid, smooth egress from commercial spray equipment. Addition of fumed silica (e.g., Cab-O-Sil, Cabot Corporation) was used.

Table 4.4: Compounds Tested with the MGA

Acetone	Formaldehyde
Acetone cyanohydrin	Gasoline
Acrylonitrile	Isoprene
Ammonium hydroxide	Isopropyl alcohol
Aniline	Kerosene
Benzaldehyde	Methanol
Benzene	Methyl ethyl ketone
Butanol	Octane (2,2,4-trimethyl pentane)
Carbon disulfide	o-Dichlorobenzene
Carbon tetrachloride	Petroleum ether
Chlorine water (saturated)	Phenol (89%)
Chloroform	Pyridine
Cyclohexane	Sulfuric acid
Cyclohexanone	Tetrahydrofuran
Ethanol	Trichloroethylene
Ethyl acetate	Water
Ethylene dichloride	Xylene
Ethylene glycol	

Source: EPA-600/2-77-151

Gelling Agent Blend Optimization: The four powdered polymers, each capable of congealing at least one class of hazardous liquid into an immobile mass, were originally selected and combined in equal parts by weight and fluidized with fumed silica to form MGA.

To achieve an optimum blend, the proper proportioning of the above components had to be pursued. A number of laboratory tests were performed to obtain a formulation which would minimize material cost without sacrificing the effectiveness against a broad spectrum of spills.

Table 4.5 shows the formulations tested. Blend D was, of the first four tested, superior in gelling efficiency. Blend D was, therefore, selected as the principal mixture to be used in experiments pertaining to control of physical characteristics of the gelling agent and testing of dissemination equipment. Orders were placed for sufficient quantities to perform these experiments. Subsequently experiments were performed on substitute gelling agent materials which led to Blend E described below.

Table 4.5: Formulation of MGA Blends (percent by weight)

Blend	Gelgard	Imbiber Beads	Hycar	Carbopol	Cab-O-Sil	Klucel H
A	20	20	20	20	20	*
B	15	30	35	15	5	*
C	5	30	35	20	10	*
D	5	30	30	25	10	*
E	5	35	20	*	10	30

*Not used.

Source: EPA-600/2-77-151

A substitute material which has potential for the immobilization of alcohols, the most difficult of all hazardous liquids to gel, was evaluated. Klucel is a hydroxypropyl cellulose, nonionic, water-soluble cellulose ether. Klucels M and H both have the same structure, with M having a higher molecular weight. These materials are superior to Carbopol 934 in gelling the low molecular weight alcohols. In addition, the gels formed with the Klucel were more elastic and possessed greater coherence.

Klucel H was incorporated into Blend D as a substitute for Carbopol later in this study (Blend E). This blend was tested for dosage requirements and gelling speed. Blend E is slightly superior to Blend D; the overall advantages, however, were not considered to be so important as to warrant repetition of packaging and dissemination experiments performed with Blend D.

It should be noted that since the start of this work, the manufacture of Gelgard M has been discontinued by the Dow Chemical Corporation. Small supplies of Gelgard are still available, and Dow will resume manufacture if an order is sufficiently large. Substitute material has also been found that works similarly to Gelgard. This material (Kelzan) is discussed with other potential substitute materials in the next section.

In order to broaden the scope and universality of the gelling agent, a search for substitute gelling agent materials was performed. Manufacturers of gelling and thickening compounds were contacted and products obtained for laboratory evaluation.

The materials were evaluated against the following criteria:

material must be a dry powder with low moisture content;

material must form viscous gels without extreme stirring or heating; and

material gelling efficiency must compare directly with current MGA components.

Each of the potential substitute materials was screened using these representative liquids from the four major chemical classes:

Class I—liquids (aqueous solutions)—water
Class II—liquids (nonpolar organics)—cyclohexane/isooctane
Class III—liquids (polar organics)—trichloroethane
Class IV—liquids (alcohols)—methanol

Specific Product Evaluations as Substitute Gelling Agent Materials: Soloid (Kelco Company) — The commercial development data sheet states that "Soloid is a new solvent thickener that is soluble in most aromatic, aliphatic and chlorinated hydrocarbon solvents. Soloid is not soluble in methylene chloride."

Soloid was found not to wet with Class I liquids even when vigorously stirred. The material floated on the surface. Class II and Class III liquids gelled extremely quickly. Soloid did not thicken Class IV liquids but tended to swell approximately three times in size.

Soloid was compared with Imbiber beads in gelling isooctane (a Class II liquid). Although the Soloid compound formed a gel very fast, the Soloid powder tended to float on the surface. Mixing was needed to get the powder into the liquid mass, with a noticeable adverse "flour-bag" effect (gelling the exterior surface of masses of agent and rendering the inner material unavailable for reaction). It was found that approximately four times the amount of Imbiber beads compared to Soloid was required to gel isooctane.

Soloid is, therefore, a candidate substitute material for the MGA components for both Class II and Class III liquids.

Kelzan (Kelco Company) – Kelzan is a xanthan gum-polysaccharide which is a thickening agent for Class I liquids. The data sheet obtained shows 1% aqueous solutions to have high viscosities. Kelzan was compared with Gelgard M. When Kelzan was added to water, the powder spread immediately over the surface and formed a surface gel of sufficient strength to float additional Kelzan on top of the surface. Upon stirring, the powder would agglomerate around the stirrer, producing a large "flour-bag" effect. Approximately twice the volume of Kelzan compared to Gelgard, was needed to gel equal volumes of water. However, the "flour-bag" effect was so great that less than this amount of Kelzan was used in the gelling effort. This product is a potential candidate substitute component in the MGA.

Klucel H and M (Hercules) – Klucel was evaluated as a substitute material for Class IV liquids. A direct comparison was made between Klucel and Carbopol 934. It took approximately twice the amount of Carbopol 934 compared to Klucel to gel methanol. Thus, Klucel is a potential substitute component for the MGA.

Storage Requirements: Samples of the powdered form of the MGA, Blend D, were tested to find the limitations that are imposed upon storage. Containers of MGA were stored at 0 to 100% humidity and at -20° to 103°C (4° to 217°F). These samples were examined weekly for two months and less frequently thereafter.

The MGA stored at room temperature in an open container (relative humidity 50 to 80%) formed a lumpy mass while MGA stored at 100% relative humidity formed a solid spongy mass.

The powdered MGA when stored at high temperature (103°C, 217°F), changed color from white to deep tan in one month, indicating a possible degradation of the organic polymers. Tests on the partially degraded material showed that gelling effectiveness was decreased but not eliminated.

Samples were also stored for six months outdoors, exposed to the elements in containers to evaluate the type of container and the stability of the agent over wide temperature ranges. Fiber drums (Fibre-Pak, Continental Can Company) were filled with MGA and placed outdoors in an open unprotected area where they were subjected to the natural environment. The weather over the test period, February through July 1975, was typical of Western New York State.

The MGA powder stored showed no indication of any deterioration and there was no change in quality when compared to laboratory stored samples. The storage container was not damaged by the weather although there was superficial rusting of the metal parts.

Requirements for storage based on our laboratory evaluation are that the MGA must be stored at temperatures less than 50°C (122°F) and be free from moisture.

Preliminary Bioassay: A bioassay screening test was performed to determine the effect of various concentrations of MGA on creek minnows. This particular test was performed to evaluate the consequences of large amounts of MGA washing into a stream or lake. Fish were exposed for 72 hours to 0, 0.01, 0.05, 0.1, 0.5 and 1% MGA (by weight) in water. A 1% solution results in a viscous solution in which the fish could barely swim. The maximum safe concentration was found to be 0.01%. The acute effects of gelling agent were due mainly to the thickening solution which caused suffocation and gelled the fish into the surface layer. After 24 to 48 hours, MGA precipitated out of solution and remained at the bottom with no effect on surviving fish. Fish exposed to 0.05% MGA solution lost equilibrium for 30 minutes to 24 hours but recovered. The results presented, however, are only preliminary indicators of the toxicity of the MGA upon the water quality when it is used in spill control.

Physical Form of Gelling Agent: In the early stages of development, the MGA consisted of a fine powdery mixture of components. That consistency was selected to provide maximum surface area and thereby achieve maximum speed of reaction with spilled liquids. Early field experiments revealed some disadvantages of the powder form. The most serious was the drift of the fine powder from the target area even under conditions of light to moderate winds. In addition, the first particulates to contact the spilled liquid reacted so quickly that a thin surface film of gel was formed in upper layers which prevented penetration of additional agent into lower layers that had not been affected. Unless sufficient turbulence existed in the stream to provide the necessary mixing of agent with the spilled liquid, flow continued at the lower levels.

It was apparent from these experiments that the optimum consistency of the gelling agent would be one that preserved the large amount of surface area to promote rapid and complete reaction, and yet contained particles of sufficient size to be relatively unaffected by wind and sink into the volume of the spilled liquid rather than react only with the surface layers. Various techniques for pelletizing and forming loose aggregates of the powder were investigated in an attempt to achieve these results.

It was recognized early in this investigation, which was performed concurrently with investigations of dispersal techniques and equipment, that the utility of a given gelling agent form was in some cases dependent on the dispersal technique used. Since different dispersal equipment may be used for different spills, depending on the size, nature and location of the spill, the objective of this investigation was to select a material form which was usable with the widest variety of potential equipment and which was effective under static conditions without artificial mixing.

Optimization of MGA Physical Form – The material forms tested included:

the original fine powder;
the powder agglomerated by a water spray;

a variety of tablets;
crushed tablets that were sieve-graded into two-mesh ranges;
webbed, dustless powder;
a mat formed from the webbed powder; and
a roll compressed form chipped into proper sizes.

The standard test procedure used consisted of gently pouring 1.0 g of MGA into a vial containing 10 ml of test liquid (the 10:1 gelling ratio) and determining the time required for complete gelation. The criterion for gelation was that no flow should be observed when the vial was gently tipped. When gelation was not observed in this test the experiment was repeated with a doubling of MGA until the liquid was gelled (i.e., 5:1, 2.5:1, 1.25:1 gelling ratio).

Each of the MGA forms was tested against three test liquids from each of the four chemical classes (Table 4.6).

Table 4.6: Test Liquids

Class I
Saturated chlorine water
Concentrated ammonium hydroxide
Water

Class II
Kerosene
Benzene
Cyclohexane

Class III
o-Dichlorobenzene
Ethylene dichloride
Carbon tetrachloride

Class IV
Methanol
Ethylene glycol
Isopropanol

Source: EPA-600/2-77-151

The results of the tests may be summarized as follows. The "roll compressed" form, produced by chipping the solid layer resulting from roller-mill compression of the powder, and the two "graded" samples, produced by grinding MGA tablets and subsequent sieving for size distribution control, appear to represent the near-optimum compromise in material form. These do not react with spilled liquids as quickly as the original powder. However, even under light wind conditions, the quantity of MGA reaching the target area may be increased by factors as great as five. These three forms behave well in wind and in some liquids sink slowly through liquid rather than reacting with surface layers only and appear to be useful with all dispersal equipment with which they were tested. However, sufficiently large samples of these forms were not available for testing with all types of equipment.

From the standpoint of ease of manufacture, it appears that the "roll compressed" form has advantages over the other two. On the other hand, the gelling efficiency of the graded forms is slightly superior. The "roll compressed" form and the "graded" forms should be retained as candidate materials for large-scale testing.

The agglomerated form of material performed admirably in the laboratory tests, but was not significantly better than powder in windy conditions. Additional manufacturing costs (which are substantial for this form) are not justified.

The only other acceptable MGA form tested in these experiments was the original powder. Because of problems discussed earlier, however, it must be considered a poor fourth choice.

Spun Polyolefin Bag Evaluation – Several advantages could be achieved if the MGA could be disseminated in porous bags. Wind drift would be eliminated; retrieval of the MGA-spilled liquid gel would be greatly simplified; manual dissemination for special purposes, such as formation of a dike perimeter to prevent flow of a spill, could also be simplified.

To test the utility of porous bag packaging, spun polyolefin bags were obtained from Gedcor Environmental Protection Corporation and filled to half capacity with several of the MGA forms discussed in the previous section. The bags were submerged for 10 minutes in one chemical from each class of test liquids and removed for measurement of weight change due to absorbed material. In all cases, the quantity of liquid collected was substantially smaller than that gelled in previous experiments with the same agents freely disseminated. Upon examining the bag contents, it was found that reaction of the test liquids with the outer layer of gelling agent had formed a gel that protected the inner portions of the agent from contact with the liquid.

In an attempt to eliminate this "flour-bag" effect, numerous wicking agents (including paper pump fibers, straw, vermiculite, sawdust and IBM card chips) were mixed with the MGA before packaging, but no significant improvement was observed. Only the "roll compressed" form could be completely wetted and here the reaction rate, apparently limited by a partial "flour-bag" effect, was greatly retarded.

As a result of these simple experiments, the concept of packaging MGA in porous bags for dissemination was abandoned.

Conclusions – It appears that the optimum "graded" form should consist of particles between 0.5 and 2 mm. This material can be produced by the same manufacturing techniques as used during this program (exercising proper selection of sieve mesh).

The "roll compressed" form used on this program was inferior to the "graded" forms in static tests, but performed well in the field. As previously discussed, the reasons for the reduction in performance are understood and can very probably be corrected by slight adjustments in manufacturing procedures. Because of potential economic advantages of these manufacturing procedures, retention of the "roll-compressed" form for testing in large scale experiments is very much recommended.

Deployment Methods: Three kinds of off-the-shelf equipment proved to be useful in the small scale field experiments performed on this program. These include the following: pressurized tanks similar to equipment manufactured by Mine Safety Appliances (Company of Pittsburgh); venturi/compressed-air types manufactured for sandblasting purposes; and auger-fed pneumatic-conveyor equipment similar to the "Rockduster" (Mine Safety Appliances).

The "Rockduster" appears to be most suitable for treatment of large spills. Because of costs and the difficulty in transporting this equipment, it is probably suitable for distribution only with large stockpiles of gelling agent. Less expensive, easily transportable type equipment, such as sandblasters, would be more appropriate for distribution with smaller local supplies of the agent, such as at firehouses. All three forms should be retained as candidates for large-scale testing. Any of these units can be used with straight pipe nozzles or nozzles in which the exit tube is bent at approximately 30°. Large-scale tests should be performed with straight pipes equipped with detachable extensions bent at 30°.

High Expansion Systems

> The information in this section is based on *Control of Hazardous Chemical Spills by Physical Barriers,* EPA-R2-73-185 prepared by J.V. Friel, R.H. Hiltz and M.D. Marshall of MSA Research Corporation for the Environmental Protection Agency.

In the majority of spill situations, the demand for control materials will exceed the supply. An alternative to providing larger portable or mobile units is to increase the expansion ratio of the original starting material.

The search for candidate starting materials focused on materials which were polymeric in nature or capable of being polymerized or gelled to a highly expandable, rigid, chemically inert foam structure. There were two candidate materials—water-soluble polymers and conventional rigid foam systems. The water-based systems were capable of high expansion, but needed development in rigidizing. The rigid foams had not been developed for high expansion. Experiments with water-based systems are discussed in this section; rigid foam systems will be discussed later in this chapter.

Polyvinyl Alcohol-Borax Gel: Primary emphasis in the water-soluble polymers focused on solutions of polyvinyl alcohol since such systems can be solidified after foaming. A 1% solution of PVA can be readily foamed, especially with the aid of a surfactant. The interaction of borax with PVA is rather striking. For example, as little as 0.1% of borax in a solution of 5% PVA causes immediate gelation. Physically, the entire solution forms a firm, solid mass. All attempts to produce a high expansion foam from solutions of PVA and borax were unsuccessful. The major difficulty with this system was the rapid rate at which gelation occurred. An examination of the PVA-borax gel prepared without foaming indicated a texture that was firm, but somewhat brittle. The PVA-borax system was not considered a good choice for further development.

PVA-Congo Red Gels: Solutions of PVA can also be gelled by the addition of Congo red. This system forms a thermally reversible gel in that below 40° to 45°C the solution is solid, and above it is liquid. The warm liquid solution foams exceptionally well with or without the addition of a foaming agent or surfactant. The properties of the PVA-Congo red gel were found to be more consistent with the needs of this program. In general, the gels were extremely adhesive, tough and elastic. Foam masses dried to an extremely tough durable cover which adhered well to smooth surfaces such as glass and metal.

Generating low expansion foam from a 40° to 50°C solution could be accomplished by a number of methods. However, the foams blown could not be considered for diking systems. In general, foams blown from solution at 45°C were of low expansion requiring about 5 to 10 minutes to develop a reasonably nonflowing structure but some several hours to develop a strong matrix. The dried masses were surprisingly strong and resistant to water and other chemicals, such as acetone.

The results of preliminary tests were encouraging but the time required to develop strength along with the need for a warm solution were recognized as major weaknesses. In addition, preliminary results were obtained at a relatively low expansion of 30:1. Studies were conducted to circumvent the need for elevated temperatures and to examine the foam quality produced at higher expansion. Hopefully, eliminating the temperature requirement would also benefit the set time.

Results of these runs showed the PVA-Congo red foams without added surfactant could not be foamed. With an added foaming agent, a high expansion foam (150:1 expansion) was formed which possessed poor structural strength and no apparent life or stability.

PVA-Metal Ion Gels: Other reagents reportedly able to gel PVA were also examined (36). These included chromium(III) and titanium(IV). The results were similar to those obtained with Congo red. Foams could not be produced using the high expansion system without the presence of a foaming agent. Even with a foaming agent, the freshly generated foam failed to develop any structural strength. At expansions greater than 150:1, which could only be achieved with the use of added foaming agents, the life of the foam was exceptionally short. There was no true gelling and total collapse occurred within one hour.

Miscellaneous PVA Studies: In additional efforts with PVA systems, several attempts were made to minimize the use of some of the water by dispersing the Congo red in a mixture of methylene chloride and Freon 12. The intent was to use the Freon as a blowing agent. The PVA and the anhydrous Congo red solution were mixed and foamed into a warm tray to volatilize the Freon. Foams produced in this manner were of low expansion (30:1) and in all but one case were relatively quick to shrink and/or collapse. In addition to the above approach, numerous support materials were added to the Congo red-PVA system in an attempt to build immediate strength into the foamed mass.

Other Water-Soluble Polymer Systems: Emphasis shifted to other water-soluble polymer systems, particularly those which could be used at higher initial concentrations. Of special interest were the Acrysols. This class of materials is acidic and when neutralized, forms soft gels at concentrations of 1 to 2% of the neutralized polymer. The Acrysols are supplied at ~28% concentration in water. ASE60 was used as representative of the Acrysols. Neutralization and subsequent gelation can be induced by the addition of a number of basic materials including sodium and ammonium hydroxide, amines, etc. Neutralization by sodium carbonate was of special interest since carbon dioxide would be liberated to enhance the foaming action. Various neutralization methods were tested, but in all cases the foam possessed a wet strength no greater than that of shaving cream.

Application to Sinking Chemicals

The information in this section is based on *A Feasibility Study of Response Techniques for Discharges of Hazardous Chemicals that Sink,* CG-D-56-78 prepared by T.D. Hand, A.W. Ford, P.G. Malone, D.W. Thompson and R.B. Mercer of the U.S. Army Engineer Waterways Experiment Station for the U.S. Coast Guard.

Background: Gels have been proposed for use on land and surface water spills, but no literature has been uncovered related to spills of hazardous chemicals that sink. Ziegler and Lafornara (37) reported on the use of gelling agents on land spills. Polymer-based, water-soluble gels were placed on soil to prevent penetration of the spilled material. Density and water solubility were keys to the success of this method. Substances less dense than water and insoluble or only slightly soluble in water may be prevented from entering the soil (37). Further, substances miscible in water could be retarded by thick layers of gelled water. Heavier than water substances that were insoluble would pass through the barrier.

Pilie et al (27) have combined four powdered polymers into a universal gelling agent. Each polymer was effective against a certain class of hazardous chemicals: aqueous liquid, low-polarity inert spirits-type liquid (e.g., cyclohexane), polar organics, and alcohol. The feeling was that a universal agent could be applied with similar effectiveness to a wide range of spills. Initial application was intended to be on land or on the surface of water to reduce spreading or improve entrapment and rehandling of the spilled liquid. Reactions including polymers were noted to be exothermic, and explosion hazards are possible, depending on the concentration of polymer and spilled substances. Catalysts used in the polymerization process could also be harmful to aquatic life.

Field tests were conducted with various blends of gelling agents. Commercially available devices, including shovels, hand-pump dusters, dry chemical fire extinguishers, and sandblasters were used to disperse the gelling agent. On water, it was noted that some gelled materials would sink after a period of time and that bioassays revealed that fish ingesting the material died. It was noted that personnel handling these materials should be protected against skin contact with and inhalation of the gelling material. After entrapment of the spilled material by the gelling agent, the hazardous chemical could be easily removed by mechanical means.

Application to Sinking Chemicals: Application of gelling agents to sinking chemicals has not been made. However, the previous background information provides a framework for possible development of methods for gel deployment in the case of sinkers. A number of the liquid chemicals under consideration could conceivably be gelled by one or more of the components of the universal gelling agent.

A pressurized delivery system, with a broad dispersion head similar to that considered for application of sodium alginate films (38) could possibly be used for dispersion of the gelling agent. By bringing the gelling agent into close contact with the spilled liquid, it might be possible to gel it in situ, thus preventing further spread of the material and promoting early recovery using conventional dredging equipment.

The major problem with such a method is finding a gelling agent with sufficient specific gravity to remain on the bottom. It might be necessary to mix the agent with a carrier material to ensure successful bottom placement. Epstein and Widman (38) considered several carrier materials in their polymer film work, including titanium dioxide, barium sulfate, iron powder, Georgia kaolin, and others. Further, impacts on benthic and pelagic biota would have to be studied in order to determine whether or not the gelling agent would result in more harm to the environment than the spilled chemical.

Application of gelling agents is probably feasible in water up to 120 feet in depth. This is based on some of the predictions for performance mentioned in the literature. Some form of pressurized delivery system would be needed. Strong currents will present some problems, though no documentation was found to indicate the significance of currents on the technique. Gelling agents are probably applicable in all of the scenarios in this study with the exception of areas with very strong currents. The gelled material could perhaps be recovered using conventional dredging techniques, although this would be dependent on postdeployment properties of the material, which are in most cases not determined at this time. The greatest potential for gelling agents lies in their use as temporary mitigation techniques for hazardous liquid spills.

Several different types of gelling agents might be needed to cover the range of different liquid chemicals in this study which would increase the cost of this technique. Extensive development work is needed on the applicability of gelling agents to hazardous chemicals that sink. For these reasons, only a low rating can be assigned to gelling agents as a response technique at this time.

Treatment of Floating Chemicals

> The information in this section is based on *Feasibility Study of Response Techniques for Discharges of Hazardous Chemicals that Float on Water,* CG-D-56-77 prepared by J.S. Greer of MSA Research Corporation for the U.S. Coast Guard.

The formation of a gel influences the evaporation rate and therefore the vapor concentration of a spilled chemical. This influence is not great, however, and the primary benefit obtained from gelation is immobilization or confinement. This technique also has good flexibility, due to a great variety of gelling agents, good shelf-life and low toxicity.

The time required for the gelling reactions to be completed is a limiting factor for this technique. There are a few gels which may be formed within minutes, but most gelling reactions are measured in terms of hours. The "Universal Gelling Agent", developed by Calspan Corporation, has a specific blend of gelling agents combining the flexibility and rapid reaction rates required for treating spills of floating, hazardous chemicals.

Gelation or emulsification techniques affect spills of hazardous chemicals within five distinct areas: (1) detection and monitoring; (2) vaporization rate reductions; (3) recoverable forms; (4) recycle or reclamation; and (5) a base for the application of other amelioration techniques.

Detection of spilled chemicals and monitoring of the surface spread is an important part of responding to spills of floating, hazardous chemicals. Immobilization and containment, by the formation of the gel or emulsion, significantly reduces the problems of distinguishing surface phases. There is little effect on the detection and monitoring of vapor concentrations, however.

Evaporation rate reduction is achieved by forming a continuous cover of gelled material to encapsulate the more volatile spilled liquid. Gelling techniques which result in the formation of a thin, surface barrier are more efficient for reducing evaporation than those gelling a thick mass of material. The ability to limit or stop the spread of hazardous chemicals over the water surface also reduces evaporation since the amount of vapor released is proportional to the surface area.

Recovery or removal of spilled chemicals often presents a problem even after steps are taken to decrease the hazards. Gelling agents interact with the chemical to form a mobile phase, which is more easily removed by mechanical means.

There is a possibility for reclamation and recycle of the hazardous chemicals after the gelling technique has been used. The toxicity of some compounds is not diminished by common disposal techniques (i.e., landfill or incineration). The combination of expensive materials, ready markets and harmful waste products for these chemicals increases the desirability for a method of reclamation. The gel produced by this technique provides a stable base of immobilized material, which could be used in conjunction with other response techniques, to provide additional protection. This compatibility with other response techniques is an asset which has not been utilized in the development of combination response techniques.

The major factors influencing the rate of evaporation from gelled masses are: (1) extent of physical and chemical interaction; (2) homogeneity of gel; and (3) environmental factors.

Gelling agents interact with the spilled chemical to considerably increase viscosity and form a semisolid (gel). These polymerization type reactions are often quite exothermic, however, and the reactants must be critically chosen to avoid excessive heat release while forming stable gels. The extent to which this compromise can be attained governs the release of spill vapors and eventual reclamation.

A homogeneous layer of gelling material (and the gel formed) also promotes the formation of a stable gel. Particle size and distribution equipment influence this parameter and must be chosen to compliment each other. Environmental factors, such as ambient temperatures and wind, influence the effectiveness of gelation. Both affect the gelling reaction and final integrity of the gelled layer as it is formed on the surface of a body of water.

Vaporization rate reduction by gels is the result of the formation of a continuous layer of chemically and physically bonded material and the reduction of convection currents and heat transfer within the spilled chemical. Experimental measurements in this investigation showed only 40% reduction of the evaporation rate of benzene.

The flammability hazard was another part of the evaluation of gelling agents as a response technique for spills of floating hazardous chemicals. Glassman et al (39) have reported that convective flow is an important parameter in determining the flammability of pools of liquid fuels. They concluded that "thickening fuels decreases the flame-spreading rate, but increases the ignitability". Therefore, gelling agents would need to have an inhibitor included in their formulation to reduce flammability hazards.

The amounts of material and rates of application are within the compatibilities of existing equipment. Costs, although significant, are not unreasonable and could possibly be reduced by judicious selections of gelling agents and dispersal equipment. The efficiency index rating of this technique is acceptable.

A major attribute of the gelling agent technique is its ability to control the spread of a hazardous chemical spill and significantly aid confinement methods. This is particularly applicable in assessing the burden to the environment. The semisolids produced by gelation are more easily removed and reclaimed from aquatic environments. The only negative aspect found during the development testing program was when the gelled material was mistaken for food by fish.

The prime benefit of gels may be as an adjunct to foam. If the spill can be immobilized, the surface area can be minimized and foam more readily applied. Further, since gels do decrease vapor release they will provide an add-on to the foam block extending the time required for vapor permeation through the foam and/or reduce the minimum foam thickness required for control.

The major part gels may play, although other materials may also be effective, is as a base for foam support. It has been shown that foam is degraded by a number of chemicals in the hazardous list. In many cases this degradation occurs only upon contact of the foam with the liquid. If the foam can bridge the liquid so that it contacts vapor only, the degradation is minimized if not eliminated. Gelation, at least, of the spill surface, may be a means of avoiding foam-liquid contact.

Immobilization of Land Spills with Universal Gelling Agent

> The information in this section is based on *Methods to Treat, Control, and Monitor Spilled Hazardous Materials,* EPA-670/-2-75-042 prepared by R.J. Pilie, R.E. Baier, R.C. Ziegler, R.P. Leonard, J.G. Michalovic, S.L. Peck and D.H. Bock of Calspan Corporation for the Environmental Protection Agency.

The extent of environmental damage produced by any spill is dependent on the nature and quantity of spilled material and the distribution of the spill within the environment. Obvious advantages can accrue whenever it is possible to minimize the areal or volumetric extent of hazardous concentrations of spilled material either by interrupting the spill before containers are empty or by preventing the spread of material that has already spilled on land to a nearby waterway. A significant fraction of this project was devoted to these ends, and a wide variety of problems was considered.

Ideally, a spill should be terminated as soon as possible by sealing leaking or split containers before they are empty. When hazardous chemicals do spill onto the ground, immobilization of the chemical to prevent flow of hazardous liquids to surface water and to minimize percolation of liquids to subterranean aquifers should be the next line of defense. If spilled material reaches a watercourse, it is desirable to minimize the spread of floating liquids across the surface or trap water immiscible liquids in streams and ditches without interrupting the flow of water. In all of the above cases, additional advantages result if the immobilization procedure leaves the spilled material in a form which can be safely and quickly removed and packaged for further treatment, shipment or disposal.

This section discusses "universal gelling agent" (also known as "multipurpose gelling agent" and discussed in detail earlier in this chapter) as used to seal narrow splits in containers, to completely immobilize liquids on land, to prevent percolation into the soil, to reduce surface spreading and improve the effectiveness of booms on water, and to facilitate the trapping of liquids floating on streams in small mesh nets or screens. The resulting gel was easily removed in convenient form for subsequent treatment or disposal.

Universal Gelling Agent: One major subclass of chemicals transported in huge volumes includes those precursor organic chemicals which by polymerization, thickening, or other chemical techniques are turned into solid end products, such as organic monomers which react to form the polymers which are the common plastics of everyday commerce. An obvious immobilization technique for this class of chemicals was to polymerize them in place at the site of the spill either on the ground or in the leaking tank.

Most polymerization reactions, however, are exothermic, giving off substantial heat as they proceed so that explosion dangers are inherent in the solidification of material by this type of uncontrolled self-polymerization. In addition, a further environmental danger would have derived from the use of polymerization catalysts of the common type such as benzoyl peroxide or lauroyl peroxide, since the catalysts, themselves, are often poisonous and explosive. More innocuous thickening, solidifying or immobilizing techniques were desired. Over the course of the project, such a countermeasure was developed to encompass a wide variety of hazardous spills, including monomers, organic solvents, feed stock chemicals, and inorganic reactants.

The mechanism by which the "universal gelling agent" acts is as follows. The individual gel components selectively interact with the appropriate chemicals themselves to create an immobile semisolid mass which is easily removed by mechanical means. This is in contrast to other spill immobilization techniques which have been described and demonstrated, which simply absorb the spilled liquid into a finely powdered mass of inexpensive, easily accessible, easily deployable materials (such as fly ash or Portland cement) with few, if any, secondary environmental hazards.

The "universal gelling agent" represents a formulation created by the mechanical blending of at least four and preferably five or more specific ingredients, each having a specific purpose. The first ingredient is a material of the highly water-soluble polyelectrolyte-type, typified by polyacrylamide. This material could be substituted by any of a number of other polymers including proteinaceous materials such as gelatin and casein. It is important that this powder, and all the other components of the blend, be particle-size controlled within a precise range for speedy interaction with the target liquid, and for ease of deployment.

It should also be manufactured or admixed with a small surface-active additive to assure rapid contact with aqueous liquids.

The second component of the blend is a loosely crosslinked copolymer of the ilk typified by polytertiary-butylstyrene copolymerized with divinylbenzene. This material is selected to interact most strongly with those liquids having low polarity (such as cyclohexane, gasoline fractions, and a variety of other inert spirits).

A third component is a material of the polyacrylonitrile-butadiene copolymer class which is chosen to be especially effective against polar organic chemicals such as acrylonitrile, ethylene dichloride, and other chlorinated liquids.

The fourth component of the gelling agent is a material to cope with the most difficult of all hazardous liquids to thicken, solidify, and immobilize in place, as typified by methyl alcohol and other chemicals of the alcoholic class. Materials suitable for this use include the polycarboxymethylcellulose polymers or the polyethylene oxide materials. For these types of polymers, one of the less expensive polysaccharide exudates produced by bacterial cultures can be substituted.

The four component blend used experimentally on this program consisted of equal parts of: (1) Dow Chemical Corporation, Gelgard, to combat spills of aqueous liquids; (2) Dow Chemical Corporation, Imbiber Beads, for spills of the inert spirits-type liquids (typified by cyclohexane); (3) BF Goodrich Corporation, Hycar 1422, to combat the polar organic chemical spills including the chlorinated hydrocarbons; and (4) BF Goodrich Corporation, Carbopol (and in some cases, Union Carbide, Polyox) to selectively thicken and control alcohol spills.

Ease of delivery of this four-polymer blend required fluidization to ensure rapid, smooth egress from commercial spray equipment. A one-fifth by proportion addition of fumed silica Cab-O-Sil, produced by Cabot Corporation was used. It is well-known in the paint and pigment industry that such finely powdered silicas can be used as thickeners for most organic vehicles. While the kinetics of such thickening action are far too slow to be useful in the manner envisaged here, the addition of Cab-O-Sil to the "universal gelling agent" for the primary purpose of fluidizing the blend for ease of field deployment had the secondary benefit of providing a stiffer, hydrolysis-resistant gel of almost all hazardous liquids tested.

Field Experiments: Most of the field experiments performed on the project were designed to investigate methods for immobilizing hazardous chemicals. These experiments were performed at a half square mile tract of abandoned farmland owned by Calspan near Bethany, New York. After a series of soil tests showed that the possibility of significant percolation of purposely-spilled chemicals into subterranean aquifers or surface drainage into nearby streams was negligible, the experimental facilities described below were constructed.

Two sets of ditches, 2 feet wide by 2 feet deep and 100 feet long, were constructed with 1 and 2% grades, respectively, to determine the effectiveness of immobilization procedures against surface flow of spilled materials. At the foot of each set of ditches, a large excavation was constructed to house two 12-foot diameter, 3-foot deep, plastic-lined swimming pools intended to capture all spilled material before escaping to the natural environment.

In accordance with the general theme of this project, the chemicals used for experimental spills in the field were selected to represent classes of materials that pose serious hazards in the real world. Pure water was used to simulate weak aqueous solutions. Cyclohexane was selected as a serious real threat which is representative of chemicals which are immiscible with water and float on the surface, while ethylene dichloride was selected to represent immiscible chemicals that are more dense than water.

Sulfuric acid, the hazardous chemical shipped (through the United States) in greatest quantity, was chosen in an attempt to solve the special problem of frothing that has produced serious equipment damage during treatment of land spills of acids and bases with neutralizing agents. In attempts to immobilize heavy metals, the eventual fate of the compound used in the experiments could not be predicted. Therefore, a relatively harmless compound, manganese sulfate, was selected for experimental use. The results of these experiments are summarized below.

Immobilization of Land Spills with Gelling Agents: To minimize cost and cleanup problems after experiments, specific gelling components were used in all large-scale field experiments. The "universal" agent was tested only in the laboratory.

Initial immobilization experiments in the field were performed with pure water spills to represent weak aqueous solutions and to develop safe and effective field test treatment procedures before using hazardous chemicals. In these experiments, Gelgard was used exclusively as the gelling agent. Ten such experiments were performed on 55-gallon water spills in the 1 and 2% ditches.

To minimize the area affected by the spill, it is necessary to inhibit flow as soon as possible. It is therefore particularly important to treat the head of the flow first. Not only does this create a dam to interrupt the flow, but the head is a region of turbulence which produces excellent mixing of the agent with the liquid, promoting efficient treatment. Typically, the first dam of congealed material reached a depth of ½ to 1 inch before overflow began. It was most effective then to move downslope to form a second dam. With the flow inhibited by the first gel dam, the second was more easily formed, and by continued treatment was built to depths sometimes exceeding 2 inches before overflow began. By progressive treatment in this way the final dam usually exceeded four inches in depth with a 55-gallon spill before flow was terminated in the ditch. Once this was accomplished, it was a simple matter to move upstream and treat the liquid trapped behind the dams that had been formed earlier.

Several dispersal methods were tested with Gelgard in these experiments. It was apparent that broadcasting the agent with shovels was a simple and effective procedure for spills with minimum dimension of four to five feet. Much of the material was lost in the narrow ditches with this procedure. When attempts were made to sprinkle the material from shovels, the distribution of the agent across the surface was usually uneven. As water flowed around the thicker regions, an impenetrable gel formed at the surface, producing a large clump with dry, unused agent at its center. For more uniform distribution, sprinklers resembling salt shakers were constructed from three-pound coffee cans to which six-foot long handles were attached. These sprinklers proved to be very effective for treating in narrow confinements and were used extensively in subsequent experiments.

Hand pump dusters normally used for insecticide application in home gardens could not deliver the agent fast enough to stop a ½-gallon per second flow in the 2-foot wide ditches. Larger hand-powered dusters intended for agricultural use delivered material at an adequate rate but were so tiring that continuous operation could not be maintained. This type of duster could be equipped with battery-powered explosion-proof motors to produce effective portable equipment for treatment of spills that have reached remote areas (e.g., where the spilled material flows into heavily wooded areas through gulleys or small streams).

Dry chemical fire extinguishers produced an airborne plume of agent that was too wide to treat spills in confined areas. Since fire extinguishers cannot hold a sufficient mass of agent for treating large spills, they do not seem practical for use against material that is already on the ground. However, with a modified nozzle design, they could be made into very effective portable devices for sealing narrow splits in containers.

High pressure dispersal devices appear to be suitable for immobilizing large spills that require large amounts of gelling agent where the use of compressor-operated equipment is appropriate. Although paint sprayers proved effective, they could not deliver the agent at a sufficient rate to treat large spills. In these experiments, sand blasters, delivering 5 to 10 pounds of agent per minute, provided an appropriate distribution for large spills even though some of the agent was always blown out of the simulated spill area by the wind.

The bulk of gelled material was easily removed mechanically by shoveling the material into 55-gallon drums. Heavy earth-moving equipment could be useful for large spills of nonflammable materials, but should be avoided where a fire hazard exists. The sticky consistency of most gelled chemicals makes pumping inappropriate at least for land spills.

Typically, 75 to 85% of the spilled material was recovered in gelled form during the experiments. It was found that some material was always inaccessible to shovels after gelling chemicals on land. But the amount was a small fraction of the mass lost in control experiments by a combination of evaporation and percolation into the soil when the ditches had not been presaturated with water before the experiment. In some cases in which Gelgard is used on highways or city streets, thorough washdown should be required to eliminate the lubricating effect of dilute mixtures of this material and water. Like most polyelectrolytes, Gelgard produces a nearly friction-free environment that could be extremely hazardous to pedestrian and automotive traffic. Manual scrubbing followed by thorough rinsing would be adequate for small spills. Vacuum scrubbers commonly used on airports for hangar floor and ramp washing would be ideal for clean-up after treatment of large spills. Before leaving the area of a spill treated with Gelgard, all affected surfaces should be tested while wet, since friction is not affected by the dry form.

In addition to the experiments with water, experiments were performed in the ditches to test immobilization procedures against cyclohexane and ethylene dichloride. In both cases, the 2% ditches were saturated with water but no standing water was present. Seventy-five pounds of Imbiber beads were used to completely immobilize the 55-gallon cyclohexane spill within 80 feet of the spill point. Forty-one and one-half pounds of Hycar arrested the 55-gallons of ethylene dichloride in slightly less than 70 feet. After the treatment was completed in each case, water was pumped through four garden hoses into the spill zone to determine if the gelled material could be dislodged. As water seeped under the cyclohexane, buoyancy dislodged approximately 50% of the material in 30 minutes, but none flowed down the ditch. The ethylene dichloride was unaffected. A surge flow of water, produced by releasing 55 gallons at the head of the ditch in 2 minutes, washed approximately 75% of the gelled cyclohexane into the pool. The consistency of the floating material was such that it could easily have been trapped with a small mesh net or screen. Again, the ethylene dichloride was unaffected. In both cases, a total of 75% of the spilled material was recovered in gelled form by shoveling and 75% of the respective control spills was flowed into drums placed at the outfall 100 feet downstream. Thus, it was demonstrated that significant benefit can be achieved by using gelling agents to immobilize land spills of these two classes of materials.

Protection of Personnel and Property: The experience acquired in preparing for and executing these experiments constituted an extremely valuable portion of the overall results. Since the chemicals used in these demonstrations were toxic by skin contact and inhalation, and represented significant fire hazards, a variety of safety precautions had to be followed. Similar and improved preparations must be made in advance for the protection of members of clean-up teams for real spills.

Gas masks and air packs were provided to team members working in areas of potential hazards from vapors of the spilled material. Oxygen was available in case of inadvertent inhalation of vapors in spite of these precautions. In addition, two vapor sensors were used to warn against potential hazards either from inhalation or fire.

Protective clothing, including boots, trousers with elastic ankle bands, slip-over jacket with hood and elastic waistbands, and gloves, were provided to all workers. When worn with face masks, the clothing seemed quite adequate as temporary protection against skin contact. Garden hoses with running water were available continuously to rinse these garments in case of minor splashing or immersion. A small plastic swimming pool filled with water was maintained on site for total immersion of any team member suffering extensive contact with the spilled chemical.

Two types of protective suits were used in these experiments, conventional reusable garments, and "Safety First" disposable two-piece plastic suits with hood, available from Protective Clothing Supply, 1243 Military Road, Buffalo, N.Y. The latter were light and fairly comfortable but were more susceptible to tears and abrasions than the heavier reusable garments available. Both types of protective gear were extremely warm when worn in sunshine and neither type offered significant protection against fire.

Significant improvements could be made in protective clothing. For example, incorporating a metallized finish with the fabric or plastic would seem appropriate to reflect heat from the sun for increased comfort and, more importantly, to offer some protection against the initial flash and subsequent radiation from fires.

With the materials used in these experiments (and many of the spills that occur in the real world), the fire hazard is by far the most serious. The potential fire hazards must be considered in each spill and suitable tactical measures must be implemented for protection of personnel and property. Additional research is required to develop countermeasures which are specifically applicable to the spill situation.

Only two concepts for minimizing the fire hazard were tested on this program. Attempts were made to mix a variety of dry chemical fire-fighting agents with the universal gelling agent in the hope that, if a fire started, the CO_2 generated by combustion of the gel would extinguish the fire or at least reduce its intensity. However, laboratory experiments showed that even with 50% dry chemical fire-fighting agents, there was no significant reduction of hazard. Dry chemicals and foam were effective, however, for extinguishing fires fueled by the gelled material.

The second approach, that of reducing evaporation of spilled material to maintain incombustible vapor-air mixtures above the spill, was more successful. To accomplish this end, the spilled material was treated with a high-density fire-fighting foam before application of the gelling agent.

These experiments suggest that in a large spill the fire hazard due to spilled liquids might first be minimized by coating the entire spill with foam. The universal gelling agent might then be used to treat relatively small portions of the spill-affected area. When gelling is complete in one area, additional foam could be applied while the gelling treatment is applied elsewhere. By such a procedure, the overall fire hazard might be kept to manageable proportions. Before firm conclusions can be drawn, however, testing would be required.

In addition, only high-density foams were tested on the program. Considering the variety of fire-fighting foams and application techniques available, substantially more testing is warranted before making specific recommendations.

Immobilization of Heavy Metal Spills on Land: Solutions of heavy metals can be immobilized on the surface by congealing the solution with the universal gelling agents or absorption with fly ash or Portland cement. Whatever fraction of the solution that has percolated into the soil is unaffected by these procedures, however. It was postulated that this material could be immobilized by application of sodium sulfide to the contaminated soil in order to convert the soluble salt to an insoluble sulfide. Heavy metal sulfide particles formed underground should be completely immobile except by erosion of the land surface. If the sodium sulfide treatment is used for treatment of land spills, it must be recognized that significant overtreatment may be necessary due to formation of sulfides of other competing metals already in the soil.

Logistics Immobilization Procedures: For this study, use of the universal gelling agent was considered and the following assumptions made. (1) To be effective, attempts to seal leaking containers should be made within half an hour of the occurrence of the spill-producing accident so that small-scale local supplies and equipment are required. (2) To be effective, immobilization procedures for materials that are already on the surface should be initiated within a few hours after the occurrence of the spill.

The only practical concept visualized for maintaining local supplies of the universal gelling agent available for sealing leaky containers is to provide such supplies and equipment to either local fire departments or local law enforcement agencies. It is anticipated that equipment similar to dry chemical fire extinguishers with specially designed nozzles suitable for ejecting a narrow stream of gelling agent five to ten feet from the operator can be developed for this purpose. Ten thousand local organizations might conceivably be equipped.

The requirements for distribution of sufficient equipment and supplies for treatment of material on the surface is substantially more complex and must be considered on a regional basis. The distribution of caches located in Nevada, for example, need not be as dense as in the industrial regions of the northeast. Delivery of equipment and supplies from the cache to a spill in metropolitan areas must almost necessarily be by truck. In rural areas, delivery and direct dispersal from helicopters would be advantageous in many cases. Packaging problems are significantly different in the two cases.

Without attempting to solve these problems, it was assumed that as many as 200 sites throughout the country might be equipped and supplied for immobilizing spilled chemicals with the universal gelling agent and that delivery would be accomplished with ground transportation.

FOAMS

The information in this section is based on *Control of Hazardous Chemical Spills by Physical Barriers,* EPA-R2-73-185 prepared by J.V. Friel, R.H. Hiltz and M.D. Marshall of MSA Research Corporation for the Environmental Protection Agency.

Control of the spill of the hazardous chemicals from transporting vehicles must be considered as one major area in the prevention of hazardous chemical materials from damaging the water ecosystem. Available technology should allow the successful evolution of both portable and mobile systems which could provide surface containment of spilled materials at the site of the accident either by the formation of dikes or flow-diverting barriers. Technology is available to provide in situ generation of surface covers to prevent vapor losses from such spills. Completely contained, the spilled materials can then be cleaned up by a number of techniques. Subsurface losses, seepage and the like, would require additional controls.

In considering the erection of dikes or diversionary barriers the initial thought is usually to natural materials. Although significant use should be made of natural barriers and depressions by diverting spills toward them, their use to construct dikes or other barriers pose a number of problems. Not all types of surface materials are amenable to the formation of dikes of any significant height. When the ground is frozen or water-soaked, movement by hand is difficult if not impossible, and under the worst conditions even heavy equipment may not be sufficient.

To provide a general purpose system operable under all of the environmental conditions to be experienced, it becomes necessary to consider structures built up out of artificial materials. Whether these materials form an enclosure or divert the flow of spilled materials, the basic requirements are the same. Materials should be resistant to chemical attack, nontoxic, disposable and fire retardant if not inflammable. Most important, maximum benefit should be realized from as small an amount of material as possible using portable or mobile equipment with small power requirements.

In assessing the requirements, the limiting value would appear to be the necessity of obtaining large quantities of material rapidly from small portable or mobile systems. These conditions appear best satisfied by materials which can be foamed or expanded. Two material classes possess the basic requirements. These are polymer foams and so-called foamed inorganic materials.

Although a large number of polymers can be produced in the foamed or expanded condition, only polyurethane appears to possess the necessary characteristics for field operable units. Formed by the exothermic reaction of two materials, the heat generated volatilizes and expands a reaction product or an added blowing agent which creates the foam. Polymerization occurs at the same time, setting the foam to a rigid mass. The only power necessary is pressure to expel and blend the two components. Other polymer foams require mechanical agitation and may be of open cell structure and thus form a porous medium.

There is a broad spectrum of inorganic materials which can be produced in lightweight form by blending them with a prefoamed detergent or protein foam. The major item in this class is foamed concrete. Foamed gypsum is also an available material, as well as sodium silicate foam. The latter, however, requires mechanical agitation, is partially open-cell and subject to hydrolysis.

Once contained, rapid cleanup of the spill is almost mandatory. Vapor release to the air and seepage into the ground constitute a signficant material loss which poses immediate problems to the surrounding environment. Groundwater movement can transport spilled material into the water system.

Large exposed surfaces of spilled chemicals also pose an immediate threat, in terms of toxicity and flammability, to the personnel attempting to control and cleanup the chemical as well as other individuals in downwind locations.

Isolation of the chemical surface along with control of boil-off would appear to be desirable. The surface covering should be such that it does not interfere with the cleanup operation. In the past year foam has been shown to be effective as a surface cover to isolate and control vaporization rate of selected chemicals such as gasoline, ammonia and vinyl chloride monomer.

Foam Technology

> The information in this and the following section is based on *Feasibility Study of Response Techniques for Discharges of Hazardous Chemicals that Float on Water,* CG-D-56-77 prepared by J.S. Greer of MSA Research Corporation for the U.S. Coast Guard.

The overall efficiency of spill amelioration techniques is controlled by their physical and chemical compatibilities with the hazardous chemicals. This compatibility can be selected, or modified to some extent, by the formulation of the foam. Soft foam systems may utilize one of two foaming agents: (1) protein extracts or natural surfactants; and (2) synthetic surfactants.

(1) Protein foams have been developed for many different applications in recent years. Special alcohol foams have been developed for fire extinguishment, but as yet they will not provide a useful life when placed upon polar solvents and reactive chemicals.

(2) Synthetic foams have somewhat greater flexibility in that there are different classes. Anionic, cationic, nonionic and amphoteric properties are used to denote the major classes in this formulation system. There is a separate class of synthetic foam formulations which employ fluorocarbon or silicone surfactants as a means of forming a surface film. These formulations (aqueous film-forming foams) offer an alternative approach to ameliorate spills. They are not considered within the scope of this section since they are nonpersistent foams.

The Foaming Process: The production of foam is facilitated by lowering the surface tension of a foaming solution. The initial height of the foam may be correlated with the surface tension of the foaming solution; greater initial foam height produced by foaming solutions will lower surface tension.

However, for foam to have any stability, the foam film produced must have some elasticity, i.e., as the film thins and stretches, there must be some restoring force generated by the stretching process which opposes this effect and prevents the stretching from continuing with eventual rupture of the film. It is for this reason that pure liquids, even with low surface tension, do not foam. In solutions of surface-active agents, this restoring force is believed to be due to the higher surface tension which a newly increased surface shows. According to this mechanism, when a portion of a foam film is stretched, the local concentration of surfactant is decreased momentarily and the surface tension at the stretched spot becomes momentarily greater than the tension in the surrounding areas. This causes a surface tension gradient, radiating out from the stretched spot, which draws liquid in from the surrounding areas, countering the stretching of the film and thickening it.

For this mechanism to act effectively, it is essential that the surface tension at the stretched portion not be reduced too rapidly; i.e., the reduction of the surface tension to its equilibrium value by surfactant migrating to the surface from the interior should not take place before sufficient material from the perimeter has been moved in to thicken the film under the influence of the surface tension gradient. Therefore, a surfactant which is a good wetting agent, i.e., rapidly reduces the surface tension at an extended interface, can act as a defoamer by preventing the operation of this film-healing mechanism.

The general concensus concerning foam collapse is that the prelude to breakage is drainage from the cell walls that leaves the bubble in a delicate or friable condition. Subsequent rupture invariably results in the generation of a mist of foam fragments. This behavior is considered similar to that observed with distorted liquid droplets in an air stream. Foams begin to collapse as soon as they are generated.

Foam Generation: There are three mechanisms used to generate foam. These are mechanical agitation, impaction of a spray against a screen or bubble formation through controlled orifices. Mechanical agitation normally results in a low expansion foam less than 80:1. It is characterized by small bubble size and good stability.

The impaction of a spray against a screen is the most common method of generating medium to high expansion foam from 100:1 to as high as 1,500:1. The air velocity required to generate high expansion foam is usually provided by a fan. However, air can be induced to obtain a good foam by using high pressure water sprays. This eliminates the need for a fan or an external power supply. Aspirating generators have limitations. Expansion is restricted to a maximum of 400:1 and individual units are restricted in size.

The generation of foam through controlled orifices has seen limited commercial utilization. These use a system termed a flooded plate generator. This is a compartmentalized unit in the form of a box. A horizontal perforated plate separates it into two halves. Foam solution flows over the top of the perforated plate and the flowing air is introduced below the plate. Bubble size (expansion) is controlled by the size and spacing of the perforations. This system has the advantage of insuring complete containment of all air in the foam with no leakage.

Its use has been limited to the generation of foam from solutions too viscous to be handled in other types of generators. Expansions have usually not gone above 200:1, but higher expansions appear possible.

Foam Collapse: Surfactant foams begin to collapse as soon as they are generated. The rate varies with the surfactant used. Drainage is a controlling factor; the faster the water drains the more rapid the natural collapse and the more susceptible the foam is to chemical or mechanical influences. Without outside influences surfactant foams exhibit linear collapse rates in the range of 0.2 to 2 m/hr (8 to 80 inches per hour).

Collapse can be slowed down in anionic foams and the best of fire fighting foams show rates in the 20 to 30.5 cm (8 to 12 inches) per hour range. This is accomplished by balancing the chemistry in the bubble wall. Further reductions are possible by the incorporation of water-soluble polymers. Mecca et al (40) report a foam formulation which collapsed at the rate of only 0.25 cm (0.1 inch) per hour.

Foams as a Response to Floating Spills

The efficiency with which foams reduce evaporation from a simulated pool was found to depend upon seven factors: (1) chemical compatibility; (2) foam layer thickness; (3) surfactant concentration; (4) water-solubility of vapors; (5) expansion ratio; (6) thermal insulating capability; and (7) environmental factors.

Without a foam cover, factors such as wind, surface turbulence, liquid convection currents and spreading coefficients would have greater cumulative effects on the rate of evaporation.

The chemical compatibility of foams with floating, hazardous materials is the most important factor influencing the formation of a foam cover. Foams are generally compatible with nonpolar compounds and with those having a pH close to water (neutral).

Developments in foam formulations have expanded their compatibility with polar compounds, but the extended lives necessary for controlling vaporization and vapor hazards limit their use to compounds having a dielectric constant less than 3. Each formulation offers slight variations of chemical compatibility which may be used to extend the number of compounds which may be treated.

High Expansion Foams: "Type L foam", manufactured by Mine Safety Appliances Company, and "Universal Foam", manufactured by National Foam Systems, Incorporated, were the only formulations found to be generally compatible with the floating, hazardous chemicals. Type L foam would be used as a 30% solution of the foam concentrate and Universal Foam would be used as a 10% solution of the foam concentrate in water. Type L foam is the more stable, having a collapse rate of 1.27 cm/day (0.5 inch per day), and was more impermeable to the chemicals used in the experimental tests done in this investigation.

Some experimental tests were performed during this investigation to measure the vapor permeability of these foams. These tests showed that a 40 cm (16 inch) layer of high expansion foam (~200:1 volume ratio) would be sufficient to reduce vaporization and ameliorate flammability hazards. A 15 cm (6 inch) layer of low expansion foam (~10:1 volume ratio) would provide similar protection.

A high expansion (~200:1) foam cover for the largest typical spill would require the generation of approximately 1,520 m^3 (53,700 ft^3) of foam. This approach would include handling approximately 7.6 m^3 (270 ft^3) or 7,600 liters (2,000 gallons) of liquid in the high-expansion foam generating equipment.

Most commercial foam generators produce foams of higher expansion ratios than the 200:1 recommended for vaporization suppression. This lower expansion is the preferable compromise for vaporization rate reduction, foam life and foam resiliency.

Foam generators may be powered by electricity, gasoline or water. Each of these generators has the flexibility to generate the desired foam requirements. The water-powered units would provide the desired flexibility while being free of any flame initiating hazards.

The MSA "Model 6000 Foamaker" is a typical example of high-expansion foam generating equipment applicable for hazardous spill response. This unit weighs approximately 102 kg (225 lb) and would require 2.5 m^2 (27 ft^2) of deck space. It normally operates at water pressures between 2.8×10^4 and 7×10^4 kg/m^2 (40 to 100 lb/in^2) with water flows between 235 and 390 l/min (62 to 103 gal/min).

Some modifications would be required to make the MSA Model 6000 standard foam generator produce foams with a nominal 200:1 expansion ratio. This unit is designed to deliver foam at a rate of 557 m^3/min (6,000 ft^3/min) with a nominal expansion ratio of 600:1. Changing the ratios of gears driving the fan would be one method of obtaining the desired foam. Similar, minor changes might also be required for skid or cart-mounted units which would make them amenable to service on ship decks. Similar modifications would be required on the other foam generating equipment referenced in this report.

This equipment would require approximately 30 minutes to generate the 1,520 m^3 (53,700 ft^3) foam layer. Multiples of this typical unit or some larger units could be used to decrease the total response time. Using multiples of smaller generators could also aid the production of a uniform foam layer.

The foam generating system would include a water supply, holding tank and ancillary equipment for transporting, metering and mixing, in addition to the foam generator. The entire system would weigh approximately 4,500 kg (9,700 lb), and require approximately 3 m^2 (32 ft^2) of deck space.

Foam generating systems would have an estimated life of approximately 10 to 20 years. It would be possible to operate several foam generator units from one large pumping station and reduce the amount of control required while increasing total foam output. The foam generating technique is considered a one-man operation for the systems described. Regular maintenance would be required to preserve the design capabilities of the equipment against the rigors of shipboard exposure.

Low Expansion Foams: Low expansion foams (10:1 volume ratio) may also be considered as a method to achieve evaporation rate reductions. The volume of foam required in response to the largest typical spill investigated in this program can be calculated from the minimum layer required (0.15 m or 6 inches) and total area 3,800 m^2 (40,700 ft^2). The amount of foam solution required to form this 620 m^3 (21,900 ft^3) of low expansion foam would be between 6,200 and 18,700 liters (1,600 and 4,800 gallons).

This layer of low expansion foam would not completely stop the evaporation of a floating, hazardous chemical. However, a layer between 7.6 and 10 cm (3 to 4 inches) would be sufficient to inhibit flammability or arrest combustion reactions from flashing back into the pool of floating chemical.

Low expansion foams may be used to reduce vapor flammability. This technique involves the generation of a volume of foam sufficient to cover the spill area with a continuous layer. This layer need not be the full 15 cm (6 inch) thickness required for vaporization rate reduction, but must be continuous (allow for wind and wave action).

Experimental tests of foams generated at various expansion ratios showed that, while all foams will reduce flame propagation rates, only those generated at expansion ratios less than 10 to 1 would not support combustion after equilibrium contact with representative chemicals. These tests also indicated that low expansion foams would act as a flame arrestor to stop any vapor flame reaction from flashing back into the liquid pool.

A 10 cm (4 inch) layer of low expansion foam would be sufficient to inhibit combustion and arrest flames before they could flash back into the pool of floating, hazardous chemical. The foam layer would require the use of 3,800 to 11,400 liters (1,000 to 3,000 gallons) of foam solution to cover a 38 m^3 (10,000 gallons) spill. MSA Type L foam would be made from a 30% solution, while Universal Foam would be made from a 10% solution; both should be generated at an expansion ratio of approximately 10:1.

Foam-Generating Equipment: Some typical foam equipment required for this response technique is tabulated in Table 4.7.

Table 4.7: Typical Foam Generating Equipment

Manufacturer	Model	Operating Pressure (psig)	Dimensions Diameter x Length (inches)	Weight (lb)	Foam Capacity (ft^3/min)
Low expansion					
Rockwood Engineering, Inc.	DM-FFF	100	–	67	500
Rockwood Engineering, Inc.	LWP-500	150	–	75	500
High expansion					
Mine Safety Appliances Co.	Mini-X	80–100	12 x 12	14	750
Rockwood Engineering, Inc.	S Jet-X	50–250	15 x 54	10	1,000
Mine Safety Appliances Co.	3000	150	19 x 35	15	2,800
Mine Safety Appliances Co.	6000	100	36 x 60	225	6,000
Rockwood Engineering, Inc.	Jet-X-2	30–120	22 x 25	48	2,000
Walter Kidde and Co., Inc.	WD-150	70	66 x 55	550	15,000
Walter Kidde and Co., Inc.	PI-135A	*	58 x 102	1,160	13,500

*Electric pump.

Source: CG-D-56-77

A low expansion foam generating system would weigh approximately 910 kg (2,000 lb), and occupy about 3.7 m^2 (40 ft^2) of deck space. The foam response technique requires one man to operate the equipment and direct the foam over the spill. The service life of foam generating systems would be expected to be in the range of 10 to 20 years, with proper operation and maintenance.

Problems Specific to Spills of Floating Chemicals: It will be necessary for the Coast Guard to monitor the EPA program to insure that it is providing the data needed for floating chemicals. There are three areas that may not receive attention in a program primarily designed for land based spills which may require special investigations. These concern the influence of flowing or otherwise moving water on foam persistence, the possibility of interaction between foam and the floating chemical to produce a soluble hazardous substance, and the inerting of foams, particularly high expansion, by controlling the composition of blowing air.

Waves and Other Water Action – It is expected that the collapse rate of foams will be one factor in the selection of an acceptable system. To provide adequate vapor control, it will be necessary to maintain a minimum foam thickness. This becomes an economic consideration in the long run, so that persistent foams become the desirable situation. Because wave action or other water motion may exaggerate foam collapse, Coast Guard requirements may be more severe in this regard than for land based spills. Thus, consideration may be warranted to develop foams of sufficient persistence to be economically as well as technically useful with spills on water.

Foam Inerting – Technology shows that all foams ultimately absorb measurable quantities of the liquids they are covering. Thus, air-gas mixtures can develop which fall within the flammable and explosive ranges. With low expansion foams, the flammability hazard is negligible due to the fact that the foam acts as an inhibitor. With higher expansion foams, explosions cannot be propagated but they do burn back at high rates.

Work sponsored by the Maritime Administration on tanker inerting has shown that high expansion foams can be blown using stack gases or gases produced by an inert gas generator. Foams so blown cannot be ignited even after assimilation of significant quantities of flammable gases due to the lack of oxygen. Ships available to the Coast Guard may have the capability of utilizing such gases to blow foam, a capability which may not be readily available in land spills.

The ability to deliver a nonflammable high expansion foam may alter the recommendations for foam in spill situations. If economic or other conditions can be shown to favor the use of high expansion foam, and the flammability which can develop is the only drawback, it may be advantageous to work out procedures to blow foam with the stack or other inert gases.

Foam Effects and Solubility – The final consideration is the possibility of interaction, such as saponification, between foam and the spilled material which would result in conversion of the floating material to one which is soluble or a sinker. The latter occurrence is probably remote but solubility is possible. Foam-chemical interaction would be confined to the interface area, but foam drainage passing through the spill into the water substrate would carry such reaction products into the water.

In land spills penetration into the substrate is to be expected and some exaggeration due to the influence of foam will probably be of little significance. With spills on water, soluble products will end up in the water. These reactions may be minor and the amount carried into the water negligible given dilution and diffusion rates. Data in this regard is not defined at present. If, as expected, foam is found to be a principal spill control mechanism, investigation of this effect will become necessary.

High Expansion of Conventional Rigid Urethane

> The information in this and the next two sections is based on *Control of Hazardous Chemical Spills by Physical Barriers,* EPA-R2-73-185 prepared by J.V. Friel, R.H. Hiltz and M.D. Marshall of MSA Research Corporation for the Environmental Protection Agency.

This phase of this report deals with the screening, selection and evaluation of highly expanded materials capable of serving as a suitable hazardous material land spill barrier.

The search for candidate starting materials focused on materials which were polymeric in nature or capable of being polymerized or gelled to a highly expandable, rigid chemically inert foam structure. There were two candidate materials, water-soluble polymers (discussed previously in the section on gels), and conventional rigid foam systems. The water based systems were capable of high expansion but needed development in rigidizing. The rigid foams had not been developed for high expansion.

The most simplified polymeric system from an application point of view is undoubtedly the rigid urethanes. The nominal working density of conventional polyurethane is about 2 pcf although densities of the blown foam can approach that of soft pine or about 22 pcf. Although there have been reports of urethane as low as 1 pcf, which offers an expansion of about 75:1, there was little if any information of lightweight material in the 0.25 to 0.50 pcf range.

An alternative to the urethane is the urea-formaldehyde type formulations. Several contacts with industrial applicators of this type system indicated that this finished foam would not retain sufficient structural strength. A sample of urea-formaldehyde foam at approximately 0.6 to 0.8 pcf was received and was found to be exceptionally weak. In fact, the material hardly survived shipment to the laboratory. A concerted effort was therefore directed toward the development of a urethane type formulation with the basic urethane properties but a density of less than 1 pcf. This work was initiated in parallel with that of the first phase on this program. Formulation investigations to yield urethanes less than 1 pcf were performed in this phase along with preliminary tests; it became clear, however, that densities less than 0.5 to 0.6 pcf were not practical.

Low Density Urethane: The chemistry of urethane foam is complex. In brief, the essential ingredients of a urethane formulation consist of a polyol, polyisocyanate, a surfactant, catalyst and blowing agent. In operation, the polyol and the isocyanate in the presence of a suitable catalyst react to form the plastic polyurethane. The reaction is accompanied by the evolution of considerable heat which tends to boil the added blowing agent, in turn causing the plastic mass to foam. The surfactant or cell control agent regulates the foaming phase so that uniform cell structures are obtained. Although these are the necessary ingredients of the foam, numerous combinations of ingredients are possible due to the availability of a selection of polyols, catalysts and blowing agents.

Blowing agents most commonly used are the Freons; however, any low boiling component could be considered. Water is not only a reactant (polyol) but can also serve as a blowing agent since the reaction of the isocyanate with water yields carbon dioxide.

Familiarizing runs were made using a base formulation and merely incorporating more blowing agent into the mix. The excess blowing agent did produce a somewhat higher expanded foam but the foam was of poor quality as evidenced by the relatively large and uneven cell sizes. Eventually, with increasing amounts of blowing agent, additional amounts of catalyst were required to initiate the reactions. Foam densities ranging from 0.8 to 1.0 pcf could be obtained by this approach; however, the solution was unwieldy with the excess blowing agent and catalyst.

Water was next added to the mix with triethylamine (TEA) found to be the most effective catalyst. Numerous attempts then followed to obtain the proper combination of polyol-water and the required amount of TEA catalyst. The work was performed by maintaining a master mix of the polyol, surfactant and fluorocarbon in one solution and varying the amount of water and/or catalyst. Excellent foam textures could be produced in the range of 0.6 to 0.7 pcf. However, all foam was found to shrink upon standing. Shrinkage was attributed to the pressure differential formed in the individual cells when the gas phase cooled. The degree of shrinkage was sufficiently serious to indicate the likelihood that the applied foam would tear itself away from the surfaces upon which it was deposited. Ultimately, it was found that nonshrinking foams could be formed by using methylene chloride in place of the fluorocarbon. With some modification in the polyols and the amount of TEA catalyst, it was possible to form a high quality foam in the range of 0.6 pcf (expansion ratio 115:1).

A secondary development of the study occurred when it was found that water with TEA could react with the isocyanate to yield a foam-like material. By optimizing the amount of TEA catalyst it was found that high quality foam down to 0.29 pcf could be routinely made in the laboratory. Visually, there were no apparent differences in the water/TEA/isocyanate foam generated at 0.3 and 0.6 pcf when compared to material with the conventional 2 pcf density. The lightweight material was, of course, weaker but considerably less brittle than the high weight material.

The results obtained with the low density urethane-type foams were quite encouraging. The expansion with the lightweight material (230:1) was undoubtedly much greater than could possibly have been achieved with the water-soluble polymer systems discussed in the section on gels.

The evolved formulations were of particular interest for another reason. Both formulations (0.6 pcf, 0.29 pcf) utilized water. It was possible, if required, to incorporate into the urethane mix the water-soluble polymers which by themselves were found ineffective. In this respect the water-isocyanate reactions were the quick set mechanisms lacking in the water-soluble polymer systems. At this point the possibilities for a high expansion polymer system appeared very promising.

Preliminary spray tests using conventional urethane spraying systems were conducted on both formulations. Both streams of reactants were fed to a conventional Binks 18 FM Mine Gun which is standard urethane hardware. Spray tests were conducted indoors at about 70°F. Optimum foam generation using this system consisted of 3.5 parts activator to 1 part water-catalyst mix.

Using the 0.29 pcf formulation, the foam mix was sprayed onto both smooth cardboard and concrete. The freshly sprayed material appeared as a free-flowing liquid and tended to run off vertical surfaces. Cream time or the time for initiation of rise was approximately 3 seconds. After the initial period, the rise of the applied foam was remarkably fast. Time to achieve full rise was about 5 to 7 seconds.

It was found that the applied foam did not adhere well to either the cardboard or concrete surfaces. In addition, considerable amounts of vapors of TEA were evolved during the actual spraying and subsequent reaction. Sections of the applied foam showed a density of 0.35 pcf which approached the laboratory data.

Further tests of foam generated with conventional hardware showed the major weakness to be the lack of adhesion to smooth surfaces. In all other respects the generated foam was structurally sound.

Spray tests were also conducted with the 0.6 pcf formulation, similar to those described with the lighter material. Optimum foam production was achieved at a ratio of 1.3 to 1.4 parts activator to 1 part resin mix. Fumes of DMEA were minimal.

The sprayed solution was free-flowing with a cream time of about 4 seconds and reaching a full rise in about 8 seconds. Sections of the foam showed a density of about 0.6 pcf and were obviously much stronger than the 0.35 pcf material sprayed earlier.

The sprayed and cured foam was found to be only weakly sealed to smooth cardboard. Whole chunks of foam could easily be freed from the cardboard surfaces. A much improved adhesion was evident on the smooth concrete. In some regions it was apparent that adhesion to concrete was stronger than the cohesive force within the foam itself. Subsequent liquid penetration tests showed no significant penetration by dyed water.

Overall, the 0.6 pcf density material showed a definite improvement over the 0.3 pcf formulation, especially in the adhesion on concrete. However, when compared to a conventional 2 pcf material, the 0.6 pcf formulation was understandably inferior, both with respect to adhesion and strength. The results of these initial spray tests, indicated that the 0.6 pcf formulation possessed the capability of substituting for the 2 pcf but in terms of strength alone was obviously less sufficient.

Unquestionably, the major weakness of the water-isocyanate foam was its apparent lack of adhesiveness. Additional laboratory studies to improve adhesion consisted of adding water-soluble polymers, especially film forming varieties, to the original water-isocyanate mix. The polymers were added to the water phase of the mix with the intent that the reaction with isocyanate would remove all water, leaving a dried film of polymer to improve adhesion and possibly cohesion. The results in general were unsatisfactory, however.

A series of spray tests were conducted on concrete in which the foam was applied 10 inches high in a 3 foot wide circle to form a container. Water was added to the inner area. With a four inch height of water, seepage at the concrete-foam interface was apparent. In addition, water was found leaking through channels which had inadvertently formed in the barrier wall. With about 5 to 6 inches of water, the barrier uplifted from the concrete. There was no evidence of fracturing or tearing of the foamed mass indicating that structurally the foam would serve well as a barrier if the problems of fixing it firmly to the substrate were solved. These results clearly indicated that, because of the buoyancy, the lightweight material probably required exceptionally good adhesion qualities even to dry surfaces. Although the higher the expansion the greater the amount of material delivered from a portable unit, the unsolved problems would appear to limit the practical urethane systems to densities of 0.6 pcf or greater.

Towards the end of the initial program, two techniques were developed which markedly improved adhesion to surfaces wet with water. It is necessary to determine if these procedures are beneficial in situations where the wet surface is due to the spilled chemical. Sufficient testing should be conducted to define a procedure covering the broadest range of wet surface conditions. The selected procedure would then be incorporated into the portable unit, either directly into the chemical package or as an auxiliary unit.

By making minor changes, in the catalyst from triethylamine (TEA) to dimethylethylamine (DMEA) and a major change in the blowing agent from a Freon type material to methylene chloride, a rigid polyurethane foam was achieved with a density between 0.5 to 0.6 pcf rather than the normal 1 to 2 pcf. The rise time of this particular form of polyurethane foam was much more rapid than that of normal materials but the cream time, that interval before the onset of foaming, was increased to some 12 to 14 seconds from the normal 2 to 4 seconds increment. It is this last feature which allowed a better penetration into vegetated surfaces and gravel.

This particular system had one apparent drawback. Because of the exceptionally fast rise and the concentrated heat release, there was a tendency of the foam to shrink slightly upon cooling from the reaction temperature. Field tests, however, showed that the bonding strength to dry surfaces of concrete, asphalt and bare ground was sufficient that the shrinkage did not adversely affect the material's ability to contain liquids. Further, the cream time was sufficiently long that penetration did occur into vegetation to the point that effective seals could be achieved against the surface as long as the vegetation was not too thick and matted. Improved penetration was also achieved on gravel surfaces, but never to more than some 2 diameters of the stone. It is doubtful if this material could be effectively utilized to control liquid spills on a gravel surface.

The benefits due to the increased expansion and the extended cream time of the formulation without sacrificing any of the desirable characteristics exhibited by the normal 1 to 2 pcf foam led to the adoption of this particular material as a standard material in this phase. The use of methylene chloride as the basic blowing agent produced some changes in materials viscosity. To maintain the 5 cfm delivery rate, and to get good mixing in the nozzle of the correct proportions of both resin and activator, minor changes were necessary. To achieve the desired 1:1 resin-activator proportions, the cross-sectional area ratio of resin and activator orifices was changed to 2.3:1. Some mixing problems were encountered but were overcome by adding 10% by weight of Freon R12 to both resin and activator. Volatilization of this material in the nozzle provided for good mixing. With the Freon-nitrogen pressure system, 200 psi had to be used as the initial tank pressure. With the methylene chloride, the pressure was increased to 220 psi.

A significant number of field studies were run using this material on a variety of substrates to block and hold water with good results except when wet or extremely cold surfaces were involved.

In an effort to improve bonding to cold surfaces, catalyst changes were made to accelerate the reaction and concentrate the heat output. Of the materials tested, an addition of 0.10% lead naphthenate had the best effect. It decreased the cream time to some 3 seconds at temperatures as low as 15°F.

The accelerated reaction improved expansion and bonding at low substrate temperatures. It was detrimental, however, with respect to the ability to penetrate into vegetated surfaces. Thus, at this state of development, a choice must be made between the ability to achieve good bonding to a cold substrate or a vegetated one.

The initial work was with hard surfaces, asphalt, concrete, and hard packed dirt, and significant success was realized in applying the urethane and successfully containing significant volumes of water 2 to 3 feet in depth. A circular form of polyurethane 36 inches high and 30 inches inside diameter was set up on concrete and maintained full of water without leakage for a period of days. Similarly, polyurethane was used to block a curb drain set in concrete. Sealing to asphalt was as good if not better than sealing to the concrete.

Similarly, effective seals could be accomplished against bare ground. The ability of the polyurethane to dike liquids in that particular situation, however, depended upon the integrity of the bare ground. A breach of the dike eventually occurred.

The breach was always in the substrate just below the level of the urethane dam and usually restricted to a small localized area. The time required to achieve the breach was a function of the head of water and the density of packing of the substrate. Even though the dike was eventually breached, several inches of water could be held for an hour or more. In the case of bare ground, where poor containment was experienced, additional downstream dikes could be erected to extend the control time of the liquid material.

Dams could be erected with this type of polyurethane material. By expanding this material into a rubber or plastic envelope it was possible to block water flowing in culverts and other similar circular drains.

The final effort in this portion of the program was an evaluation of the ability of the polyurethane materials to block or divert already flowing streams. These efforts were completely unsuccessful. The primary difficulty was the inability to achieve good adhesion between urethane and a surface wet by water or one of the hazardous chemicals. Several mechanical aids were tried, such as the use of hold-down stakes; heavy filler, such as rock; or a support system such as screen or netting. None of these were successful in improving the ability of the polyurethane to combat the flowing stream situation. In addition, they complicated the construction of the dam appreciably.

Chemical Compatibility: The chemicals which can be controlled by urethane foams need to be clarified further. Data on specific chemicals is limited. Water-based liquids with the exception of strong acids are contained as are nonpolar organics. Data is also positive for chlorine and ammonia. Polar solvents are a question. Acetone is controlled but methyl alcohol is not. If the foam is applied and rigidized prior to contact with the spilled chemical, it is doubtful that any catastrophic reaction will occur. Reacting chemicals can break bonds destroying the adhesion to the surface or they may even seep through the foam mass as is the case with methyl alcohol. Thus in most cases urethane will exercise some control if applied correctly. Even in those cases where control is not affected it is not expected that the spill situation would be exaggerated. Guidelines in this general area are almost completely lacking, however.

The foam behavior with methanol is not understood. It is not a destructive reaction but rather one of absorption. The absorbed methanol does destroy foam rigidity, however. Tests run with acetone, a polar compound like methanol and acrylonitrile which is also polar in nature did not produce this effect. Further work is necessary to explain the phenomenon and to define what other materials might behave as methanol. Polyurethane is not recommended for containing methanol.

Storage Tests: Three short term storage cycles were initiated. Two storage tests were at room temperature (75°F) and one at 40°F. As expected, the lower temperature storage adversely affected the expansion of the foam as well as increasing the cream time.

The initial tests at ambient were conducted over a six week period. No change was noted in the materials or the foam generated from them until the sixth week. At that time separation had occurred in the resin mix with two liquids formed. They could be reblended by agitation but separated again within 24 hours. Since the only major change that had been made in formulation was the methylene chloride, that was assumed to be the contributing factor.

A second test package was made in which the methylene chloride was added to the activator instead of the resin. Eight weeks of storage with this arrangement showed no problems.

Conclusions: Polyurethane foam will serve a number of functions such as construction of barriers, diverters, plugging of sewer drains, covering storm drains and damming of drainage ditches, etc. The single factor which will exert the most influence in determining the usefulness of foam will be the type of surface. On dry, firm surfaces such as concrete and asphalt and at temperatures above 30° to 35°F no difficulties should be experienced. On this type of substrate tests have shown that 4 foot high barriers can easily be constructed in which at least a 3 foot head of water can be contained. Surfaces which are cold and/or wet (water, solvents, etc.) will impair the quality of the initial deposit of foam. This layer of low quality foam can serve as a buffer zone upon which additional foam can be successfully applied.

Unfortunately, cold, wet surfaces impair the bonding or adhesion and the low quality foam which results is prone to leakage. Grassy and weedy surfaces also prevent the penetration of the foam and thus the formation of a good seal. Leak-plugging and strengthening can be achieved to a certain degree by selected addition of more foam, but in general the performance of barriers built on this type of surface is unpredictable and most frequently merely limits the flow of liquid.

In many cases spilled material ends up following existing courses to water systems. The foam is able to effectively seal sewer openings and storm drains to prevent the chemical from entering the water system in that way. On major highways and railroad right-of-ways drainage paths including ditches, culverts and conduit are used to divert and drain water. The foam is capable of sealing these paths to effectively contain the spill. There are examples of chemical spills which have entered urban sewer systems or have reached rivers by flowing through railroad drainage systems.

High Expansion Foam Covers

Spilled material contained either by artificial or natural barriers can still pose a major threat. Release of dangerous vapor can cause a toxicity hazard not only in the proximate location but in downwind areas as well. A potential fire and explosive threat may exist perhaps accentuated by the movement of men, equipment and existing circumstances. The danger to the cleanup crew may slow down that operation allowing time for groundwater movement or other means of moving the contaminant into the water system. This section deals with the evolution of soft surfactant type foams which when applied to a spilled contaminant would form a protective covering.

Foam Resistance to Wind: Prevailing wind speeds are a major consideration in the application of detergent foam in open areas. Since foam covers will perform partially as a function of their thickness, it is necessary to have information on the effect of wind speeds. Preliminary tests were, therefore, conducted to determine the resistance of foam heights to various wind speeds. Tests were conducted in a wind tunnel in which foam was applied onto a liquid substrate. Trichloroethylene was contained in a tray measuring 1.5 feet wide by 2.0 feet long by 1.5 inches high. Various heights of high expansion foam were applied by conventional means onto the surface of this simulated contaminant.

The foam formulation consisted of polyethyleneimine, polyvinyl alcohol and MSA long lasting foam agent. This particular formulation was selected since it reasonably approached the type of ingredients which were expected ultimately to be used as a vapor barrier. The expansion of the foam used in these tests was about 200:1.

The test results were not able to establish clear-cut boundary conditions. Sufficient experience was gained to indicate that at 9 mph wind speeds all high expansion foam masses greater than 4 inches could be uplifted and completely displaced from a liquid surface. The maximum allowable height in a 5 mph wind was about 4 to 6 inches. Even in this latter case it was strongly suspected that the ability of the foam to anchor onto the sides of the tray and the outside surfaces contributed some stability to the total foam mass. A foam height of 6 to 8 inches under the same 5 mph wind speed was obviously displaced from the liquid surface.

These tests indicated vulnerability of high expansion foams to wind in the 5 to 10 mph range. Thus, maximum vapor suppression and protection should be accomplished with foam heights no greater than 3 to 4 inches.

Conventional Foams: A conventional fire-fighting foam was evaluated to determine the covering capability of material which would be readily available. MSA standard fire-fighting foam was selected as a typical material and toluene as the contaminant. The selection of toluene was based on its relative chemical inertness and absence of reactive functional groups. Toluene is lighter than water and is also a very poor solvent for water; hence it would not be affected by the foam drainage.

A 6.5 inch height of foam was applied to the toluene from a 3% solution of MSA foam agent. The collapse of the foam was evident almost immediately. After about 15 minutes of contact with the liquid toluene, the initial 6.5 inches of foam cover had completely collapsed exposing the liquid surface. The normal collapse rate for this foam in the absence of environmental effects is 8 to 10 inches per hour.

An additional test was performed using a 6% solution of foam agent. The results were similar, complete collapse of the foam within 20 minutes. The rate at which this foam collapsed was surprising especially in view of the presumed inertness of the selected contaminant. These data must not be construed as demonstrating the complete ineffectiveness of existing foam agents. Actually, the limited data collected showed a very significant reduction in vapor evolution up until the foam collapsed. In cases of emergency where other facilities and materials may not be available, the use of existing foam agents could provide a useful service. These laboratory data indicate that the beneficial effects would be short-lived, but repeated applications could be made.

Long Lasting Foams: Since vapor suppression was obviously associated with the life of the foam, a proprietary preparation of MSA was of special interest. This formulation consists of polyethyleneimine, polyvinyl alcohol and MSA foam agent and in the past had been the source of foams with lives of over 3 months or more. An 8 inch layer of this foam generated at an expansion of about 200:1 showed excellent vapor suppression for up to 16 hours which was the effective life of this covering. Ordinarily, the collapse of this foam during a similar time period but without the presence of a contaminant would have been less than 0.1 inch.

New Foam Development: The results of these preliminary tests showed: (1) the limitations of existing fire-fighting foam agents; (2) the effectiveness of thin but densely packed beds of foam; and (3) the adverse effects on foam life posed by the contaminant.

Emphasis at this time shifted to a foam cover which could cope with the more chemically reactive species such as alcohols, ketones, etc. These materials are not only produced in large quantities but were known to be effective poisons for surfactant foams.

The most difficult class of compounds to smother effectively are the low molecular weight, volatile, highly polar solvents with methanol as an outstanding example. Most foam preparations collapse instantly on contact with the alcohol. With continued addition of foam solution the subsequent foam drainage alters the alcohol surface tending to decrease the adverse effects giving a false sense of foam stability. On a practical basis, however, where large spills were involved, such dilution would not occur.

For the purpose of further materials screening, methyl alcohol was selected as the test contaminant. Three approaches were taken which showed considerable promise. These included: solution of silicates which formed a relatively hard layer of solid silicates on the surface of dry methanol, solutions of pectin, which formed a gel on the surface and solutions of Acrysols which also formed a thick rubber-like coating on top of the methanol. Both these latter types of cover eventually dried to a hard impervious crust. All three materials were best applied from a low expansion system.

Silicate System – The silicate solution employed was a commercial grade water solution containing approximately 30% of dissolved sodium silicate. This solution was diluted with approximately an equal portion of water and a foaming agent added to maintain its concentration at about 4%. Foams blown from this formulation collapsed on contact with dry methanol; however, a dense, crystalline structure resulted. If applied at a high expansion ratio, insufficient material was available to form a continuous blanket. In that case the crystallized segments sank. At low expansion, the generated foam froze into a solid mass with sufficient entrapped air to remain buoyant. Subsequent tests showed that the silicate formulation was effective for methanol but was not suitable for other nonalcohol type materials, polar or otherwise.

Acrysols – The Acrysols are a class of acid type polymers which when neutralized by a base form a solid gel. A foam of this variety is difficult to prepare since gelation occurs immediately upon contact with the base. Low density foams could be produced by using solutions of sodium bicarbonate. The CO_2 liberated from the mix served as a blowing agent to expand the foam.

In contact with alcohol the gelled foam formed a thick coating which tended to float on the surface of the alcohol. Some dissolution of the gel occurred but the rate was slow. This type of polymer was also an effective cover for ammonium and amine type compounds where gelation occurred upon contact. Like the silicate, it was limited to a small group of materials.

Pectin Type Foams – The most widely applicable foam was prepared from citrus pectin. Pure citrus pectin is a powder which dissolves readily in water provided sufficient mixing is available. It is commercially available in preparations offering a range of set times and thickening power. Chemically, citrus pectin is a weak acid and maximum thickening is achieved at a pH of about 3.1. Once the added powder is adequately dispersed, wetted-out and dissolved, the process of thickening begins. Normally, a 1.5 to 2.0% solution of pectin in water becomes viscous within 30 to 60 seconds. On a practical basis some thickening of the solution begins immediately following the introduction of the powder.

Foams generated immediately after dissolving the pectin showed exceptional capabilities of minimizing the evaporation of a number of substances. Similar to the above systems, maximum results were obtained by applying the pectin in a low expansion (30:1) form. It was successful in part because the contaminant was swamped with the pectin solution. Immediately upon contact, the foam collapsed slightly but entrapped finely divided air bubbles, forming a tough impermeable buoyant skin on the surface of polar contaminants. With the nonpolar type, contaminant skin did not form but the foam remained intact behaving somewhat similar to conventional type foam agents. In both cases, however, the foam eventually thickened and dried to a permanent structure.

Effective reduction in vapor evolution was achieved by these foams for a variety of compounds with different chemical entities. The results achieved with these formulations and especially with solutions of pectin were exceptionally promising. They were able to be floated on a variety of materials, even methanol and acetone with quite low liquid densities, with good control of vapor release.

There are some inherent disadvantages to the use of pectin. Water is the only solvent and, because of the viscosity problems, the maximum concentration of pectin in water is limited to about 1%. Further, once the pectin is added to the water the properties of the resulting solution begin to change. Thus, the characteristics of the solution exiting at a foam nozzle depend upon the pectin dwell time in the line. Fortunately, dwell time can be controlled by inserting the pectin into the water line near the nozzle, at the pumper or at intermediate locations.

Introduction of the pectin into standard systems posed some problems, however. A powder eductor was examined as the first means of introducing pectin into a hose line. This device consists of a large funnel attached to the intake of an eductor. Sufficient pectin could readily be injected into a 10 gpm water line to maintain a pectin concentration of well over 2%. The exiting water solution was collected and showed the pectin to be finely dispersed with no obvious lumps or aggregates of undissolved powder. To add both foam concentrate and pectin into the line at the same time, two approaches were examined: (1) a completely powdered system containing a solid surfactant blended into the pectin; and (2) a stabilized suspension of pectin in a liquid surfactant. The latter system would be more consistent with existing foam dispensing equipment.

Subsequent laboratory studies showed powdered amphoteric surfactants to be excellent foaming agents for pectin. A powdered mix was prepared by blending one part surfactant to one part pectin. The problem of preparing a free-flowing, stabilized suspension of pectin was only partially solved. The best formulation consisted of equal parts by weight pectin, a liquid alkylaryl surfactant and propyl alcohol as a viscosity modifier. The resulting suspension was only temporary and complete settling of the pectin occurred within 24 hours.

Several attempts were made to spray the two pectin concentrates using conventional fire-fighting type generators. The solid blend of foam concentrate and pectin was evaluated in a Feecon type low expansion nozzle. Materials were preblended in a 1:1 weight ratio. Two tests were completed with the funnel eductor installed at two different locations along the hose line to provide a dwell time of approximately 5 and 50 seconds. Output foam of the low expansion nozzle was directed onto 2 feet wide by 4 feet long trays containing about a 2 inch depth of dry methanol. The output foam in both cases was exceptionally poor when compared to a conventional foam agent used alone.

The pectin foam was dispensed as a heavy, wet foam which caused excessive splashing and dilution of the alcohol. Some foam cover did develop with the shorter dwell time. However, the solution runoff was exceptionally gritty indicating the pectin had not completely dissolved. In both cases, dilution of the alcohol had occurred to render the test results inconclusive. It was obvious that a major problem existed in duplicating the quality of the foam used in the laboratory tests.

An attempt was made to produce a high expansion foam starting with the liquid suspension or slurry of pectin and surfactant described earlier. The generator was of conventional design and capable of expansion of 200 to 300:1. The liquidized pectin was uplifted into the line via an eductor at a location yielding a dwell time of abut 20 seconds. A reasonably high expansion foam was produced but was found to collapse readily on contact with the alcohol. Samples of the runoff and drainage again indicated that the added pectin was not completely dissolved.

Small scale tests were next performed in the laboratory using a miniature spray system. The device consisted of a 0.5 gpm spray nozzle and a flat screen of about 1.5 inch diameter. The entire spray head or generator was housed in a cylinder measuring 1.5 inch diameter by 3 inches long. The system operated batchwise; that is, a foam solution was prepared and charged to a pot. Pressurized nitrogen or air (100 psi) fed the solution to the generator. The study portion of the effort consisted of charging the pot reservoir with water, then when all was in readiness, quickly adding the pectin mixes followed by foaming.

Various time delays were examined between adding the pectin and the actual foaming. Results showed that optimum foam cover was developed with a 30 second dwell time. Foams produced at this time were thick creamy mixtures which formed excellent covers on methanol and acetone. With holding times greater than 30 seconds, the expansion ratio decreased until at one minute the foam output was only slightly higher than 1 to 1. Even at the lowest expansion the foam (or aerated solution) formed an excellent cover. If dropped from any height, this material sank deeply into the alcohol but immediately surfaced and remained buoyant.

It is clear that, although the medium expansion imine modified system can be generated with commercially available generators, the low expansion pectin-fortified system does not lend itself readily to the common low expansion generators. It may be necessary to devise a generating system for application of that material. Foam pumps are a new item on the market that have the potential of handling the pectin system and should be evaluated before undertaking development of a new generating mechanism.

It should be cautioned that the use of a foam cover should in no way change the procedures which would be employed to handle the chemical without it. Its purpose is as an extra precaution. Further, whenever a foam cover is being applied, either initially or as a recover, those people applying the foam should wear full protective gear. Body protection such as the Acid King Clothing of Wheeler Protective Clothing Company with bottled oxygen or the butyl suit of MSA with a self-contained Chemox rebreather is recommended.

Foamed Concrete

The objective of the foamed concrete task of the program was to demonstrate the feasibility of erecting a fast-set foamed concrete dike and to outline the general requirements for an emergency field unit. Both objectives have been achieved.

Foamed concretes and foamed gypsums are common materials of construction, with well developed formulation and application technology. To use these foams for spill containment, it was necessary to provide them with instant setting capabilities which allow barriers to be built up without forms. To accomplish this, it was considered necessary to impart a gelling action to the foamed material as it was delivered to the barrier location. Thus, the basic mechanism to be evolved was a gellant system to provide a uniform, semirigid foamed mass to standard foamed concrete and foamed gypsum mixes.

Foamed concrete is one type of lightweight concrete. It is an established construction material with a density range usually given as 25 to 100 pcf. [The American Concrete Institute (ACI) has defined the density range of 50 pcf or less to be low density concrete.]

Foamed concrete is produced by introducing controlled quantities of air, water and foaming agent under pressure into a foam nozzle and blending the resultant foam with cement slurry or cement-aggregate slurry in a variety of mixing devices, either batch or continuous. The foam must have sufficient stability to maintain a structure until the cement sets to form a matrix of low density concrete. Foam generation and proportioning into the cement slurry are regulated to achieve control of final strength and density. Up to 80% of the final mix volume may be air, added as preformed foam. The resultant foamed product can be pumped to distances up to 1,000 feet, to a height of 300 feet. The pumping parameters are simply a function of the slurry viscosity, pump specifications and slurry set time.

Existing foamed concretes have similar fluidity and set times as normal concretes and thus require forms. There are, however, a number of fast set gypsum and concrete formulations which have never been formulated in the cellular, or foamed mode. Reg-Set, a commercial fast set concrete, for example, can reportedly be made to take its set within 5 minutes after mixing with water. Hardstem and Hardstop, both gypsum formulations, were developed as rapid set formulations by the British for pour-in-place mine bulkheads (41). Accordingly, Halliburton (42) developed both fast set gypsum and concrete-based formulations upon investigation of materials suitable for sealing abandoned mines to prevent acidic mine drainage. Various additives and sodium silicate were employed to achieve extremely rapid sets. This latter work, although employing normal density materials, provided a suitable basis for the study of foamed materials. The program consisted of: (1) preliminary formulation evaluation; (2) small-scale barrier pours; and (3) field testing.

The objective of this laboratory study was to develop leads for possible formulations or techniques that could be developed into an instant-set pour system with slurry mixing equipment. Therefore, commercially available slurry mixing, preformed foam and pumping equipment were employed without modification in order to expend more effort upon the formulation development and the feasibility of the technique as a spill control measure.

Preliminary Evaluations: The evaluation studies were conducted on a small scale, on the order of ¼ to ½ lb of material per experiment. Studies with gypsum formulations considered accelerators and gel-set formulations employing the water-soluble polymer polyvinyl alcohol (PVA) and a gelling agent were made.

Similar studies were also conducted on two fast-set cement formulations developed by Halliburton Company. One of the Halliburton formulations consisted of a mixture of cement, bentonite, fly ash and gypsum. It employed bentonite to obtain high viscosity and gypsum to obtain a fast set. A second, more rapid-set formulation, employed cement and bentonite for structural strength and sodium silicate to impart the rapid set to the mixture. A specialty gypsum (Hydroperm) and a regulated-set cement (Huron Cement Company) were also investigated.

When using an accepted accelerating agent (Na_2SO_4) at several concentration levels, set times were reasonably rapid when compared to normal plaster formulations. However, when using larger amounts of sodium sulfate, the time to set was not nearly fast enough for this intended application.

The gel time for polyvinyl-alcohol-containing gypsum formulations was also slower than desired. Polyvinyl alcohol solutions can be gelled to a rigid structure with basic materials such as borax, and various dyes such as Congo red. Although some highly viscous mixtures were obtained by adding small amounts of saturated solutions of $Na_2B_4O_7$ to slurries containing polyvinyl alcohol, at no time was it felt that the time to set was short enough for building barriers of practical heights.

Evaluation of the cement-gypsum formulation developed by Halliburton Company was initially encouraging, but failed to develop into a practical fast-set system. The formulation, as developed, is a two-slurry system; one containing cement, fly ash and water; and the other, gypsum and bentonite. In Halliburton's application, the bentonite-containing slurry is premixed to allow the bentonite to hydrate before mixing with the cement-fly ash slurry. Initial experiments conducted in this manner were encouraging; the formulation resisted pouring after only one minute and took a hard set after five. For the intended emergency spill control application, however, the two-slurry system was complicated and time consuming. Thus, efforts were directed toward using this formulation compounded as a single slurry system, without prehydration of the bentonite. Using this method, the time to a nonpouring formulation increased from one minute to four and the hard set time from five minutes to nine.

No particular improvement was noted when the fast setting Reg-Set cement (Huron Cement Company, Alpena, Michigan) was substituted for the Type I Portland cement. This was not too surprising since Reg-Set cement, although much faster setting than the Type I cements, will not normally set before five minutes. Adding potassium sulfate as an accelerator to the Reg-Set cement did not appear to materially aid the initial set; however, once initiated, a hard set rapidly developed.

The last experiment was somewhat encouraging in relation to the ultimate goal. Protein-based foam was added to the slurry to achieve a final wet set density of 35 pcf without significantly altering the set time of the mix. Unfortunately, none of the above formulations set fast enough for the desired application, but this experiment gave hope that if a fast set mix could be developed, the addition of preformed foam to achieve the desired low-density barrier material would not affect the set time.

The most encouraging result came from a modification of another of Halliburton's specialty formulations developed for sealing mine entries (42). This system was also a two-slurry formulation containing water and cement in one slurry, and a water, bentonite, sodium silicate mixture in the other. As mentioned previously, it was felt that the two-slurry system was not practical for this application, but that the addition of a solution of sodium silicate to the foamed cement formulation as it exited the nozzle was entirely feasible. The plan involved forming a lightweight slurry of cement, bentonite, water and preformed foam, and adding the silicate solution by means of a water ring at the nozzle to achieve a fast set.

Initial laboratory scale studies were discouraging until it was discovered that gel formation occurred within seconds after adding the sodium silicate, and that any additional agitation only served to break up the exiting gel. Once formed, the gelled material resisted flow and appeared to have reasonable strength. In addition, preformed foam made from MSA salt water foam concentrate was compatible with the cement-bentonite mixture and encouraging samples of foamed concrete of 37 and 45 pcf were obtained.

A final experiment was an attempt to determine if the cement-fly ash-gypsum formulation would also gel with sodium silicate. This was found to be the case.

Small Scale Barrier Pours: Small scale barrier pours using field-scale equipment were conducted using a Mason Flow Mixer, a continuous slurry generator developed by Hoge-Warren-Zimmerman of Cincinnati, Ohio and a Model 2000 Slurry Rock Dust Distributor, a batch slurry mixer manufactured by MSA for use in coal mines. The formulation development work and pouring techniques were developed with the Mason Flow Mixer.

The Mason Flow Mixer consists essentially of a hopper for powdered solids, a screw feed for solids metering, a slurry chamber and a slurry pump. For producing foamed products, preformed foam is added either in the slurry pump or immediately downstream of the slurry in the delivery hose.

The unit is capable of producing slurry in the range of 5 to 10 gal/min. It requires 220 V power and a water supply, but can also be powered by a gasoline engine. This particular unit is adapted for laboratory experimentation. Controls are available to vary the water-to-dry powder ratio, the rate of slurry production, and the speed of the slurry pump. On-off controls for the water and powder feed as well as automatic wash out capabilities are available on a remote control, hand-carried push-button device.

In operation, water is started through the mixing chamber and slurry pump, followed by the activation of the dry powder feed. Approximately 20 seconds is required to blend the dry material into a slurry and deliver it through the slurry pump into the delivery hose.

Preformed foam was produced from commercial foam units using both protein and detergent-based concentrates. The essential elements of a foam unit are a tank for foam solution, a proportioner, and a refiner column. A 4 to 6% concentration of the foam concentrate in water is fed under pressure to the proportioner where it is mixed with air and fed to the foam refiner column. The proportioner may consist of either metering valves for both air and foam solution, or an orifice plate for the solution with a valve to regulate the air. The refiner consists usually of a 1 to 2 inch diameter copper column 1½ to 2 feet in length, full of a suitable packing material (e.g., stainless steel wool, Raschig rings, Burrell saddles, etc.).

The proportion of air to dilute concentrate solution determines the foam density. The total pressure on the system (usually 45 to 90 psig) determines the rate of foam production.

Protein and detergent-based foam were each used initially in the studies. The protein foams were produced by Mearl Corporation, Mearlcel for gypsum-based and Mearlcrete for cement-based mixtures. MSA's detergent-based foam concentrate was later found to be at least equivalent to the protein-based concentrates and compatible with both gypsum and concrete-based mixtures. Therefore, it was used almost exclusively in the test pourings and field trial studies.

The initial unsupported barriers were produced with gypsum formulations. Set times of the order of minutes were achieved by adding accelerators, but the set was not fast enough to produce a nonflowing slurry.

Development work with commercial quick-set cement formulations was carried out with Huron Cement Company's Regulated Set Portland Cement (RSPC). This is a special Portland cement supplied with an accelerator that produces an initial set in 5 to 15 minutes, depending primarily on water content and temperature. With added foam, the set time increased considerably and the cement was of poor strength and quality. The retarded set was likely due to the protein-based foam since some organics (e.g., proteins, sugars, starches, etc.) are known to retard the set of Portland cement.

The preliminary development studies on Halliburton-based formulations also set too slowly for this application. Following the lead of the laboratory studies, however, formulations containing sodium silicate as a fast gelling agent showed promise, and were slowly developed into barrier forming materials.

Because of the extremely fast gelling action of the sodium silicate, it was necessary to modify the generating and delivery system. The cement-bentonite slurries were formed in the Mason Flow Mixer and combined with preformed low expansion detergent foam downstream of the slurry pump. The silicate was subsequently blended into the foam slurry by the use of a shotcrete nozzle as it exited the delivery hose. A water ring, consisting of two circular rows of holes in the shotcrete nozzle, was used to inject the silicate solution.

In the initial formulations the development work stressed variations in the silicate concentration, bentonite addition, type and quantity of detergent foam formulation and technique. Quick gelling mixtures were obtained having densities of 32 to 93 pcf. In these runs two problems persisted: one was the tendency of the unsupported barrier base to be displaced by subsequent layers; and the second was the difficulty in obtaining a uniform foam delivery rate. Both of these problems are related; the first is a function of foam density and silicate concentration while the second is caused by variations in downstream slurry pressures. Air pressure was used to produce the foam and to introduce it into the cement slurry mix.

The rate of foam production and its addition is a function of the pressure drop across the metering system. Thus, as the viscosity of the cement slurry varied, the back pressure in the feeder line varied which affected the foam addition rate. This condition produced concrete of varying densities and support strength. In the first three runs, bentonite added to the slurry mix to increase its viscosity produced a high back pressure, resulting in foamed concrete in excess of 90 pcf. In subsequent runs, therefore, the bentonite was not used. Although some improvement was noted, considerable density variations still persisted indicative of a foam feed-rate problem.

Although the presence of problems was recognized, it was apparent that the sodium silicate-gelled foamed cement offered promise for unsupported barrier formations. Accordingly, the remaining research and development effort centered on this formulation, and detailed data was gathered in an effort to delineate the conditions for best barrier production.

The basic cement formulation for these studies contained 5% by weight bentonite. The slurry feed without bentonite was too thin to effect a reasonable barrier buildup. A concentration of 5% bentonite, however, materially aided in increasing the viscosity of the slurry without causing undue pumping problems. In later runs, additional lime was also added. Free lime is present in Portland cement. With high sodium silicate addition, and resultant calcium silicate formation, however, it was felt that the final cement blend would be deficient in lime. This additive proved to be beneficial.

Consistent dikes of over 16 inches were poured with concrete densities of 32 to 37 pcf. Gel structure breakdown, with a resultant slurry slide, limited the pour height to approximately 2 feet on a 12 to 16 inch wide base. Once the initial gel structure was broken, slides continued to occur along the same break line.

Short asbestos fibers (5% by weight) were introduced into the powder feed in order to mitigate the slide problem. Back pressure surges, however, as determined from a pressure gauge in the delivery line, resulted in foam feed problems and a necessity for adding increased water. The total effect was to lower the dike height.

At the recommendation of Hoge-Warren-Zimmerman, the manufacture of the Mason Flow Mixer, the preformed foam was added to the slurry in the Moyno pump chamber to overcome the varying feed rate of the foam generating unit. In order to carry the increased capacity of the slurry and the foam, it was necessary to increase the pumping speed of the Moyno, by a factor of four, to its maximum. Driving the Moyno pump at such high speeds caused troubles not previously encountered. Pumping the water-cement slurry and the preformed foam resulted in electrical overload and breaker cutout. In addition, it was very difficult to balance the flow of water-cement slurry and preformed foam delivery so as not to overflow the Moyno pump. These runs were very short lived; no sooner were flows established than the Moyno pump would overload and spill foamed concrete onto the floor.

Although homogeneous foamed concrete barriers of excellent height and strength were occasionally produced, the electrical overload problems made this procedure impractical. The slurry mixer automatically shut down several times during each run and was restarted only with difficulty to continue the run.

To avoid the Moyno feed and electrical overload problems, a pump was installed on the preformed foam unit to meter and pump the foam into the concrete downstream of the Moyno pump. This enabled the speed of the Moyno to be reduced and larger quantities of preformed foam to be added consistently in order to produce lower density foamed concrete. At the same time a larger diameter delivery line was installed to alleviate high feed line pressures. The change enabled production of foam concrete as low as 23 pcf. The resulting concrete structure lacked good integrity, however, which may have been due to a poorly blended foam slurry mix. This method of foam introduction greatly improved the operation in that it eliminated the electrical overload problem.

The one inch (i.d.) spiral mixer-delivery hose system was installed in the above system to provide additional blending action of foam and slurry. This action produced reliable barriers of significant height.

Field Tests: Field tests were conducted with the Mason Flow Mixer on the ability of the quick-set foamed concrete formulation to block and impound flowing liquids. The powder feed for the tests consisted of: 90 parts by weight cement, 5 parts by weight bentonite and 5 parts by weight lime. The unit was placed on a flat bed truck for the tests and connected to water, air and power supplies at the test site. A mobile compressor provided the necessary air.

A problem with the unit, as far as its utility in an emergency system, was found during these tests. In order for proper water mixng to occur, the unit must be nearly level. A slight slant rearward allows the water to flow backwards out of the mixing chamber without contacting the powder feed.

At ambient temperatures below 32°F, further field test work was adversely affected. At these temperatures, the foam solution lines and various water lines on the slurry mixer tended to freeze. The incorporation of either aliphatic or polyhydric alcohols in the foam solution alleviated the freezing problem but unfortunately these antifreeze materials retarded the cement set and the foam concrete tended to be easily dissociated by flowing water.

The use of inorganic solutes, $CaCl_2$ and NaCl, in high concentration to lower the freezing point of the detergent foam solution resulted in poor concrete quality ($CaCl_2$) or poor foam quality (NaCl). A barrier to a 15 gal/min flow of water using MSA salt water foam and 5% $CaCl_2$ solution produced a 13 inch high barrier of 47 to 48.5 pcf concrete. The concrete was of poor integrity, slow to gel and tended to dissociate.

A later run, conducted on an above-freezing day without added antifreeze material in the foam, resulted in a barrier height of 27 inches which was one of the best dikes produced.

The field tests using the Mason Flow Mixer established the utility of foamed concrete as a barrier for hazardous material spills and the feasibility of erecting barriers under field conditions. The two problems encountered, however, orientation sensitivity and cold weather dike formation, need further investigation.

Following the field tests discussed above, an MSA Model 2000 slurry rock dust distributor was evaluated as an example of a batch operation foam concrete generator. The unit consists of a slurry tank of approximately 200 gallons capacity, and a Moyno pump for slurry delivery. Both the Moyno and the slurry tank agitators are powered with variable speed hydraulic motors, which are in turn powered by an electrically driven hydraulic pump. A pump was employed to introduce the detergent foam downstream of the Moyno pump. The hose delivery system and silicate blending nozzle used on the Mason Flow Mixer were also employed on this system.

The advantages of using a batch system, rather than the continuous unit like the Mason Flow Mixer, are simplicity and ease of field operation. A batch unit eliminates the need for metering the powdered solids and water feed. This unit also would operate in off-level positions.

The ability of the foamed concrete to form barriers throughout the earlier experiments was generally inferior to that finally developed from the continuous mixer. Problems were experienced initially with extremely rapid slurry feeds and balancing the preformed foam and silicate solution feeds to the slurry. In order to obtain good powder mixing within the slurry tank with the small agitators normally used to slurry the rock dust, an abnormal amount of water had to be employed. Whether due to this or to unbalanced preformed foam or silicate feeds, barrier heights exceeding 14 inches were never realized. Poor gel formation, resulting in slides, inevitably held batch processed dikes to the above height limit.

Summary: The feasibility of foamed concrete to produce a dike which will hold back or impound a typical spill of liquid has been demonstrated. Tests on such varied substrates as clay, shale, chipped limestone, grass and weed-covered ground have been successful. Screening tests have shown that such hazardous liquids as methanol, 1,1,1-trichloroethane, phenol, acetone cyanohydrin and acrylonitrile do not affect freshly poured foam concrete.

The rate of buildup of hydraulic head on the barrier system is probably the single most important factor to successfully impounding a spill. An extremely high liquid velocity would probably not allow barrier formation in the flowing stream. In such a spill, however, most of the contents would likely be lost and the damage done before emergency equipment could be brought to bear. Where low flow velocities are present, or a large impoundment area available to allow filling over a period of 15 minutes or more, a foam concrete dike could be successfully employed.

It seems that the batch type operation offers the most simple approach to an emergency field unit. Such a unit could be trailer mounted, suitable for a pick-up truck operation. It would produce approximately 50 ft^3 of foamed concrete at 45 pcf per batch within 30 minutes of delivery to the site. Repeat cycles would take approximately 25 minutes each.

Three such batches (150 ft^3) could be produced in approximately 80 minutes to build barriers of such sample dimensions as: 2 ft x 2 ft x 38 ft; 1.5 ft x 2 ft x 50 ft; and 1.5 ft x 3 ft x 33 ft.

Raw materials for one batch are: 10 bags of cement (940 lb); 30 gallons of sodium-silicate solution; and 1.5 quarts of foam concentrate.

Material for one run would be transported on the unit. Material for three batches would weigh approximately 4,900 lb and require about 1,800 lb of water (215 gallons) at the site. Additional material could be brought in if needed.

The sodium silicate and foam concentrate would be the only material required to be stored in quantity for multiple batches. Enough Type I cement could be stored for 1 to 3 batches, with additional cement picked up as needed.

REFERENCES

(1) Mattson, J.S. and Merck, H.B., Jr., *Activated Carbon: Surface Chemistry and Adsorption from Solution,* Marcel Dekker, Inc., New York (1971).

(2) Condon, P.A., *Activated Charcoal, 1953–1964, A List of Selected References,* Library List No. 82, U.S. National Agricultural Library, Washington, D.C. (1966).

(3) Harrison, E.A., *Activated Carbon, A Bibliography with Abstracts,* NTIS Doc. No. COM-73-11383/9, Springfield, Virginia (July 1973).

(4) Wheaton, R.M. and Seamster, A.H., "Ion Exchange," *Encycl. Chem. Technol.,* 11. 871-899 (1966).

(5) Kunin, R., *Ion Exchange Resins,* 2nd ed., R.E. Krieger Publishing Co., Huntington, New York (1972).

(6) Wolfe, F. and Schaaf, R., "Molecular Adsorption on Microporous Ion Exchange Resins," *Plaste Kaut.,* 18 (1), 21-26 (1971).

(7) Yamabe, T., Yamagata, K. and Seno, M., *Properties of Macroreticular Ion Exchange Resins,* Lawrence Radiation Lab., University of California at Livermore, NTIS Doc. No. UCRL-TRANS-10411, Springfield, Virginia (November 1969).

(8) Hunter, R.E., *Methods for Separating Oil from Water Surface,* U.S. Patent 3,723,307, assigned to Ocean Design Engineering Corporation (March 27, 1973).

(9) Singewald, A., *Flotation of Ion Exchangers,* German Patents 1,767,360 (December 30, 1971) and 1,302,084 (December 18, 1969) both assigned to Winterschall, A.G.

(10) Sopkova, A., Flotability of Ion Exchangers, *Rudy 17* (1), 6-10 (1969).

(11) Bhapper, R.B., *Froth Flotation of Ion-Exchange Resins and Its Application,* New Mexico Inst. Mining and Technol. State Bureau of Mines and Mineral Resources Circular 53, Socorro (1961).

(12) Mason, D.G., Gupta, M.K. and Scholz, R.C., "A Mobile Multipurpose Treatment System for Processing Hazardous Material Contaminated Waters," *Proceedings of the 1972 Conference on Control of Hazardous Material Spills,* Houston, Texas (March 21-23, 1972).

(13) Bennett, G.F., ed., *Proceedings of the 1974 Conference on the Control of Hazardous Material Spills,* San Francisco, California (August 25-28, 1974).

(14) U.S. Environmental Protection Agency, American Petroleum Inst., and U.S. Coast Guard, *Proc. Cong. Prevention and Control of Oil Pollution,* San Francisco, California (March 25-27, 1975).

(15) Onishi, T., *Recovery of Oil-Soluble Compounds from Their Aqueous Solutions in Emulsions,* Japanese Kokai 70-18,978 (June 29, 1970).

(16) Bowen, H.J.M., "Absorption by Polyurethane Foams: New Method of Separation," *J. Chem. Soc.* A1970 (7), 1082-1085.

(17) Yashida, N., Imonaka, Y. and Katayana, K., *Polyurethane Foam Filter Material,* German Patent 2,244,685, assigned to Teijin, Ltd. (April 5, 1973).

(18) Miller, E., Stephens, L. and Richlis, J., *Development and Preliminary Design of a Sorbent-Oil Recovery System,* Environmental Protection Technology Series Report W73-11071, EPA-R2-73-156, Contract No. EPA-68-01-0066, Hydronautics, Inc., Laurel, Maryland, (January 1973), NTIS Doc. No. PB 221497/1 (microfiche), Springfield, Virginia (January 1973) and EP1.23/2:73-156 (hard copy), USGPO, Washington, D.C. (January 1973).

(19) Cochran, R.A., Jones, W.T. and Oxenham, J.P., *A Feasibility Study of the Use of the Oleophilic Belt Oil Scrubber,* Report No. TPR-14-70, Final Report USCG-714103/A/002, Contract No. DOT-CG-00593-A, Shell Pipe Line Corporation Research and Development Laboratory, Houston, Texas (October 1970), NTIS Doc. No. AD 723598, Springfield, Virginia (October 1970).

(20) Cochran, R.A., Fraser, J.P., Hemphill, D.P., Oxenham, J.P. and Scott, P.R., *Oil Recovery System Utilizing Polyurethane Foam,* Feasibility Study, Edison Water Quality Research Laboratory, Edison, New Jersey (1973), NTIS Doc. No. PB 231838/4GA, Springfield, Virginia (1973).

(21) Shuckrow, A.J., Mercer, B.W. and Dawson, G.W., "The Application of Sorption Processes for In-Situ Treatment of Hazardous Materials Spills," *Proceedings of the 1972 National Conference on the Control of Hazardous Material Spills,* Houston, Texas (March 21-23, 1972).

(22) Suggs, J.D. et al., *Mercury Pollution Control in Stream and Lake Sediments,* Water Pollution Control Research Series, 16080 HTD-03/72, Office of Research and Monitoring, U.S. Environmental Protection Agency (March 1972).

(23) Christianson, D., Battelle Pacific Northwest Laboratories, Richland, Washington, personal communication.

(24) Bauer, W.H., Borton, D.N., Bulloff, J.J., *Agents, Methods and Devices for Amelioration of Discharges of Hazardous Chemicals on Water,* CG-D-38-76 (August 1975).

(25) Unpublished Technical and Promotional Data, ICI United States, Inc., Wilmington, Delaware.

(26) Harding, J.F., Calgon Corporation, Bridgewater, New Jersey, personal communication.

(27) Pilie, R.J., Baier, R.E., Ziegler, R.C., Leonard, R.P., Michalovic, J.G., Peck, S.L. and Bock, D.H., *Methods to Treat, Control and Monitor Spilled Hazardous Materials,* EPA-670/2-75-042 (June 1975).

(28) Dawson, G.W., Shuckrow, A.J. and Swift, W.H., *Control of Spillage of Hazardous Polluting Substances,* Water Pollution Control Research Series, FWQA Report 15090FOZ (November 1970).

(29) Boyd, G.E., Anderson, A.W. and Meyers, L.S., Jr., "The Exchange Adsorption of Ions from Aqueous Solutions by Organic Zeolites II Kinetics," *JACS* 69 (November 1947).

(30) Kressman, T.R.E. and Kitchner, J.A., "Cation Exchange with a Synthetic Phenolsulphonate Resin," *Faraday Society Discussion* No. 7, 194.

(31) Helfferich, F., *Ion Exchange,* McGraw-Hill, San Francisco (1962).

(32) Dawson, G.W. and Stradley, M.W., *A Methodology for Quantifying the Environmental Risks from Spills of Hazardous Material,* presented at the AIChE Conference Boston-Sheraton (September 8, 1975).

(33) Kirsch, M., Vrolyk, J.J., Melvold, R.W. and Lafornara, J.P., "A Hazardous Material Spills Warning System" in *Control of Hazardous Material Spills, Proceedings of the 1976 National Conference on the Control of Hazardous Material Spills,* Information Transfer, Inc., Rockville, Maryland (April 1976).

(34) Silvestri, A., Goodman, A., McCormack, L.M., Razulis, M., Jones, A.R., Jr. and Davis, M.E.P., "Detection of Hazardous Substances," *Proceedings of the 1976 National Conference on Control of Hazardous Material Spills,* Information Transfer, Inc., Rockville, Maryland (April 1976).

(35) Silvestri, A., Goodman, A., McCormack, L.M., Razulis, M., Jones, A.R., Jr., and Davis, M.E.P., *Development of a Kit for Detection of Hazardous Material Spills Into Waterways,* Department of the Army, Edgewood Arsenal Special Publication ED-SP-76023 (August 1976).

(36) Dreyrup, A.S., U.S. Patent 3,492,250, assigned to E.I. DuPont de Nemours and Company (January 27, 1970).

(37) Ziegler, R.C. and Lafornara, J.P., "In Situ Treatment Methods for Hazardous Material Spills," in *Proceedings of the 1972 National Conference on Control of Hazardous Material Spills,* Houston, Texas (March 21-23, 1972).

(38) Epstein, M. and Widman, M., *Coatings for Ocean Bottom Stabilization,* paper presented at 158th Meeting, American Chemical Society, New York, New York (1969).

(39) Glassman, I., Hansel, J. and Eklund, T., "Hydrodynamic Effects in the Flame Spreading, Ignitability and Steady Burning of Liquid Fuels," *Combustion and Flame 13,* 99 (1969).

(40) Mecca, J.E., Jensen, H.F. and Ludwick, J.D., *Noble Gas Confinement Study. I. Development of a Long-Lived, High-Expansion Foam for Entrapping Air-Bearing Noble Gases,* U.S.A.E.C. DUN-7221 (1971).

(41) United Kingdom Mines Rescue Service, Nottinghamshire, Darbyshire and Leicestershire District, *Use of Gypsum Powders in the Construction of Stoppings,* National Coal Board, Hobart House, London (1969).

(42) Halliburton Company, *New Mine Sealing Techniques for Water Pollution Abatement,* FWPCA Contract No. 14-12-453 (March 1970).

MOBILE UNITS

EMERGENCY COLLECTION BAG SYSTEM

Introduction

The information in these sections is based on a paper presented at Conference 1976 by F. Roehlich, Jr. and R.H. Hiltz of MSA Research Corporation and J.E. Brugger of the U.S. Environmental Protection Agency. The title of the paper was "An Emergency Collection System for Spilled Hazardous Materials."

A news story which has become increasingly familiar is that of the accident involving a tank truck or railway tank car. Almost invariably such an incident results in the spillage of the material within the damaged tanker. Usually that material is one which is potentially harmful to the environment into which it is released. In order to keep such spills from streams, lakes, sewer systems, etc., several control methods have been devised. These include foamed-in-place plugs to stop the flow from the damaged tank, foamed dams to contain the spill on the ground, and the method which is the subject of this paper—a collection bag system. This system has the advantage of fast response time. Spilled materials are collected before they can breach dikes, be absorbed into the ground, or otherwise dissipated into the environment.

At the August 1974 Conference a paper was presented (1) which described the development and fabrication of the original system. In this paper the authors review that work and present the results of the program continuation in which a modified version of the Collection System was designed, constructed and tested.

System Design

The system is designed to collect and temporarily store diked or pooled spills of hazardous fluids until appropriate removal equipment can be brought to the scene. Its size and weight are such that it can be mounted as a unit on the bed of a half-ton pickup truck, van, dual wheel railroad vehicle, or airlifted by a small helicopter and be deployable at a spill site over all types of terrain by two persons. The collection bag to contain the hazardous material has a 7,000-gallon capacity (950 cf). Burst pressure is 11 psig; working pressure is 3.3 psig, equivalent to a water fill on a 30° slope; maximum working pressure is 5.7 psig, equivalent to a water fill on a 60° slope. The bag fabric and transfer hoses are selected to withstand temporary storage of various types of hazardous materials. The system pump has a nominal throughput of 60 gallons per minute from pits up to 15 ft deep. This gives a fill time of approximately two hours. The ancillary components of the system are specified to be available commercially.

Two models of the system were designed. The major difference between them is the type of power utilized by the pump. The original model has a rechargeable battery power supply whereas the second model operates from an explosion resistant gasoline engine.

Other design considerations include:

- Proper grounding to decrease accumulation of electrical charges during transfer of low-conductivity fluids.
- Reduction of operating complexity.
- Provision for application flexibility.

First Model: The collection bag actually consists of four bags fabricated from a polyester-reinforced PVC known as Shelterlite and held together by a series of nylon fabric strips. Three of the bags are each connected to the fourth bag. The latter serves as a manifold or "header" bag. Each bag is 20 ft long and 3 ft in diameter. The bags are vacuum folded into a steel-reinforced aluminum housing which is approximately 3 ft high, 4 ft wide and 1 ft deep with a spring-loaded cover or lid. The bag housing, together with the pump, motor, batteries, motor starter, hoses, hose reels and ancillary equipment is mounted on a 4 ft x 4 ft plastic pallet.

The pump is a centrifugal type, driven by a 1 hp explosion proof 24-V dc motor. The pump has a rated zero head capacity of 80 gallons per minute and a related suction lift of 25 ft. The motor is powered by twelve 12-V gelled lead-acid batteries connected in series-parallel which were sized to run the pump 2.5 hours under load without recharging.

The hose reel has a 15½-inch diameter and is separated into two compartments by a bulkhead. The reel carries a 50-ft length of 2-inch diameter reinforced chemical hose and a 50-ft length of 2-inch diameter stainless steel braided hose with a Teflon liner. The latter is attached to a rotary seal through the hose reel axis, to a selector ball valve. Stainless steel pipe connects the selector valve to the pump inlet. Another length of stainless steel pipe leads from the pump outlet to a pair of quick-connect hose couplings. These fittings are accessible through the bag housing.

A 10-ft length of 1.5-inch diameter braided stainless steel hose with a conductive Teflon liner is used to connect the header bag to the pump via one of the quick-connect couplings. The purpose of the conductive Teflon lining is to maintain static charge-free transfer of low-conductivity liquids. This hose is stored in the bag housing. The second quick-connect coupling is provided to accommodate a second set of bags or other holding vessel. The 50-ft section of chemical hose carried in one compartment of the hose reel can be removed from the reel and attached to a second inlet port of the selector ball valve.

System Operation: The following is a description of the anticipated collection procedure which would follow a spill incident. The collection system and two operator personnel arrive on the scene in a pickup truck. One operator depresses a latch on the bag housing door and the lid drops away. The interconnected folded bags are then placed on the ground and unfolded. The 10-ft hose section is then attached to an inlet port of the header bag and to one of the quick-connect couplings of the system.

Deployment procedure for the hoses depends on (1) distance from spill pool sump and/or tanker and (2) whether one or two points are to be evacuated. When pumping from a single point (sump, diked area, or tanker) located within 50 ft of the system, only the hose connected to the rotary seal is deployed with the appropriate fitting on the hose end. Fittings are provided aboard the system for pumping and for connection to tanker drain ports.

When pumping from a point located up to 100 ft away, both hoses are unreeled simultaneously and coupled together. For two point intake sources, such as a ruptured tanker and a ground spill or two spill pools, the unattached hose is connected from the second source to the second inlet port of the selector ball valve.

The pump frame is then grounded by cable and spike and the pump motor actuated. (The pump cavity should have been primed before the system arrived on the scene.) Several minutes time may elapse before actual transfer of material into the bag begins. During this time some adjustments may be made in placement of the bags on the ground. If a faster pump rate is desired, the wiring harness on the batteries can be changed to produce a 36-V output which will increase pump rotation from 2,250 to 2,900 rpm.

When the bags have been filled, the inlet lines may be rearranged to pump material from the bags into a second tanker or other storage vessel.

Second Model: In designing the second model of the collection system, efforts were made to improve upon the operating efficiency of the first model. These efforts resulted in the following changes:

(1) A 2-ply urethane-coated nylon was specified for the bag material to achieve greater abrasion resistance and better low temperature capability.

(2) The bags were equipped with screw-type ports with internal sleeves to facilitate back transfer of material from the bags to another vessel subsequent to spill cleanup.

(3) Bag strappings were replaced by belts and tie patches in order to reduce stresses resulting when the bags are filled on uneven terrain.

(4) Accordion-type bag interconnecting tubes were used to prevent pinching during bag filling which occurred with the straight tubes in the first model.

(5) Bag housing was increased approximately 90% to 4 ft x 4 ft x 1.5 ft in order to accommodate these changes.

(6) A 3-hp explosion resistant gasoline engine was substituted for the electric motor. This provided greater pump speed and reduced the weight of the system considerably by eliminating the batteries, motor starter, and extensive support structure. The battery charger was also eliminated and replaced by an explosion-proof fuel container mounted in a sheath.

(7) Further improvements in pumping efficiency were attained by increasing the size of the pump outlet quick-connect couplings and eliminating bends in the pump inlet piping.

(8) The pallet size was increased to 4 ft x 5.5 ft to accommodate the new bag housing and piping. The pallet was constructed of ribbed aluminum which provided more rigidity and stability than the slightly lighter weight plastic used in the original model.

(9) The compartmentalized hose reel of the first model was replaced by two separate reels to facilitate independent removal of each hose. Two chemical hoses were used instead of one chemical and one stainless steel. The steel hose is more flexible but can be crushed more easily and is considerably more costly.

Tests of Components: Several tests were run to evaluate bag construction materials and pumping capabilities with a variety of fluids and power sources. Two field demonstrations were also conducted.

The first bag material test consisted of exposure to various chemicals and subsequent tear-strength testing using swatches of vinyl fabric. Results are presented in Table 5.1.

A second bag material test involved filling the bags of the first model with water to near-capacity to determine seam integrity and effectiveness of a bag patching kit. Results showed that seams held but that patches could not be applied to a wet bag surface.

Pump performance curves were determined for the battery powered model at both 24 and 36 V and for the gasoline powered model. Pump capacity was measured as a function of viscosity for both models. Pump capacity as a function of vapor pressure was also determined.

Table 5.1: Tear Strength of Vinyl Fabric After 24 Hours Exposure to Chemicals

	Pounds
Phenol	312
Methyl alcohol	203
Acrylonitrile	276
Benzene	281
Acetone cyanohydrin	43
Xylene	298
Sulfuric acid	13
Aldrin toxaphene group	216
Acetone	254

(continued)

Table 5.1: (continued)

	Pounds
Nitric acid	247
Ethyl acetate	222
Sodium hydroxide (30% solution)	202
Methyl ethyl ketone (MEK)	196
Hydrofluoric acid	197
Hydrogen peroxide (30% solution)	223
N,N-dimethylformamide	263
Butyl ether	296

Source: Conference 1976

Demonstrations

The information in this and the following section is based on *Emergency Collection System for Spilled Hazardous Materials*, EPA-600/2-77-162 prepared by R.H. Hiltz and F.R. Roehlich, Jr. of MSA Research Corporation for the Environmental Protection Agency.

Upon completion of component assembly, the first model was field tested to verify its performance characteristics. The site selected for testing was a pond which serves as a chemical disposal facility located on the MSA Evans City plant property. The pond contains about 200,000 liters (50,000 gal) of water with a pH of about 9. It was planned to set the unit up at the edge of the pond and go through all of the procedures involved in normal operations to demonstrate operability and to note any problem areas not anticipated previously.

The collection system was moved by fork lift vehicle to a point adjacent to the pond. In the initial test, the bag and housing were removed from the system and not used. The 15 m (50 ft) length of stainless steel /Teflon hose was unreeled and its nozzle dropped into the pond. The pump was started and water was pumped from the pond, through the system, and back to the pond. In this series of tests, the pump was located about 2.5 m (7.5 ft) higher in elevation than the surface of the pond water. Pumping time required for initial flow from the system exit port was about six minutes. Flow rate was 163 l/min (43 gpm).

At the conclusion of the initial test, the batteries were recharged at 27.3 V. Charge current was maintained at 1.0 A. The bag and housing were then reinstalled on the pallet for a field demonstration of the system. The bag was removed from its housing, deployed and the pump started. Within six minutes the header segment of the bag began to fill. Pumping was stopped after 35 minutes with approximately 5,700 liters (1,500 gal) in the bag segments. About 40% of this total was in the header segment. The bag was then emptied to the pond, cleaned, dried, evacuated and refolded. Batteries were again recharged.

The final demonstration of the series was similar to the previous one, but involved two hours of pumping time. No measurement was made of the volume pumped, but the bag system was about three-quarters full. The demonstration was witnessed by EPA personnel and recorded on film.

In a full-scale demonstration, the second model collection system was mounted on the bed of a ¾-ton pickup truck. A 26,500-liter (7,000 gal) capacity tank of the type used in road tanker trucks was filled with water to which an innocuous red dye had been added. In the demonstration run, a tanker leak was simulated by actuating a motor-operated ball valve located on the underside of the tanker. Urethane foam dikes were then established nearby to collect the spilled material and to divert portions of the spill into the primary dike. The pickup truck bearing the collection system was then driven to a point adjacent to the diked area. Hoses were deployed to the dike and to the tanker. The bag was then removed from its housing and unfolded on the ground behind the truck. The engine was then started and the spilled material pumped from the dike area trench to the bags via one of the 15-m (50 ft) hoses. The second 15-m hose was connected to a fitting located at one end of the tanker.

When the tanker leak subsided, the 3-way valve in the system was rotated to admit liquid from the tanker to the pump which in turn continued to transfer it into the bags. The hoses were then reconnected and liquid was pumped from the bags back into the tanker. Events of the demonstration were recorded on film and witnessed by EPA personnel.

Trouble Shooting and Maintenance

Improper operation of the Emergency Collection System can occur due to several causes. The following instructions will be helpful in tracing the cause of a failure or improper system operation. Possible difficulties are listed in the sequence in which they might be encountered while operating the system.

Difficulty	Instructions
Hose reel jammed	Check crank locking wheel located at side of reel mount
	Check for binding of hose(s) against a part of the system
Bag housing door will not open	Probably due to twisting of the bag housing, use screw driver or similar object to pry out top of door
Pump will not start	Check battery connections and voltage
	Check for foreign objects in shaft and coupling area
	Open motor starter control box and check fuses and overload heaters
	Check for possible locked rotor by removing coupling guard and turning shaft by hand
No charging current	Check fuses in starter control box, selector switch must be in OFF position until input power is applied to the charging unit
	Check for open circuits in battery wiring
	Check power cord from 115 V ac line
Starter rope jammed	Check for binding in rope recoil reel
	Check for foreign objects in pump
Starter rope will not rewind	Check spring in recoil reel

No general maintenance should be required other than to clean the hoses and pump lines after use. A repair kit for the bag is included with the system which contains various types of hose sealers. Maintenance on specific system components should be performed as directed by the supplier, or by returning the unit to the vendor for servicing.

MUDCAT/PROCESSING SYSTEM

Introduction

> The information in these sections is based on *Removal and Separation of Spilled Hazardous Materials from Impoundment Bottoms,* EPA-600/2-76-245 prepared by M.A. Nawrocki of Hittman Associates, Inc., for the Environmental Protection Agency.

Practical methods for the removal and processing of hazardous or semihazardous materials from the bottoms of water bodies are receiving relatively high priority as targets for environmental action. Not only must the offending material be removed from the bottom sediments in an efficient and safe manner, but the removed sediment and hazardous material mixture must also be processed and disposed of in an environmentally acceptable and safe manner.

Consequently, Hittman Associates, Inc., under contract to the U.S. Environmental Protection Agency, conducted a demonstration of a system for the removal and processing of hazardous and semihazardous materials from the bottom of a shallow pond. Since it is difficult to justify the spilling of hazardous materials even under research conditions, a number of simulated hazardous materials were spilled onto a pond bottom for the demonstration. These simulated hazardous materials were relatively innocuous substances whose physical properties were chosen to represent a range of properties which might be displayed by real hazardous materials.

The purpose of the demonstration project was twofold. The first was to demonstrate a technique for removing hazardous materials from bottoms of water bodies at a high rate but with minimal adverse effects on the surrounding water body. The second purpose of the program was to evaluate a portable system which could be set up to process the sediment and hazardous materials mixture and return clean water to the pond.

The removal system used was a Mudcat dredge manufactured by National Car Rental System, Inc. It is specially designed for use on small bodies of water, and to impart minimum turbidity to the water while dredging. It can discharge approximately 1,500 gpm of slurry with a solids concentration of 20 to 30%. Processing was performed by a system consisting of a pair of elevated settling bins, a bank of hydrocyclones, a standard cartridge-type water filter unit, and a bag-type filter known as a Uni-Flow filter. Basically, the Uni-Flow filter consists of a number of hanging hoses. Dirty water is pumped into the inside of the hoses and is allowed to filter through them. The suspended matter is trapped on the inside of the hoses. Periodically, the collected sludge is flushed from the inside of the hoses.

Removal System

The system utilized for removing the spilled simulated hazardous materials from the pond bottom was an approximately 30-ft long Mudcat dredge. The dredge moves in straight line directions by winching itself along a taut, fixed cable. Bottom sediment removal equipment on the dredge consists of an 8-ft long, horizontally opposed, adjustable depth, power driven auger and a pump rated at about 1,500 gpm with a 10 to 30% solids concentration of the dredged slurry. A retractable mud shield over the auger minimized mixing of the disturbed bottom deposits with the surrounding pond water.

The dredge also comes equipped with a rock box into which objects greater than 8 inches in diameter (the diameter of the discharge line) are automatically discarded before the dredged slurry is pumped into the discharge line.

Processing System

The system used for processing the suspended sediment and simulated hazardous material dredged slurry was set up on a 50-ft high knoll, approximately 600 ft from the edge of the pond. It included, in order of processing of material:

(1) Two steep-sided elevated bins, each with an initial capacity of 36 yd^3, installed in series. They are of the type typically used in concrete batch plant operations. The discharge from the dredge was pumped directly to the first bin where settling of suspended solids occurred. The slurry was then allowed to overflow into the second bin, where additional settling occurred. From the second bin, the flow was split to either a temporary holding/settling basin or the feed pump for the secondary separation phase. Each of the elevated bins provided about 144 ft^2 of surface area for settling.

(2) A bank of hydrocyclones manufactured by Demco Incorporated, and consisting of six 4-inch style H cones with 3-gal silt pots, a closed underflow header, and automatic solids unloading.

(3) A commercially available cartridge-type water filter manufactured by Crall Products, Inc. The unit consisted of four model 16-17-51 filters, operating in parallel, with an automatic back-flushing mechanism. Each of the filters contained 51 permanent sand cartridges with filter openings rated at 25 μ.

(4) One Uni-Flow bag-type fabric filter consisting of 720 1-inch diameter, 10-ft long, woven polypropylene hoses. The hoses were arranged in six banks of 120 hoses each. This enabled the shutting down of one bank for hose maintenance or replacement while the other five banks could be kept on line. The slurry was pumped into a top header which distributed the influent to each bank of hoses. The filtrate from the hoses is collected in a bottom tray and allowed to flow by gravity back to the pond. Normally, every 5½ minutes the sludge within the hoses is drained for 30 seconds into a collection trough and allowed to flow by gravity into a sludge disposal basin.

Figure 5.1 is a schematic diagram of the overall processing and sludge disposal system.

Figure 5.1: Schematic of Processing and Sludge Disposal System

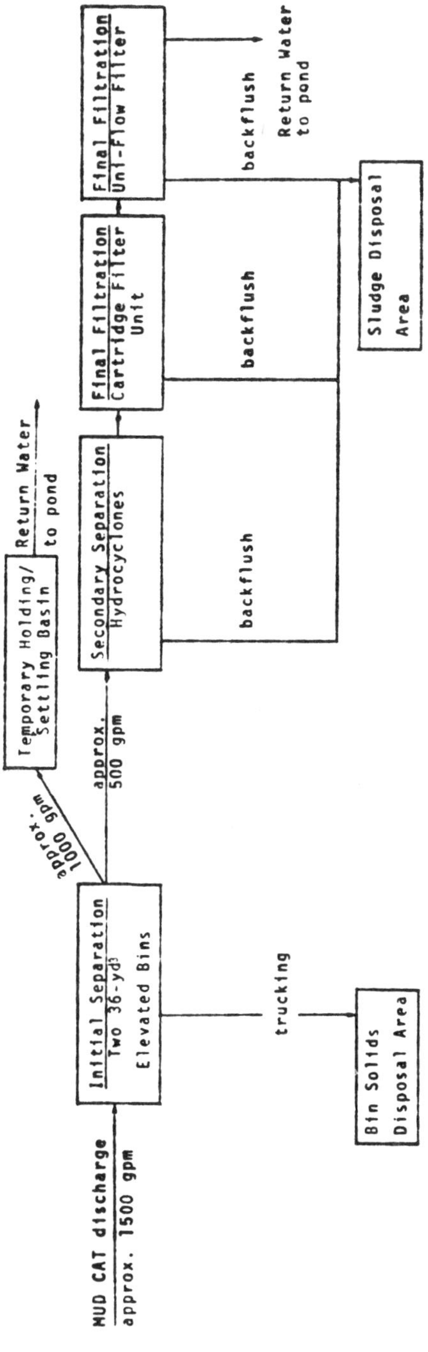

Source: EPA-600/2-76-245

Removal and Processing of Simulated Hazardous Material from a Pond Bottom

A test area 3 ft wide and 28 ft long, with lead-in and lead-out areas of 4 ft at each end, was prepared on the pond bottom such that the center line of the test area corresponded to the Mudcat's positioning cable. Marker poles were placed at the corners of the test area, and the Mudcat positioning cable over the test area was marked at 4-ft intervals to provide reference points for bottom sampling. Each simulated hazardous material was placed on the bottom of the test area in a uniformly distributed pattern using a 6-ft length of 6-inch diameter PVC pipe.

With the Mudcat positioned well outside the test area, pond water only was pumped into the bins until they were full and the process system was balanced and back-flushed. During this pumping, the process system was charged with clean pond water and all pressure controls were adjusted to proper balance. The auger of the Mudcat was lowered to the pond bottom as it reached the lead-in area at a rate of 10 fpm. After all dredge operations were stabilized during the traverse of the lead-in area, the test area was traversed in a smooth and continuous manner under normal operating parameters.

Auger rotation was stopped after the complete traverse of the test area and lead-out area and pond water only was pumped in order to clear the dredge discharge line of all simulated hazardous material. During this time the dredge advanced to a position remote from the test area.

A survey of the pond bottom was conducted before and after each simulated hazardous material test to determine the amount of pond bottom material removed by the Mudcat during each test. A La Motte Chemical Co. Code 1061 Bottom Sampling Dredge was used to collect samples of the bottom after each simulated hazardous material removal test. The bottom sampling pattern was on alternate, 4-ft centers beginning 2 ft inside the test area.

A total of seven bottom samples were collected after each simulated hazardous material removal test, except for the first test using fine iron powder when each 4-ft square area was sampled and a total of fourteen samples were collected.

The process system was balanced and back-flushed during the initial pumping of the pond water only. All process system components were operated without back-flushing during the conduct of the test. The 2-minute test run time was less than the time between normal automatic back-flushing cycles.

Laboratory Analyses: Analysis for the amount of suspended solids in the effluents of the elements of the processing system was performed by standard analytical methods (2).

Table 5.2 presents the results of the removal of the simulated hazardous materials from the pond bottom. The variability in the thickness of sediment removed for the different tests reflects the inability of the dredge operators to determine precisely the depth to the pond bottom in the very soft sediments present in the test area.

Table 5.2: Removal of Simulated Hazardous Materials from Pond Bottom

Material	Pounds Placed on Bottom	Pounds Remaining on Bottom	Percent Removed	Average Thickness of Sediment Removed (ft)	Number of Passes by Dredge
Iron powder	800	8.0	99.0	1.12	1 (forward)
Glass beads	500	<0.1	99.9+	0.61	2 (forward and backward)
Iron filings	500	0.4	99.9	0.22	1 (backward)
Coal	500	2.1	99.6	1.14	1 (backward)

Source: EPA-600/2-76-245

Removal of Simulated Hazardous Materials by Processing System: The amount of simulated hazardous material entering/leaving each system component was determined so that each component of the system could be evaluated as to its ability to remove the various test materials along with the normal suspended solids. The main interest was in the processing system units downstream from the elevated bins. Table 5.3 is a summary of the removal efficiencies of the hydrocyclones, cartridge filter unit, and Uni-Flow filter. Here, the three types of removal equipment can be compared on a step-by-step basis in terms of their ability to remove the simulated hazardous materials which remained in the dredged slurry after passing through the clarifier bins.

Table 5.3: Summary of System Component Removal of Simulated Hazardous Material

	.... Percent Removal*............			
Component	Iron Powder	Glass Beads	Iron Filings	Coal
Hydrocyclones	42.9	87.8	83.8	55.2
Cartridge filters	22.0	6.1	7.5	29.9
Uni-Flow	35.0	6.1	8.7	14.9
Return water to pond	0.1	–	–	–

*Of simulated hazardous materials reaching the processing system after the elevated bins.

Source: EPA-600/2-76-245

Removal of Latex Paint

Testing the ability of the processing system to remove a real hazardous material, in this case latex paint, was done after dredging of the simulated hazardous materials was complete. Twenty-one gallons of paint were dumped directly into the dredge discharge plume into the first elevated bin. The paint was dumped within a two-minute period as dredging of normal bottom sediments was occurring at a reduced rate of 500 gpm. Samples were collected throughout the processing system during the dumping of the paint for an additional two minutes afterward. Table 5.4 presents the results of this processing of latex paint.

Table 5.4: Results of Processing Latex Paint

System Unit	Effluent Suspended Solids Concentration (mg/l)	Adsorbed Pigment Removed	Pigment in Solution Removed	Total Pigment Removed
		............(lb)............		
Hydrocyclones	140,500	0.00	-0.03*	-0.03*
Cartridge filters	103,040	4.02	0.00	4.02
Uni-Flow	94,430	5.83	0.03	5.86
Return water to pond	1,770	0.47	0.00	0.47

*The negative sign indicates that, in the hydrocyclones, the net effect was an increase in the pigment in solution. The source of pigment was that pigment adsorbed to silt particles which was agitated into solution by the action of the hydrocyclones.

Source: EPA-600/2-76-245

Discussion of Field Results

Removal System: It is evident that the Mudcat dredge is a relatively effective method for removing undesirable particulate materials from pond bottoms. It removed from 99 to over 99.9% of the material placed on the bottom within the test area. During some of the tests, however, up to approximately 1 ft of bottom sediments were removed along with the simulated hazardous material. Even though relatively brief tests were performed, the degree to which the test materials were removed by the dredge was observed to be influenced by two factors.

First is the fact that during a backward cut, the Mudcat dredge has a greater efficiency for removal of sediment, and consequently, of spilled hazardous material, than during a forward cut. This is due to the fact that during a backward cut the mud shield is fully extended over the cutting auger and helps to prevent the resuspension of any bottom sediments into the surrounding water. This influence was first observed during the test for removal of glass beads. As the dredge traversed the area in a forward cut, the pond bottom was quickly sampled behind the dredge. The samples revealed that a large amount of glass beads had been left on the bottom. Therefore, the dredge traversed the area again, but in a backward cutting mode. The remaining tests were both performed using backward cuts by the dredge and showed higher removals than the first test when a forward cut was used.

Secondly, the specific gravity, and possibly, the relative softness of the material being picked up was observed to influence the removal efficiency. During the two tests which were made with the same direction of cut, i.e., backward only, the lighter and softer material (coal) was observed to have a lower removal rate and also be dispersed into the surrounding water body more easily than the heavier and harder iron filings.

Overall, the Mudcat performed very satisfactorily in removing undesirable particulate matter which had a wide range of specific gravities and particle sizes from a pond bottom. It is probable, therefore, that such a removal system would be satisfactory for the removal of most particulate hazardous materials spills which have settled to the bottom of a water body. The bottom sediments would have to be relatively soft to ensure a high percentage of pick-up by the Mudcat dredge. The minimum depth of cut by the dredge into the pond bottom which produced efficient removal of the test material was approximately 0.3 ft.

It has been postulated that the Mudcat dredge could be used somewhat as a floating pump platform to remove lighter hazardous materials because the Mudcat intake can be elevated to the water surface. However, this is a relatively inefficient means of removing floating or suspended hazardous materials from a water body since large volumes of water would also be pumped and have to be processed. However, the Mudcat might be used to divert small streams around a spill site.

Processing System: The ability of the processing system to remove a variety of undesirable particulate material along with suspended solids from a dredged slurry and return a relatively high quality water to a pond was documented during the field demonstration. Essentially all of the glass beads, iron filings, and coal, and 99.9% of the fine iron powder were removed from the dredged slurry before the water was returned to the pond.

Through the test on latex paint, there was also an indication that the system is applicable to removing some components of liquid hazardous materials present in dredged slurry. During this test, 95.5% of the pigment reaching the system downstream of the clarifier bins was removed along with the suspended solids. Leakage of the Uni-Flow hoses prevented a greater removal efficiency, as evidenced by the high suspended solids concentration shown in Table 5.4.

The overall quality of the effluent from the processing system was adversely affected by leaks in the Uni-Flow hoses, which allowed unfiltered water to enter the effluent. Leakage was especially bad during the testing of latex paint, when an overall effluent suspended solids concentration of 1,770 mg/l was measured.

Originally, the processing system was designed to be operated at a flow rate of 500 gpm. However, two conditions were found to prevent operation at this optimum processing rate. First, the back pressure in the cartridge filters, at flow rates approaching 400 gpm and at the dredge slurry solids concentration encountered, caused a reduction in flow rate. Secondly, sediment rapidly builds up in the Uni-Flow hoses. This buildup contributes to the back pressure in the system. Flow rates must be adjusted downward to compensate for the increase in pressure on the hoses in order to prevent their bursting. Expedient flow rates for the Uni-Flow filter were found to be between 100 and 300 gpm after the hoses had been coated by sediment.

The elements in the processing system downstream of the elevated bins operate as a closed system, the processing rate of which is dependent upon the operating characteristics of the most sensitive element. The Uni-Flow filter is the most sensitive to pressure and therefore is the element in the processing system upon which the total system is dependent.

The hydrocyclone unit has a specified operating pressure range in which removal of suspended solids is optimized. This pressure range corresponds to a system flow rate of between 420 and 540 gpm. Therefore, the maximum flow rate of approximately 300 gpm for the Uni-Flow filter after the hoses become blocked with sediment limits the efficiency of the hydrocyclone unit to a suboptimal range.

Efficiency of removal by the cartridge filters is not significantly affected by changes in flow rate or pressure. However, the cartridge filters can become the limiting element in the processing system if the filter cartridges become restricted by accumulated solids. When this happens the cartridge filter unit governs the processing rate of the system.

During the processing of the simulated hazardous materials, the hydrocyclones consistently removed the greatest amount of simulated hazardous material which entered the processing system after the bins. In three out of four tests, the Uni-Flow filter removed greater amounts of simulated hazardous material than the cartridge filter unit, even though it was downstream of the cartridge filters. After the test program was complete, the cartridge filters were opened and cracks were discovered in a number of the cartridges. These cracks were postulated to have occurred during back-flushing of the filters and would allow unfiltered dredged slurry to leak through this unit.

Very little pigment in the water base paint processed as a hazardous material was found in solution in any of the system effluents; practically all of the pigment in the effluent samples was adsorbed to silt particles. In the processing of the paint, the hydrocyclones removed no pigment. In fact, the hydrocyclone unit resuspended some pigment that had entered it adsorbed to silt particles, so that the unit made a negative contribution to the removal system. The cartridge filter unit removed no pigment from solution and removed less adsorbed pigment than the Uni-Flow filter. The Uni-Flow filter was the only element which removed pigment from solution.

The complete sediment processing system may not be applicable in all situations. Some components of the system are more suited for removing certain particle size ranges and specific gravities, and thus in some instances some of the components could possibly be eliminated. For example, as evidenced by the results, the hydrocyclones are most efficient in removing sand size or larger particles of relatively high specific gravity. If the undesirable material is composed mainly of fines, little advantage will be gained by processing the material through hydrocyclones. This was especially evident during the latex paint test.

A degree of redundancy is provided in the system by utilizing both the cartridge filter unit and the Uni-Flow filter in the final filtration step. Both of these filters are not required for final filtration. From these tests with simulated hazardous materials and previous experiments with dredged slurries, the Uni-Flow filter proved to be more useful in the processing system than the cartridge filter unit; it can efficiently remove a variety of materials with different particle sizes and specific gravities.

Sizing and selection of the individual components of the processing system should ideally be performed on a case-by-case basis. The clarifier bins, hydrocyclones, and Uni-Flow filter are all applicable to the processing of undesirable particulate material along with dredged sediments. Utilization of the cartridge water filter is limited in this application becaause of operational difficulties encountered while using it on slurries with high suspended solids contents, even though it did produce some removal of the simulated hazardous materials and the latex paint tested.

Conceptual Portable System

General Considerations: The experience gained during the field demonstration was utilized to develop a concept for a system which could be utilized to remove and process hazardous materials spills which had settled to the bottoms of water bodies. The basic criteria which were followed in defining this system were:

(1) The system had to be portable, preferably to be able to fit and be transported on one or more semitrailers.

(2) The system had to have the capability to be easily and rapidly set up and disassembled.

(3) The components of the system had to be reliable with the minimum possible chance of breakdown or breakage during processing.

(4) Maintenance should be simple and preferably be performed while the system was on line with a minimum of hazard to personnel.

(5) Capability should be present to process a wide range of anticipated hazardous materials.

(6) Capital and operating costs should be reasonable.

(7) The system had to have the capability of processing the full 1,500 gpm flow from the Mudcat at the expected solids loadings rates.

Overall System: After the data from the field demonstration were analyzed, additional data were reviewed for possible application to a portable hazardous materials processing system. These data consisted of manufacturer's literature, published reports, discussions with equipment manufacturer's representatives, and previous experience with and analysis of equipment which might be applicable for use in the processing system. The requirements of the system were balanced with the available state-of-the-art equipment and possible modifications or refinements to the equipment which were deemed feasible.

Figure 5.2 is a schematic of the overall conceptual system for the removal and processing of hazardous materials from pond bottoms. It has been assumed that the Mudcat dredge will be utilized for the removal operation. The dredge proved efficient in removing materials from a pond bottom as well as providing a minimum of contamination to the surrounding water, even under simulated failure conditions.

Initial Separation: From the results of the field demonstration and the subsequent review of additional data, it was apparent that initial separation of dredged solids and hazardous materials could best be accomplished by some form of settling basin. The elevated bins used in the field demonstration were judged to be too cumbersome in terms of erection and solids unloading, so that an alternative settling device was used in the conceptual system.

The equipment for the initial separation phase of the conceptual system consists of a portable hydraulic scalping-classifying tank combined with a spiral classifier. Basically the scalping-classifying tank is a metal tank which is used in the sand and gravel processing industry to hydraulically separate sand and gravel from slurries, and to automatically meter the release of solids to collecting and blending flumes. The surface area of the tank determines the size of particles that will be removed as a function of flow rate.

Scalping-classifying tanks have V-shaped bottoms to collect the settled solids and are equipped with valves on their bottoms which discharge the solids. Motor-driven vanes sense the level of solids in the bottom of the tank and automatically open the valves as the solids accumulate. The solids discharged through the valves drop into flumes which transfer the solids to spiral classifiers.

Spiral classifiers, also called screw classifiers or sand screws, are basically rectangular tanks with parallel sides, a vertical wall at one end, and a sloping bottom which extends to a height above the top of the other end wall. A rotating screw which operates on the incline conveys the settled solids up the sloped bottom and deposits them outside the tank.

For hazardous materials processing system, the tank would have a water surface of 40 ft x 10 ft. Two spiral classifiers supplied with the unit each has a screw diameter of 44 in and a length of 32 ft.

Figure 5.2: Schematic of Conceptual System

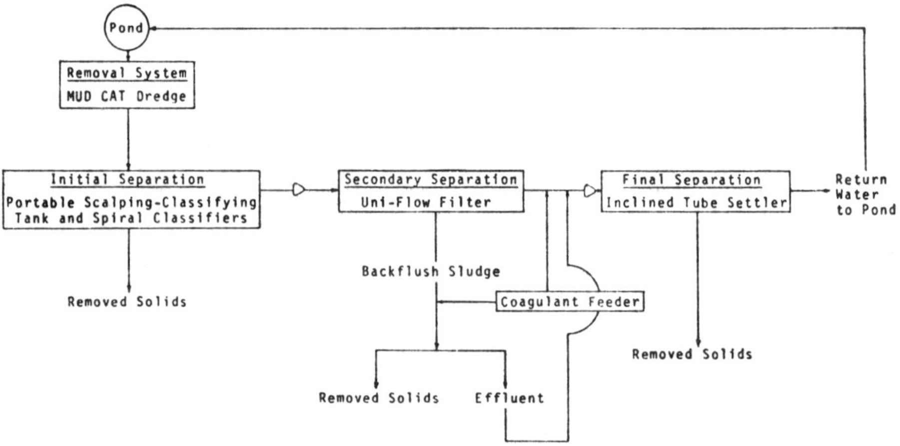

Source: EPA-600/2-76-245

Secondary Separation: The overflow from the scalping-classifying tank will go to the secondary separation portion of the system where the fine-grained materials (less than 74 μ in diameter) will be removed. In the conceptual system, hydrocyclones or cartridge filters will not be used. Hydrocyclones are most effective for removing sand size (74 μ) particles or larger. Because of operational and maintenance problems encountered with the cartridge filter unit during the processing of dredged slurries, this equipment was also deleted from the conceptual system.

The most promising piece of equipment for removing fine-grained material from a dredged slurry and returning a high quality water to the pond still appears to be some type of Uni-Flow filter. Experiments with 5-inch diameter hoses showed that they would be better than 1-inch hoses for this application. Mainly, fewer problems with hose blockage were encountered with the larger diameter hoses.

The hoses will be constructed of a polypropylene fabric and have wire cages on both the inside and outside of the hoses. The wire cages on the inside of the hoses prevent the collapse of the hoses during the draining cycle, thus producing a more effective cleaning of the hoses. Wire cages on the outside of the hoses prevent excessive bowing of the hoses during operation of the filter. This enables the Uni-Flow filter to be operated at a higher pressure and thus a higher flow rate. The influent would enter through the bottom of the hoses.

Drained solids would fall into a collection trough beneath the unit. Coagulants will be added to the back-flush sludge from the Uni-Flow filter and the effluent from this process will be recycled back into the system for final solids removal.

Final Separation: An inclined tube settler will be used as a final solids separation step in the portable system. This unit was included in the system to ensure that water of the highest quality is returned to the pond. Typically, the Uni-Flow filter has an effluent which averages a few hundred mg/l of suspended solids. An inclined tube settler would be ideal as a downstream addition after the Uni-Flow filter since settlers are used to clarify wastewater which usually have less than 1,500 mg/l of suspended solids. The settler would also serve as insurance should a Uni-Flow hose burst and release the trapped hazardous materials into the effluent before the hose can be turned off.

Inclined tube settlers are basically composed of a bank of inclined tubes which may be circular, hexagonal, square, rectangular, or chevron-shaped in cross section. Wastewater influents flowing up through the tubes tend to drop their suspended solids due to the force of gravity on the suspended particles. The steep inclination of the tubes causes the settled sludge to counterflow along the side of the tubes after it accumulates. It then falls into a sediment storage sump below the tube assembly.

In the conceptual system a coagulant will be added to the influent to the inclined tube settler. Systems employing coagulation in conjunction with an inclined tube settler can remove particles 6 μ in diameter or smaller. Typical flow rates through inclined tube settlers are on the order of 3 to 5 gpm/ft^2 of tube cross section.

Transport Requirements: The conceptual removal and separation system will require a total of five semitrailers for transport of the complete 1,500 gpm system, including auxiliary equipment. Breakdown of the transportation requirements is as follows:

Removal System—1 custom semitrailer to hold Mudcat dredge and pipe
Initial Separation—1 portable scalping-classifying tank and spiral classifying tank and spiral classifiers, size of one semitrailer
Secondary Separation—1 Uni-Flow filter on low boy trailer
Final Separation—1 inclined tube settler on standard size semitrailer
Auxiliary Equipment—2 pumps, piping, coagulant feeder, etc., to fit on one semitrailer

DYNACTOR

Introduction

The information in this section is based on a paper presented at

Conference 1974 by S.R. Rich and T.G. Pantazelos of RP Industries, Inc. and R.G. Sanders of Industrial Bio-Test Laboratories, Inc. The title of the paper was "The Short Contact Time Physical-Chemical Treatment System for Hazardous Material Contaminated Waters."

This paper describes a full-scale mobile system to decontaminate spilled hazardous materials by short contact time treatment in a continuous flow configuration at one-third million gallons per day. The work is being conducted through the coordinated efforts of scientists, engineers, biologists, and chemists of RP Industries, Inc., Hudson, Massachusetts, and Industrial Bio-Test, Inc., Northbrook, Illinois, supported by Contract No. 68-03-0228 from the Environmental Protection agency. Under earlier EPA Contract No. 68-01-0123, the feasibility of the short contact time physical-chemical treatment system was established (3).

As a result of the successful conclusion of the feasibility studies, the present program was started in April 1973. The small-scale feasibility studies and measurements had demonstrated that the dynactor/separator system was indeed capable of decontaminating substantially all of the hazardous polluting substances listed in the report prepared by Battelle Memorial Institute (4).

The dynactor, a continuous dynamic reactor, was shown to be capable of continuous flow decontamination of hazardous substances in which a variety of chemical reactions are completed in less than 0.2 sec. The reactions include activated carbon adsorption, neutralization, precipitation, ozonation and chlorination, and aeration.

In the case of carbon adsorption, the liquid being decontaminated is pumped into the dynactor. The resultant shower of thin films and fine particles of liquid aspirates large volumes of air into which dry, powdered activated carbon is deliberately introduced. The powdered activated carbon is completely scrubbed, picked up, by the liquid. The powder, now captured by the liquid, adsorbs the hazardous polluting substances in the water. Both processes take place in less than 0.2 sec.

In addition, powdered magnetic iron oxide is added to the dry activated carbon before introduction into the dynactor, permitting magnetic separation of both substances after a suitable binding agent has been added to the treated solid/liquid suspension. The binding agent (a flocculant) permits the solids to settle rapidly in a group of thin-layer, high-rate settling columns which are operated in parallel with one another, after which a continuous flow magnetic separator/dewatering system increases solids concentration to between 10 and 15% solids.

If the polluting substance is an acid, powdered, dry sodium bicarbonate is aerosolized into the dynactor. The liquid shower aspirates air into which the powdered sodium bicarbonate has been introduced. The liquid scrubs the powder, which neutralizes the acid. The proper quantity of powdered sodium bicarbonate is metered into the system to produce the desired output pH. If the contaminating chemical is alkaline, a dry powdered acid such as sulfamic acid is introduced into the air that is aspirated by the liquid shower. Again, the acid powder is scrubbed by the liquid and neutralizes the alkaline material to a desired pH.

Where the spilled hazardous substance is a heavy metal salt, a powdered precipitant is introduced into the air stream input to the dynactor reaction chamber. Most toxic heavy metals are reduced to innocuous levels in the liquid by precipitation and subsequent removal of a highly insoluble salt. After precipitation, a suitable flocculant is added so that the salt precipitate is removed in the thin-layer, high-rate settling columns which are part of the mobile treatment system.

Ozonation and chlorination are accomplished by introducing either ozone or chlorine into the air input to the dynactor reaction chamber. Substantially complete adsorption reactions take place in the dynactor within 0.2 sec. These materials (chlorine or ozone) are metered into the system at the level required for decontamination of the particular substance being treated in the system (at the concentration found in the spill-contaminated water).

Where simple aeration will decontaminate a hazardous polluting substance, the dynactor is employed without any additives other than the air that is aspirated by the liquid under treatment. The intimate contact by the liquid and gas in the dynactor reaction chamber raises dissolved oxygen concentration to full saturation during passage through the reaction chamber. Aeration alone is effective against certain contaminants and this is readily accomplished in the dynactor/separator treatment system.

In addition to the small-scale feasibility studies and measurements conducted under EPA Contract No. 68-01-0123, the dynactor and magnetic separation system for carbon adsorption of a spilled hazardous substance was tested in a field demonstration on the Little Menomonee River. This work was completed under EPA Contract No. 68-03-0181.

At the Little Menomonee River, a dynactor and a pair of thin-layer, high-rate settling columns, together with automatic controls and chemical feed systems, were placed aboard a small van truck and driven from Marlboro, Massachusetts, to Milwaukee, Wisconsin. River water, plus river bottom mud contaminated with creosote, were pumped from the riverbed into the dynactor system. Here, at 12 gpm, the technology demonstrated in the feasibility phase of the present program proved highly successful in decontamination of the river water which contained a high concentration of both suspended and dissolved organic components of creosote.

First, the river bottom mud and creosote solids were removed in one of the high-rate settling columns. The partially clarified river water containing both dissolved and suspended organic materials was then treated in the dynactor, where activated carbon (plus magnetic iron oxide) was added to adsorb the organic materials. The resulting suspension was then clarified magnetically, settling and dewatering substantially all of the solids. The system effluent contained 0.1 ppm of organic materials. The initial input to the system had been approximately 200 ppm. This dynactor system operated for approximately 80 hr over a one-month period, demonstrating that the dynactor technology was not only technically feasible but practical for use in actual field operating conditions.

Equipment Description

The information in this section is based on a paper presented at Conference 1976 by R.G. Sanders of Industrial Bio-Test Laboratories, Inc. and S.R. Rich and T.G. Pantazelos of RP Industries,

Inc. The title of the paper was "A Mobile Physical-Chemical Treatment System for Hazardous Material Contaminated Waters."

The dynactor hazardous spills treatment vehicle is self-contained, self-powered and capable of being taken over any standard road to the site of a spill of hazardous chemical without exceeding any state or federal regulation regarding vehicle weight or size.

Aboard the dynactor hazardous spills vehicle there are:

(1) A 125 kVA diesel electric generator and power supply.

(2) A dynactor—a dynamic thin-film gas/liquid/solid reactor for direct gas/liquid/solid contact.

(3) Motive pumps that pump the liquid from a spill into the dynactor, and provide the necessary flow and pressure required by the dynactor to detoxify hazardous chemicals.

(4) Reagent powder feeding equipment to feed appropriate reagent powders into the dynactor to treat liquids for which these powdered materials are appropriate.

(5) Liquid reagent feeding equipment to supply the dynactor with those reagents that are more appropriately fed in liquid form.

(6) Coagulating and flocculating chemical feed equipment to furnish those chemicals necessary to effect settling and concentration of solids either present, or produced in the liquid in passage through the dynactor system.

(7) A high-rate lamellar settling system for the rapid settling, concentration, and removal of solids either present or produced in the liquid under treatment.

(8) Sludge-voiding automatic valves for removal of sludges and solids that settle in the high-rate lamellar settlers.

(9) A gravity feed system to return the decontaminated and clarified water to the environment.

(10) A chemical analysis bench so that operating personnel can test influent, effluent, and intermediate samples of the water that is being treated in order to control the various parameters that govern the performance of the system.

(11) A logic system incorporated in the design which continually senses flow rates, levels, pressures, so that the system, once set in operation and adjusted, functions automatically.

Generator: The diesel electric generator is housed in the front portion of the trailer. It is mounted on a vibration absorption system and furnished with a "super-critical" muffler system in order to reduce the noise level that it would otherwise generate in the interior of the trailer. The generator is 125 kVA, approximately twice the actual power requirement of the dynactor equipment in order to be able to function during the high power demand periods produced by starting and stopping various pumps and components of the dynactor system. Thus, it accommodates starting current surges without undue fluctuation of generated voltage or frequency. The appropriateness of this selection has been proved out during the tests of the system conducted, starting in July 1975.

A fuel tank for the diesel electric generator is located inside the trailer and contains enough diesel fuel for forty-eight hours of operation. This permits continued performance without constant concern by the operating crew over power outage, etc. A power panel and logic system are mounted just aft of the diesel generator housing, accessible to operating personnel. Substantially all controls are located on the panel.

Transport Vehicle: The dynactor that has been incorporated in the hazardous spills vehicle is constructed of specially selected lightweight materials in order to be capable of mobile operation. Its reservoir is constructed of aluminum and protected against corrosion by a coal-tar epoxy coating system that has proved to be capable of resisting substantially any chemical with which the system may come in contact. The dynactor plenum chamber is raised and lowered by lightweight winches and an erection mechanism. For travel over the road, the plenum chamber is lowered to a level below the height of the standard "Great Dane" 45-ft trailer that houses the system. When the trailer arrives at the site of a spill, the plenum chamber is erected to its full height of 26 ft above the road.

The reaction column and return duct are constructed of fiber glass reinforced neoprene rubber. The total design permits full and proper operation of the dynamic reactor when erected on site and ready transportability for travel purposes. The dynamic reactor occupies a little less than one-third of the total volume of the vehicle. The nozzle system of the dynactor is fed by stainless steel tubing, designed to contain the 200 psi of pump pressure at the rated treatment liquid flow of 250 gpm.

Pumps: The pumps that operate the dynactor system are (1) a low pressure submersible pump that is lowered into the spill itself and (2) the dynactor motive pump that furnishes this liquid at 200 psi to the dynactor nozzle system. The submersible pump is a low power (5 hp) pump which is dropped by the operating crew into a given spill. Having "flooded suction," it readily pumps the liquid in the spill up to the dynactor vehicle, directly to the input of the 50 hp dynactor motive pump. This is a "fire engine pumper" pump, which furnishes 250 gpm of the spill to the dynactor nozzle system. Pump (1) is portable. Pump (2) is fixed and mounted directly on the floor of the spills vehicle.

Chemical Feed. The reagents which are used to treat hazardous spills are either in powder or liquid form. In order to feed reagent powders into the dynactor, a powder feed mechanism has been developed which is capable of introducing controlled quantities of solid reagents directly into the elevated plenum chamber of the dynactor. For example, powdered caustics, such as sodium carbonate, are used to neutralize spills of acids. Powdered activated charcoal is fed into the dynactor for the purpose of adsorbing a wide variety of adsorbable chemicals.

It is necessary to control the rate of introduction of these materials. For this purpose, a powder feed mechanism was developed, because no universal powder feed system was available commercially that would satisfy the engineering requirements of the system. These are (a) operation at ground level for ready introduction and mixing of chemicals, and (b) clog-proof delivery of powder to a point 25 ft above the ground, where the powder is introduced into the dynactor plenum chamber. The powder feeder comprises three basic components: (1) a hopper with a slowly rotating "tree" with metal "branches" that rotate in such a manner as to prevent clogging or bridging; (2) an auger system, whose speed is adjustable. The auger geometry and speed uniquely determine the flow rate of powder which continuously fills the auger due to the mixing action described above; and (3) a fluidizing system that fluidizes and blows the powder up into the plenum chamber without the use of any conveyors or other moving parts. The powder feed system is capable of delivering controlled and measured quantities of chemical powders into nearly any environment.

The liquid reagent feed system is comprised of a variable flow rate chemical feed pump system, which introduces reagent liquids at the suction side of the dynactor motive pump. Thus the dynactor pump acts as a mixer, and the subsequent division of the liquid into a cloud of films and droplets produces extremely high rates of reaction.

Coagulant and flocculant chemicals are introduced for the purpose of speeding up the settling of solids and sludges either contained or produced in the liquid being treated. Reagent vessels are installed below the floor level of the vehicle, and a series of peristaltic pumps introduce measured quantities of coagulants and flocculants into the flowing liquid. Various coagulant/flocculant combinations are required as appropriate to the liquid and the solids contained therein. For example, when sodium carbonate has been utilized for treatment (as in reaction with dissolved lead nitrate to produce the insoluble precipitate, lead carbonate), an excess of sodium carbonate is treated with ferric chloride to produce an iron carbonate floc. The liquid is then further treated with an anionic flocculant, Nalco No. 8174, to produce a rapidly settling solid floc.

Solids Removal: Solids are settled and removed from the liquid in a set of ten high-rate lamellar settling columns, built into one wall of the hazardous spills vehicle. The columns are mounted at an angle of 30° from the vertical and are 30 in^2 in cross section and 12 ft high. The lamella are stainless steel sheets, arranged for easy removal (for cleaning purposes). The sheets are spaced 1 inch apart. The vertical velocity of liquid is approximately 6 in/min in passage through each settler.

Solids settle and concentrate on the surface of each of the lamella. The conical bottom of each of the settling columns receives the sludge which is concentrated in the settling process. Each settling column is operated by a separate 25 gpm centrifugal pump so that liquid flow is evenly divided. Overflow of clarified liquid is automatic and gravity alone returns clarified and detoxified liquid to the environment. The conical bottoms of each of the ten settling columns are sequentially voided into a sludge collection manifold that is incorporated on the underside of the vehicle.

Operated by a timer, valves open and close at a rate which is appropriate to the conditions experienced in each spill. In this manner, clarified liquid and sludge are separated so that the clarified liquid returns itself to the environment and the toxic sludge is collected in a vessel for subsequent treatment or burial.

Analysis: A chemical bench is included in the vehicle; it can be equipped with the analytic equipment appropriate for a particular spill. Upon notification that a given type of spill has occurred, portable test equipment can be taken aboard to enable a chemist or chemical technician to conduct appropriate tests on both the system influent and effluent in order to establish appropriate reagent flow rates to effect optimum detoxification of the particular spill. (For example, if an acid spill is to be neutralized with a caustic powder, output pH must be measured in order to control the flow rate of caustic reagent.)

Control: The logic system, which interconnects and controls the hazardous spills system, is so arranged that the dynactor hazardous spills vehicle operates automatically. For example, excess liquid accumulation in the dynactor reservoir will automatically shut down the input pumps until the proper operating level is reestablished. Pressure and flow are monitored continuously in appropriate portions of the system in order to control valves, pumps, and the operation of the equipment to prevent overload, burnout, etc. These automatic controls are located just aft of the diesel electric generator for ready accessibility to operating personnel.

Decontamination studies involving aeration, oxidation, neutralization and carbon adsorption of selected hazardous chemicals in water are in progress.

MOBILE UNIT FOR IN-PLACE DETOXIFICATION OF SPILLS IN SOIL

Introduction

> The information in this section is based on a paper presented at Conference 1978 by K.R. Huibregtse of Envirex Inc. and J.P. Lafornara of the U.S. Environmental Protection Agency and K.H. Kastman of Soil Testing Services Inc. The title of the paper was "In-Place Detoxification of Hazardous Materials Spills in Soil."

Spill incidents can occur in almost any known geographic area, contaminating air, water and/or soil. Containment and treatment technology for water spills has received the most attention and is the furthest advanced. However, in many instances, both water and soil are contaminated when land spill threatens a nearby water body or the groundwater table. The state-of-the-art of land spill cleanup has consisted mainly of excavation and hauling to an approved landfill site or possibly flushing of the affected area with water. These methods are appropriate in certain circumstances. However, when the groundwater is threatened, when a large soil mass is contaminated or when no suitable disposal site is available, other approaches may be needed.

It is the purpose of this effort, funded by the U.S. Environmental Protection Agency under contract number 68-03-2508, to develop a mobile treatment system which allows in situ detoxification of hazardous materials spilled on soil. Detoxification in this context refers to amelioration of a spill's effect by chemical reaction. The project's goals were to design and demonstrate a mobile vehicle capable of encapsulating a 10,000 gal land spill in grout and treating the spilled chemicals in place by either oxidation/reduction, neutralization, precipitation or polymerization. The approach to achieving the design goals was to use direct injection of grouting material into the soil around the contaminated area to envelop the spill and isolate it from the groundwater, followed by detoxification by injection of treatment agents. This paper documents the results of the laboratory and pilot tests and the resulting preliminary system design. The vehicle was to be fabricated and demonstrated during 1978 and was expected to be a part of the EPA spill response arsenal by 1979.

The work was divided into five phases: (1) laboratory study, (2) pilot testing and design, (3) fabrication, (4) testing and demonstration , and (5) reports. The information obtained during the laboratory and pilot tests was used to develop the final system design and, as anticipated, the end-product design was modified from that originally envisioned.

Laboratory Testing

The laboratory tests had two main objectives:

(1) To determine if in situ treatment techniques could effectively detoxify chemicals present in various soil systems and

(2) To evaluate, choose and test various grout types for their potential use in spill containment.

Treatment Evaluation: Various reagents and soil types were chosen for testing the four types of chemical reactions: oxidation/reduction, neutralization, precipitation and polymerization. Chemical compounds studied as contaminants were chosen based on the following criteria: efficiency of the chemical reaction, common use of the chemical, and potential risk of spillage. Treatment agent choices were based on: the hazardous nature of the treatment chemical, its availability, its handling difficulties, and the volume needed for detoxification of the contaminant. Contaminant concentrations were established by common shipment concentrations, and the strength of the reactant was established to keep the detoxification controllable. The chemical systems are shown in Table 5.5.

Four soil types were also included in the laboratory study. It was determined that classification of soils by grain size would be most advantageous, since this characteristic often controls the soil's permeability and therefore its amenability to injection of treatment agents. The four soil types considered were clay, silt, sand and gravel.

Table 5.5: Chemical Reaction Systems Investigated

	Contaminant		Reactant	
Reaction Type	**Compound**	**Concentration**	**Compound**	**Concentration**
Oxidation-reduction	Sodium hypochlorite	12-15 % Cl	Sodium bisulfite	7.5%
Neutralization	Sulfuric acid	36 N	Sodium hydroxide	1-5 N
Precipitation	Copper sulfate	75 g/l	Sodium sulfide- sodium hydroxide	1.0 g/l 0.1 g/l
Polymerization	Styrene	100 %	Persulfate	

Source: Conference 1978

The testing procedure involved mixing specified amounts of soil and water and packing this mixture to incremental heights to achieve a specified soil bulk density. These soil columns were then contaminated with liquid to fill a certain soil void volume, the treatment agent added and samples collected at the underdrain. If sealed tests were performed on a system the contaminant/reactant/soil mixture was allowed to stand for a given time and soil core samples were taken and analyzed.

The data collected from all laboratory testing were evaluated and the percent of contaminant treated was calculated along with the residual concentration in the treated soil. Statistical analyses of these results using ANOVA design and F tests were used to identify which of the variables had significant effects on the efficiency of the reaction. The results indicate that both soil type and reaction type significantly affect the degree of detoxification, along with the three internal variables (soil conditions, detention time and loading).

The efficiency of in situ treatment in gravel was much lower than with other soils. This is a result of most of the contaminant rapidly percolating through the gravel prior to treatment. However, for the contaminants entrained on the gravel, the reaction efficiency ranged from 95 to 99%. The overall efficiency of the neutralization reactions was also lower since a prereactant water rinse was required in order to reduce the heat of reaction. Precipitation reactions were more efficient than anticipated. This may be due to the blocking effect of the precipitate which clogs some of the voids and forces the treatment agent to flow into other contaminated areas. Redox reactions were generally quite efficient under all conditions. The detention time was critical for sand detoxification indicating that too high a pumping rate will be detrimental in final treatment.

The effectiveness of sealed detoxification (surface treatment) was not anticipated. As long as void saturation was not exceeded, the treatment agent entered the fine grained soils and mixed to a degree which detoxified most of the contaminant. This apparent mixing in the small void sizes was not expected. Reduced reaction efficiencies were apparent for precipitation because the precipitate did block the reactant's path into the soil. Overall, even this reaction was quite effective. The main problem with a sealed system is that the volumes which can be treated are limited to voids available for the reactant.

Grout Evaluation: The second objective of the laboratory testing was to evaluate the grout which could be used for encapsulation of a spill. The main types of grout available include particulate grouts such as cement and bentonite and chemical grouts which are mainly acrylamide (AM-9), urea-formaldehyde resin, lignin or silicate based materials. Particulate grouts are generally used in coarse grained solids since they have a relatively high viscosity due to their suspended particles in a water base. Chemical grouts are generally in solution form and can be used to grout finer grained soils. One of the most commonly used chemical grouts is AM-9 which can be used to grout both clays and silts. However, AM-9 has an acrylamide base which is toxic to groundwaters. Therefore it was not considered suitable for the spill containment application.

Evaluation indicated that bentonite/cement or silicate grouts would be most feasible for spill containment. Depending on both the soil and chemical characteristics, one may be more applicable than the other. Both systems are environmentally acceptable, since the bentonite is a natural clay product and may eventually resorb into the soil and the silicate grout may break down with time; thus long term adverse effects will be minimized.

There are several silicate grout formulas in general usage. The silicate grout used in this survey was formed using a mixture of sodium silicate, sodium bicarbonate and a copper sulfate catalyst. Extensive laboratory testing was performed to establish the most feasible dosages. It is anticipated that this type of presentation will be included in the final systems operation and maintenance manual with instructions for choosing an appropriate mix. Chemical tests to determine the grout's resistance to treatment chemicals were also performed. The results indicated that the silicate grout while resistant to bisulfite, hypochlorite, sodium sulfide and copper sulfate, had very low resistance to acids and relatively low resistance to bases. This was expected because the silicate is an alkaline material and the gel is affected by pH. When a high pH occurs, a bentonite grout would be recommended.

The final output of this effort was to develop an approach for establishing a specific chemical's treatability by in situ techniques. This involved determining if neutralization, oxidation/reduction or precipitation would detoxify the hazardous material and establishing which type of grout would be most resistant to chemical penetration. These results will be presented in the final report and Operation and Maintenance Manual in tabular form for quick reference.

Pilot Testing

Based on the results of the laboratory tests, two reaction types and two soil types were chosen for pilot scale evaluation. Precipitation and redox reactions were selected to further define effect of solids formation. Sand and clay soils were chosen so that both flow-through and sealed procedures could be tested on a larger scale. The main objectives of the pilot testing were: to determine if the detoxification procedure was feasible on a larger than laboratory scale and to establish critical parameters such as pumping rate, injector placement and back pressure, for consideration in the development of the final system design.

Testing Equipment and Procedures: Special test cells were constructed for the two tests: a clay test box 2' x 2' x 1' 6½" and a sand test box 8' x 4' x 4' ¾" with a 1" thick Plexiglass cover. Both were made from coated plywood, the larger box having heavy reinforcing. Additional tanks, pumps, tubing and mixers were procured and used during the test operations, as needed. The test procedures for the surface and injection treatments were quite different. The surface testing was basically similar to the laboratory tests. The soil and water were compacted in the box to a given bulk density and the specified amount of contaminant was sprinkled over the surface and allowed to migrate. After 24 hours, the reactant was sprinkled on the soil surface and allowed to detoxify the soil for 48 hours. Core samples were taken at specified locations in the box and analyzed for contaminant concentration.

The flow-through testing required that the box be filled with 5,600 to 5,800 lb of sand which was placed and compacted in 3.5 cm layers to achieve the desired bulk density. Water was added to yield a 5% water content. The contaminant was again placed on the surface and the reaction was performed the same day as contamination. An injector and wet well were placed on opposite ends of the box and then the specified volume of reactant was forced through the injector into the soil. After the reactant was pumped into the system, a volume of water was injected to rinse the soil of excess reactant. Throughout the pumping period, the wet well was continuously emptied into a separate holding tank. After all liquids were pumped into the soil, core samples were collected and analyzed for moisture content and contaminant concentration.

Two pilot grouting tests were also performed to aid in choosing injector types and establishing anticipated pumping pressures and to define some of the problems associated with grouting. Various mixes of grout were pumped and the resultant grout wall observed and tested, where possible.

Results of the Pilot Tests: The effectiveness of detoxification for all of the pilot tests was quite high. As expected, the geometry of reactant injection and the shape of the pilot study box affected the detoxification. When evaluating the results of flow-through testing, it was apparent that the detoxification was most effective within a radius of 1.5 ft from the injector. However, detoxification effects did extend beyond this radius. The surface treatment results reflected those predicted from the laboratory testing. The redox reactions were very effective, removing most of the contaminant which was entrained in the surface layers. Precipitation reactions were less efficient than the redox reaction. This can be attributed to the blocking of voids by precipitate formation. Shrinkage cracks which formed when the surface dried allowed more effective reaction in some of the lower layers. However, as with the redox system, the majority of the contaminant entrained in the surface layer was detoxified.

Evaluation of the grout test results indicated that injection of chemical grout on an angle was possible, while grouting near the soil surface was not feasible because of short circuiting caused by grouting pressures being larger than the soil overburden weight. The particulate grout was difficult to handle in the shallow testing box and the only injection device which proved feasible was one with a single outlet hole.

The pilot tests also indicated:

(1) the importance of driving an injector directly into the soil as opposed to boring and then placing the injector,

(2) the necessity of a wet well equipped with a self-priming pump for liquid removal,

(3) the need for pumping systems equipped for pressures up to 80 psi,

(4) the requirement for volumes of rinse water was not as critical as originally anticipated,

(5) the back-pressure caused by higher void volume loadings of contaminant reduced the forward flow rate significantly, and

(6) the neutralization chemicals could be added using a multiholed injector (which allowed for much faster treatment).

It was determined that pilot test grout gel times were shorter than in the lab and that the chemical grout injection could be controlled by the volume added while the particulate grout addition was best regulated by pressure in the injection lines.

Prototype Design

Preliminary Design: After the pilot tests were completed, the design of the prototype system was begun. Much of the information obtained throughout both the laboratory and pilot tests significantly influenced the design. The system provides much flexibility for spill cleanup. The grout or chemicals are to be mixed in alternate batches in the two 1,500-gal fiber glass tanks. Batching eliminates potential problems associated with exact mixing of grout constituents at the point of injection and thereby allows closer system control.

Two pump types were included. For grouting, positive displacement pumps will provide the most control and the simplest operation, however, they were not sufficiently chemically resistant for chemical injection which will be accomplished by the air pumps, available in Hastelloy C. It was also determined that multiple pumps instead of extensive manifolding of injectors would allow more control of the volumes pumped into the soil. If necessary, the injectors can be manifolded in pairs to allow higher pumping rates, however, this approach may not always be feasible when difficult soil conditions are encountered. The volume of liquid added is to be metered and totalized, since in most instances the chemical solutions will be added until a calculated amount is pumped into a specified area. The injector will then be withdrawn a certain distance and the pumping process repeated.

The vehicle will be equipped with a diesel-electric generator and an air compressor. An "air-hammer" type device will be used to drive the injectors (1½" o.d., 1" i.d.) into the ground. Separate multiholed injectors will be used for chemical addition. Since the cost of chemical resistant injectors would be excessive, standard steel pipe injectors will be replaced when they corrode to the point where they are no longer usable. All components would be accessible either on the vehicle or from the side. The controls will be centralized on a panel permanently mounted on the truck. Accessory equipment will include standard test apparatus to measure soil conditions and chemical concentrations, well points for use as wet wells, some small air pumps to empty wet wells, and a surface holding tank.

Costs are presently being developed and this design may be modified depending upon the complete economic considerations.

Design Limitations and Decision Matrix: The limitations of in situ detoxification techniques either through surface treatment or direct injection of grout and chemicals must be understood before the prototype equipment is used. When a land spill occurs, alternative approaches should be evaluated and the most time- and cost-effective approach for the specific situation chosen. In order to determine if in situ detoxification is most efficient, a decision matrix will be prepared. This matrix will present an approach for evaluating the feasibility of grouting and chemical injection, as well as surface containment and treatment. Among the critical variables are type of chemical spilled, interaction with the soil, the soil's "groutability" (permeability, void loading, geometry, water table level, etc.), soil volume contaminated, feasibility of excavation and availability of treatment supplies and manpower.

This equipment will not be applicable to all land spills. However, there are many situations in which it will be a feasible technique. The surface treatment approach may be desirable in many cases even if the contaminated soil is to be removed and transported to a landfill. This pretreatment will protect equipment and may even allow redefinition of the removed soil as nonhazardous. Grouting in and of itself will be feasible even when direct chemical treatment is not possible. Construction of a grout layer will protect the groundwater if excavation is incomplete or if rain rinses the area. Although grouting will be limited to relatively coarse grained sand and gravel materials, it is these soils that allow permeation of the contaminant through the soil structure and into the groundwater.

Design Changes: Several changes have been made in the initial design concept. Most significant is the addition of a surface treatment technique for fine grained soils. Polymerization was limited to a few possible materials and was determined to be too dangerous to implement in a field situation. The pilot tests indicated that it was critical to meter liquid flows individually so the original design which included a high capacity pump with extensive manifolding of injectors was changed to include a larger number of lower capacity pumps with much less manifolding.

It was also determined that the pumping rates for chemical injection should be relatively low to allow effective reaction. Therefore, the overall time required for treatment will be longer than anticipated.

Conclusions

(1) An in-place treatment technique has been shown to be an effective land spill cleanup on a laboratory and pilot scale basis.

(2) Grouting technology appears to be an effective method to contain spills and thereby minimize potential groundwater contamination.

(3) Where small grained soils (silts and clay) preclude the use of injection equipment, a surface treatment using a diluted reactant provides an efficient way to detoxify land spills of applicable hazardous materials.

(4) In order to establish the most time- and cost-effective method for land spill cleanup, the limitations of the in-place detoxification as well as specific spill variables must be considered.

(5) A stepwise approach to containment by grout injection, followed by chemical treatment seems to provide the most flexible treatment system.

MOBILE UNIT FOR WATER-SOLUBLE ORGANICS SPILLED IN WATER

The information in this section is based on *Development of a Mobile Treatment System for Handling Spilled Hazardous Materials,* EPA-600/2-76-109 prepared by M.K. Gupta of Envirex for the Environmental Protection Agency.

Introduction

Handling and transport of hazardous materials is necessary because of the widespread usage of such materials in an industrial society. Although the essential feature of any overall program designed to minimize the risks involved in producing, handling, transporting, and using hazardous substances is prevention of spills, it should be recognized that spills will occur despite the most comprehensive precautionary measures. When a spill does occur, the immediate primary concern is containment of the spilled material—isolating the contaminating substance and preventing, insofar as possible, its encroachment on the surrounding environment. Ideally such containment will prevent the spilled material from reaching any water resources. However, such an ideal situation seldom exists in practice. Generally, containment is not possible, and a water body may be contaminated by the spilled hazardous materials which must therefore be purified.

Thus, it is reasonable to expect that a development effort needs to be directed to minimize the effects of spilled hazardous materials that have reached a watercourse of some type (ditch, pond, stream, lake, river, etc.).

This concern for the accidental entrance of hazardous materials into a watercourse has initiated several programs by the Environmental Protection Agency for the control of this problem. The program described here was undertaken to develop a transportable treatment system for on-site removal and treatment of spilled hazardous materials in aqueous solutions. This program has the following objectives:

(1) Develop bench scale design data for the treatment of various hazardous materials.

(2) Design and fabricate a mobile treatment system for spilled hazardous materials.

A listing of the hazardous materials evaluated during this study is shown in Table 5.6. This list was selected based on the priority ranking system for hazardous materials developed by EPA (5). As can be seen, the materials have been divided into two categories, but overlap in several cases. Initially, only the nine materials listed in the first section of the table were included for evaluation in the original contract 68-01-0099. However, later the above contract was amended to include a laboratory scale evaluation of the applicability of reverse osmosis for the preconcentration of materials listed in the second section of the table. The intent of the laboratory evaluations was to screen the listed hazardous materials based on a brief review of available literature and to develop bench scale treatment feasibility data for those materials for which available information was lacking.

Table 5.6: List of Hazardous Materials

Materials Considered in Selection of the Mobile Treatment System
Acetone cyanohydrin
Acrylonitrile
Ammonia
Chlorinated hydrocarbon pesticides
Chlorine
Methanol
Phenol
Tetraethyllead (TEL)
Tetramethyllead (TML)
Materials Considered for Reverse Osmosis Treatment Feasibility
Acetone
Acetone cyanohydrin
Acrylonitrile
Alum (aluminum sulfate)
Ammonium salts
Benzene
Chlorine
Chlorinated hydrocarbon pesticides
Chlorosulfonic acid
Copper sulfate
Formaldehyde

(continued)

Table 5.6: (continued)

Materials Considered for Reverse Osmosis Treatment Feasibility

Lead as in TEL and TML
Methanol
Mercuric chloride
Phenol
Phosphorus pentasulfide
Styrene

Source: EPA-600/2-76-109

This report summarizes the laboratory, design and fabrication studies performed during the course of this project. Detailed engineering drawings and equipment specifications suitable for duplicating the mobile treatment system are in the possession of the Environmental Protection Agency as are a set of manuals with complete instructions for operating, repairing and maintaining the component devices. This comprehensive material can be made available to interested parties through EPA's Industrial Environmental Research Laboratory, Edison, New Jersey.

Evaluation of Activated Carbon and Chemical Treatment Methods

The unit treatment processes investigated for the hazardous materials listed earlier were chemical treatment, clarification via settling and/or filtration, activated carbon adsorption and reverse osmosis. These processes were selected based on the results of an earlier EPA study (4) which indicated that many of the hazardous materials could be precipitated by chemical treatment or oxidized to a more innocuous state by chemical treatment followed by sedimentation and/or filtration to remove the precipitated and other particulate matter such as soil or debris. In addition, activated carbon was found to be one of the most versatile and affirmative processes for the treatment of a wide variety of water-soluble materials.

Reverse osmosis treatment feasibility evaluations were based on the premise of utilizing RO in conjunction with activated carbon. It was thought that this combination could provide tremendous versatility of treatment for both inorganic and organic hazardous materials. Reverse osmosis was promising as a preconcentration step prior to activated carbon treatment for various organic materials because of its compact size, high volume and low weight in a mobile treatment system. Reverse osmosis treatment could also be utilized for the treatment of various inorganics, such as toxic heavy metals, that otherwise could not be treated via activated carbon.

The test results and data have been divided into two separate sections as listed earlier in Table 5.6. The nine materials considered in the selection of the mobile treatment system are discussed first for chemical treatment, clarification and carbon adsorption treatment. A second section presents the results of the RO feasibility tests.

Acetone Cyanohydrin: It was indicated in the literature (4) that acetone cyanohydrin may be detoxified at high pH levels by the precipitation of cyanides via chlorination or carbon adsorption treatment. Therefore, chlorination and carbon adsorption tests were conducted for this material.

Acrylonitrile: Literature (4) indicated that this compound must also be handled at elevated pH values of above 9.0 to prevent the formation of hydrogen cyanide gas. Chlorination and activated carbon adsorption were again indicated to be viable methods of treatment for this compound. Since this material was considered to be quite volatile, tests were also conducted to study the volatilization of this solution prior to chemical treatment tests.

This testing demonstrated that activated carbon could be utilized effectively for the treatment of acrylonitrile solutions. Minimum contact time was found to be approximately 40 to 50 minutes. Because of the volatile nature of acrylonitrile, a significant amount will evaporate to the atmosphere and create an air pollution hazard. It is, therefore, especially imperative that rapid action be taken by personnel protected by adequate safety equipment.

Ammonia: Dawson et al (4) indicated that the most applicable technique for the handling of ammonia spills is dilution. It is also indicated that stripping of ammonia at high pH values may be applicable in certain situations. Chlorination can also be used for the oxidation of ammonium ion to nitrogen. Information concerning the oxidation of dilute concentrations of ammonia with chlorine is available in the literature. Generally, wt/wt ratios of $Cl_2:NH_3$ of 4:1 to 10:1 are recommended for complete destruction of ammonia. Lower concentrations of chlorine result in chloramine formation.

Hence, ammonia removal by breakpoint chlorination is chemically feasible for relatively dilute concentrations of ammonia spills, but such treatment does not appear to be logistically possible owing to the large amounts of chlorine required and the secondary pollution it would cause.

Chlorine: Activated carbon may be utilized for the dechlorination of chlorine-contaminated solutions.

Chlorinated Hydrocarbons: While activated carbon adsorption was anticipated to be among the most effective methods for the removal of most spilled chlorinated hydrocarbons, pesticides and herbicides, a brief review of the literature showed that a great deal of work has also been done on the adsorption of pesticides on various other materials. The data of King, Yeh, Warren and Randall show substantial removals of lindane from aqueous solutions when concentration of soils was high (6). In contrast, Lotse, Graetz, Chesters, Lee and Newland found much lower removal efficiency at a low concentration (372 mg/l) of soil (7). Efficiency was reported to be even lower when soils containing less organic materials was used.

The adsorption of 2,4-D on montmorillonite clay was found to be minimal by Schwartz (8). On the other hand, Huang and Liao have reported removal of DDT, heptachlor and dieldrin using clay (9). It appears from this data that the removal of suspended solids in contaminated water at a spill site will remove the chlorinated organics which have been adsorbed on the suspended particles of a suitable soil.

Carbon isotherm tests have been reported for several pesticides and herbicides using a variety of carbons. Morris and Weber (10) reported results of carbon adsorption on several herbicides including 2,4-D, 2,4,5-T, Silvex and parathion.

King, et al also performed a carbon isotherm on parathion. In this test, acetone was used as a bridge solvent (6). Aly and Faust performed carbon isotherms on several herbicides including 2,4-D and the isooctyl and butyl esters of 2,4-D (11). Carbon adsorption tests were performed on 2,4,5-T butoxyethanol ester, dieldrin, endrin, lindane, and parathion by Robeck, et al (12).

The effect of other organics in the river water on the adsorption of pesticides (as on other compounds) is clearly illustrated by this data. It is quite likely, of course, that the organic content of the contaminated water at a spill site may be quite high and that these organics will compete with the hazardous material for the activated carbon, and that the pesticides and herbicides will not be well adsorbed.

Very little information was found in the literature concerning the adsorption of DDT. Although Huang and Liao reported on the adsorption of DDT on various clays, no reference was found concerning the adsorption on activated carbon (9). Because of the scarcity of data regarding DDT, a carbon isotherm test was performed in the laboratory.

Based on the DDT testing and the literature search, it is evident that activated carbon has applicability in removing water-soluble herbicides and pesticides of various varieties. The lethal nature of these compounds makes their control at spill sites essential. The relative removal capacity of activated carbon varies widely. Because of the low solubilities of pesticides and herbicides, carbon capacity for adsorbing them should not be a major problem, as long as adequate contact time is available.

Methanol: Dawson, et al, indicated that the only method of removing a methanol spill was to allow dilution and subsequent biodegradation to occur (4). Although the literature indicated poor removal of methanol by activated carbon, no concrete data could be found. Therefore, carbon tests were conducted to establish the extent of removals. Breakthrough of the carbon column was almost instantaneous. Further testing confirmed that activated carbon is not suitable for the treatment of methanol spills.

Phenol: Dawson, et al (4) indicated that phenol spills may be successfully handled both by chemical treatment with chlorine and activated carbon adsorption. It was also indicated that if the phenol concentration was excessively high, dilution prior to treatment may be necessary. Laboratory tests on phenol were conducted both with chlorine and activated carbon.

It was concluded that chlorination of phenolic spills is unattractive not only because of the large amounts of chemical requirements but also because of possible chlorinated reaction products. It was apparent, however, that activated carbon can effectively treat phenol spills.

Organolead Compounds (TML and TEL): Both tetraethyllead (TEL) and tetramethyllead (TML) are heavy liquids having specific gravities of 1.65 and 2.00 respectively. Both these compounds are considered highly toxic and have limited solubilities in water. According to Shapiro and Frey (13), the solubility of these compounds is about 0.2 to 0.3 mg/l in water. However, higher concentrations of these compounds can occur in contaminated waters at a spill site due to emulsification or adsorption on solids. TML is also an unstable compound and generally contains 20 to 30% toluene for stability. TEL is generally difficult to recover, but can be complexed with a calcium salt or EDTA or dimercaprol (4). Laboratory tests were conducted to establish the expected concentrations of these materials at a spill site. Chemical treatment and carbon adsorption tests were then conducted on solutions containing the expected concentrations of TEL and TML. From the test results, it was evident that conventional coagulation and suspended solids removal treatment probably would not be effective for satisfactory removals of residual TML concentrations at a spill site. However, a combination of potassium permanganate treatment and activated carbon may provide more satisfactory treatment.

Reverse Osmosis Tests

The hazardous materials listed earlier in Table 5.6 were screened for reverse osmosis application feasibility based on available literature. Of the various materials listed in this table, data indicated that seven could not be treated suitably with reverse osmosis and no such treatment tests were conducted for these materials. These materials were: benzene, chlorine, chlorinated hydrocarbon pesticides, chlorosulfonic acid, organolead compounds, phosphorus pentasulfide and styrene. Generally, the reasons for elimination pertained to low solubility of the hazardous materials or expected adverse effects on the RO membranes. Among the remaining ten of the listed materials, it was indicated that sufficient application feasibility data was available from the RO membrane manufacturer for two materials and, therefore, bench scale feasibility tests were conducted on only the remaining eight compounds.

The materials for which feasibility tests were conducted were: acetone, acetone cyanohydrin, acrylonitrile, ammonium salts, copper sulfate, formaldehyde, methanol and mercuric chloride. Generally, the inorganic compounds showed good rejection capabilities except mercuric chloride. The rejection of mercuric ion at 72% was found to be lower than expected since generally most divalent and trivalent ions can be rejected in excess of 90%. Also during the brief test duration (24 hours) for mercuric chloride, it was noticed that the head loss across the membrane increased rapidly, suggesting high membrane fouling potential. Therefore, it was concluded that RO would not be suitable for the treatment of mercuric chloride.

Rejections for ammonium ion varied between 88 and 98% depending upon the type of anion associated with the ammonium ion. The higher rejection for ammonium ion when associated with sulfate ion was expected because of the generally better rejection characteristics of RO membranes in the presence of a divalent ion. Copper and aluminum ions were rejected in excess of 95% as expected.

The various organic compounds evaluated in this study showed considerably lower membrane rejection capabilities compared to inorganic materials. Methanol, formaldehyde and acrylonitrile showed poor rejection capabilities (13 to 20%) while acetone, acetone cyanohydrin and phenol showed medium separation capabilities (55 to 70%). All of the abovementioned organic compounds are generally expected to exhibit lower membrane rejection characteristics. However, it was found that the change in pH for some of these compounds can provide a marked improvement in the rejection capabilities. The effect of pH change was most pronounced for phenol where the conversion of phenol to sodium phenylate enhanced the rejection capability from 55 to 95% at comparable operating conditions.

From the test results, it is obvious that reverse osmosis has limited applicability for the treatment of only a selected few hazardous materials such as soluble inorganics and high molecular weight organics (MW >100). Out of the various hazardous materials screened only eight showed medium to good removal effectiveness. Many of these spilled hazardous materials will require extensive pretreatment for the removal of particulate solids and/or other treatment such as pH adjustment, softening, etc., prior to processing by RO. Furthermore, the reverse osmosis process, being a high pressure process, requires relatively higher amounts of power. Both the additional pretreatment and power requirements may significantly influence the logistics support requirements in a spill situation. Therefore, it was concluded that reverse osmosis has only limited application feasibility for the treatment of hazardous materials.

Design Criteria for a Mobile Treatment System

Based on the laboratory evaluation of the various hazardous materials described in the preceding sections of this report, it was concluded that a flow schematic consisting of chemical reaction, flocculation, sedimentation, filtration and activated carbon adsorption would provide the most flexible mobile spills treatment system. These unit treatment processes were shown to remove the majority of the pollutants evaluated in this study. It was concluded that RO would not be a suitable treatment method for many spilled hazardous materials because of their low water solubility and adverse effect on membranes. RO could, however, be utilized on a selective basis such as for the treatment of water-soluble inorganics. For such selective uses, reverse osmosis equipment may be mounted on a separate mobile unit that could then be utilized in conjunction with another pretreatment unit capable of providing chemical and filtration treatments.

Figure 5.3 shows a schematic flow diagram of the recommended treatment system that was utilized for the design of the mobile spill response vehicle constructed under EPA Contract No. 68-01-0099. The raw waste pumping system provided on the spill response vehicle, consists of a submersible pump, an in-line booster pump and sufficient hoses to allow the deployment of the treatment system as far as 92 m (300 ft) from the spill site. The raw flow can be controlled manually by an in-line indicating flow meter and control valve. The maximum hydraulic design flow capacity of the pumping system is 12.6 l/sec (200 gpm). The vehicle is equipped with two other pumps (filter influent and backwash pumps). Any of the pumps provided on the vehicle can be used interchangeably during emergencies.

The electrical power to operate the raw feed system as well as all of the other electrical requirements for the treatment system are provided by a gasoline fueled generator on the trailer. Various chemicals such as acid, lime, ferric chloride, chlorine or potassium permanganate (depending upon what substance is being treated) can be added in the reaction/flocculation tank to precipitate and/or flocculate the waste. An in-line pH indicator follows an in-line mixer and can be used for the adjustment of chemical addition for controlling pH. Provisions are also made for addition of a polymer to the suction side of the filter pump as a filtration conditioner.

Figure 5.3: Waste Treatment Flow Diagram for a Mobile Spill Response Vehicle

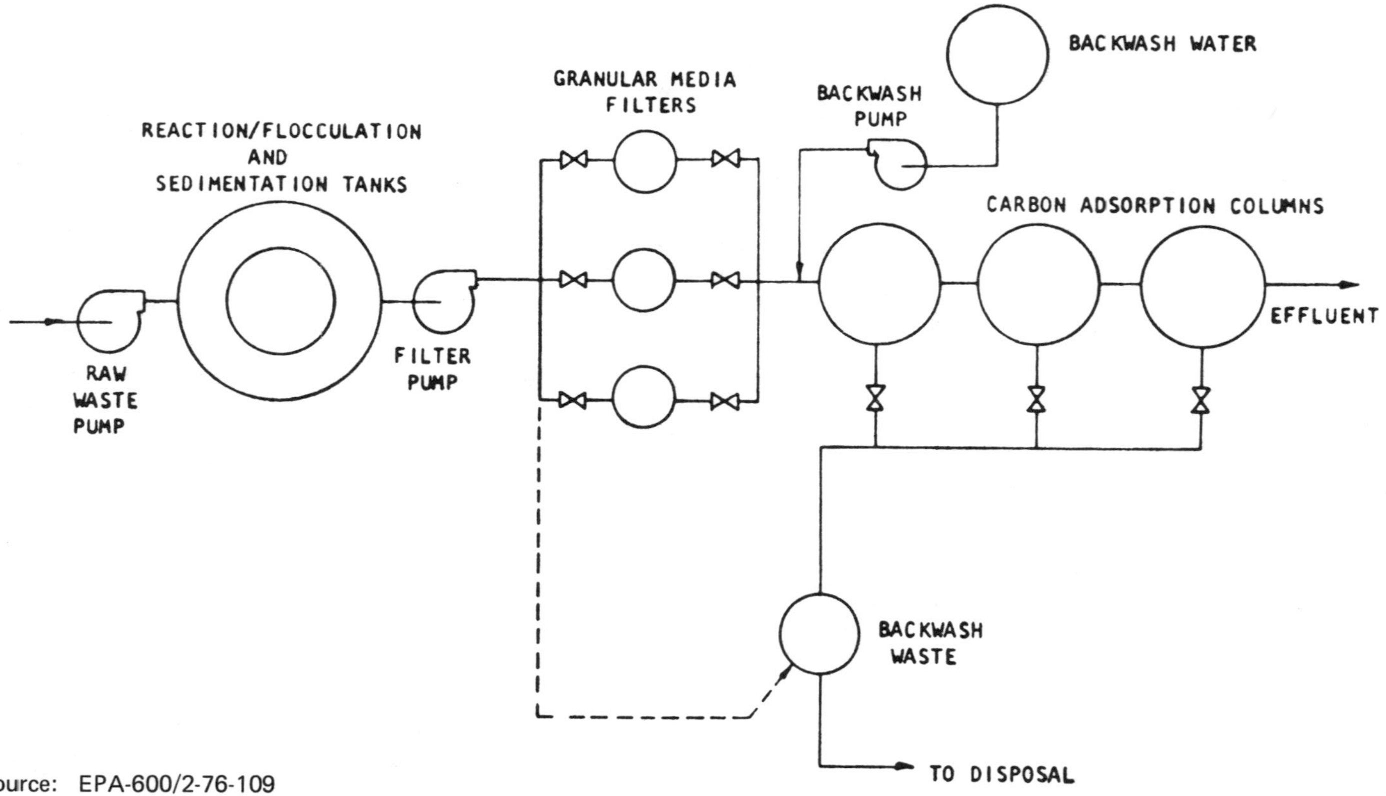

Source: EPA-600/2-76-109

The waste then flows to a settling tank. The overflow from the settling tank is filtered through granular media filters followed by granular activated carbon adsorption. Sludge is removed from the sedimentation tank and stored for ultimate disposal. An additional storage tank is provided for filter backwashing operations. Each unit process is capable of being bypassed if not required for a specific spill.

A portable treatment system must also meet the various limitations imposed upon transportable equipment. The most significant limitations involved in the design of this mobile treatment system were size and weight. A 13.7 m (45 ft) trailer in conjunction with a snub nose tractor met the required 16.8 m (55 ft) total vehicle length imposed by many states. Because a significant amount of time is required for wetting activated carbon (12 to 48 hr), the carbon should be transported wet and this increases the weight of the trailer considerably, since 0.45 kg (1 lb) of dry carbon absorbs about 0.45 kg of water. The total treatment vehicle over-the-road weight selected for this system was 47,600 kg (105,000 lb), which required special operational permits in most states.

Generally, 33,200 to 36,400 kg (73,000 to 80,000 lb) gross weight is the maximum allowable in most states under normal circumstances. However, higher weights are permissible for emergency vehicles. The maximum concentration of hazardous material in the waste was set at 1% for design purposes. However, it was recognized that concentrations would in all likelihood be much lower. The ambient temperature range was defined at 0° to 100°F.

Based on the severe corrosion anticipated for the wide variety of hazardous materials under study, 316 stainless steel was selected as the optimum material based on strength, price, ruggedness and durability. Having established the process flow sheet, the equipment had to be designed to provide maximum possible hydraulic capacity within the space and weight constraints of over-the-road equipment mentioned previously.

The importance of high flow rate in treating a dilute spill can be shown by the following example: a 19,000 liter (5,000 gal) chemical spill which empties into a stream and is diluted to 1,000 ppm creates 19 million liters (5 million gal) of contaminated water. A treatment system operating at 12.6 l/sec (200 gpm) on an around the clock basis will require 17 days to clean the water provided that the clean effluent water is not reintroduced into the spill contaminated area.

Because of the various physical constraints, it was concluded that a maximum hydraulic capacity of 12.6 l/sec (200 gpm) could be utilized for the mobile treatment system. However, flexibility was provided in the selection of the pumping system so as to operate the treatment system at lower or higher flow rates when desirable.

Flocculation/Sedimentation System: Large rubber, portable tanks are set up next to the trailer with one tank inside the other to provide the necessary detention times for flocculation and sedimentation. A minimum chemical reaction/flocculation time of 15 minutes and sedimentation time of 60 minutes were considered necessary for the design of the treatment system. This led to the selection of a 11,350 liter (3,000 gal) reaction/flocculation and 56,775 liter (15,000 gal) sedimentation tank. Longer contact periods could be provided at reduced flow rates. In order to conserve valuable space on a mobile treatment system, specially built collapsible rubber tanks were selected in place of conventional rigid tanks. A flocculation/sedimentation system was developed whereby the two 11,350 and 56,775 liter tanks could be utilized concentrically. This arrangement allowed a flocculation time of 15 minutes and a detention time of 60 minutes at a flow rate of 12.6 l/sec (200 gpm). The flocculation tank is placed at the center of the sedimentation tank. The raw waste is introduced at the bottom of the flocculation tank. Flocculation is achieved by hydraulic mixing via the use of two ejectors placed opposite to each other.

The flocculated wastewater flows out of the flocculation reaction tank into the 45,400 liter (12,000 gal) annular space of the sedimentation tank through a series of submerged orifices located around the periphery of the reaction tank. Both these tanks are cylindrical open top tanks and are supported by staves anchored into the ground. The open top tanks also permit easy accessibility for manual removal of sludge and floating materials.

A 11.9 l/sec (50 gpm) sludge pump and special suction fittings are provided for removal of settled and floating contaminants and a 11,350 liter (3,000 gal) rubber stave tank is provided for storage. The sludge can subsequently be pumped out of the stave tank by the sludge pump into a tank truck or other container for further treatment and/or final disposal.

Dual Media Filters: The supernatant from the sedimentation tank is drawn off by the filter pump through a submerged orifice header ring at the outside tank wall. A pneumatic level sensor in the sedimentation tank controls filter pump flow to match raw flow.

After the addition of a filtration conditioner, the sedimented effluent is pumped through three dual media filters in parallel for removal of residual suspended solids by the sand-anthracite filter media. Two in-line turbidimeters monitor the turbidity of the total filter influent and effluent from each tank. A differential pressure gauge indicates the degree of filter clogging. The filters may be taken off line individually and backwashed with air and clean system effluent stored in a 11,350 liter pillow tank. The tanks are located just behind the filter and backwash pumps in the center of the trailer. The filters consist of 61 cm (24 in) anthracite coal (0.9 mm effective size) over 45 cm (18 in) fine sand (0.45 mm effective size). The filters are designed for a maximum hydraulic loading of 4.8 l/sec/m^2 (7.0 gpm/ft^2). Additional pertinent filter data for each of the three filters is given below:

Filter diameter	1.17 m (3.5 ft)
Filter area	0.893 m^2 (9.62 ft^2)
Design filtration rate	2.11 l/sec (33.5 gpm)
Maximum filtration rate	4.22 l/sec (67 gpm)
Design backwash rate	6.3 l/sec (100 gpm)
Maximum allowable differential pressure	1.06 kg/m^2 (15 psi)
Maximum tank pressure	4.93 kg/cm^2 (70 psi)
Depth of sand	45 cm (18 in)
Effective size	0.5 mm
Uniformity coefficient	1.5

(continued)

Quantity of sand	860 kg (1,900 lb)
Depth of coal	61 cm (24 in)
Effective size	0.85-0.95 mm
Uniformity coefficient	1.7
Quantity of coal	0.425 m^3 (15 ft^3) (0.39 ton)

The dual media portion of the trailer is a completely assembled system ready for operation. A backwash system is provided to flush away the captured solids and thus the filter is continuously regenerated without the need for media replacement. The backwash waste is pumped to a 11,350 liter (3,000 gal) stave tank. From here it can be pumped out by the sludge pump either into a tank truck for disposal or it may be reintroduced into the raw feed line and recycled through the system.

The air-backwash system is arranged so that two filters may be left on line while one it taken off line for backflushing. The flow through the system will have to be decreased during backwashing if reducing from three to two filters on-line causes a differential pressure rise. The amount of throttling-back will depend on how dirty the filters are. It is recommended to backflush before the filters get too clogged, because when the clean filter is brought back on line it will take a high percentage of the flow at an increased filtration rate. This condition could lead to unusually high effluent turbidity. One way to cause balanced filter flows with a new filter on line is to throttle the filter inlet valve somewhat to cause an artificial head loss across the filter which will tend to balance the flows.

Carbon Columns: The filtered effluent flows through three pressure carbon columns which may be used in parallel or in series. Altogether, they contain 19.6 m^3 (700 ft^3) of carbon. This volume represents a dry carbon weight of 8,172 kg (18,000 lb) of carbon which is the maximum possible weight that can be accommodated on the trailer because of the overall weight constraints on the mobile system. The carbon columns are designed for a hydraulic loading rate of 3.4 l/sec/m^2 (5 gpm/ft^2). Three 2.1 m (7 ft) diameter carbon columns with carbon bed depth of 1.8 m (6 ft) are provided on the mobile treatment system.

The selected carbon volume of 19.6 m^3 (700 ft^3) on the trailer provides a maximum contact time of 27 minutes for the three columns at a flow rate of 12.6 l/sec (200 gpm). This carbon contact time was found to be suitable for many of the hazardous materials evaluated in this study. However, when high contact times are required, these may be provided by reducing the hydraulic flow rates through the carbon columns proportionately.

The carbon tanks occupy the back one-half of the trailer. The carbon columns are completely plumbed so that only valve adjustments are necessary to control the various modes of operation.

Whether done on the site or back at the home base, the carbon will need to be recharged either because of exhaustion, to prevent undesirable effects of mixing contaminants, or because storage with contaminants in the carbon could be hazardous. Spent carbon may be removed using the backwash pump and processed clean water stored in one of the rubber tanks for this purpose. The clear effluent water is pumped through the underdrain system of the carbon column to fluidize the bed and cause the slurry to drain out of the tank drain fitting.

When the adsorptive capacity of the carbon for the processed hazardous material is depleted, new carbon may be installed into the tanks in the field by a slurry pumping system. The slurry is pumped into the carbon column where it is dewatered by the carbon column underdrains. The water is then returned to the effluent storage tank completing the closed loop slurry pumping system.

A manual sampling valve in the carbon column inlet line and in each tank effluent line permits analysis of process water to measure removal effectiveness and to sense carbon column breakthrough.

Treatment System Layout: System layout will be indicated in many cases by the space available and terrain conditions at the spill site as well as the accessability of driving the trailer close to the site. A raw feed pumping system on the trailer includes 92 m (300 ft) of hose to transport the contaminated water to the treatment system. Hose lengths totalling 92 m are also available for discharging the treatment system effluent into the receiving stream.

It is essential that a plan for the various required locations for the raw feed submersible pump be devised before the final treatment system location is selected. In that way a central location can be selected which will allow pumping from most if not all of the planned locations without moving the treatment system. Much time and effort is lost through needless moving of the system.

The trailer location must also take into account the hazardous conditions in the immediate vicinity of the spill site. These precautions should include provisions for operator safety from contamination emitting directly from the spill site and from the treatment system itself since it functions as a reconcentrator of the spill and as such is a hazardous location.

In addition, the electrical generator is powered by a gasoline fueled internal combustion engine. The separator should not be operated in areas where there is danger from fires or explosions.

A recommended plan view layout of the treatment system is shown in Figure 5.4. The required site is approximately 15 m x 30 m (50' x 100'). There are sufficient hose provisions on the trailer to accommodate the layout shown. It is preferable that the entire site be level and free from obstructions. In particular, the reaction/sedimentation tanks rely on being level for proper function. In this regard it is recommended that, if feasible, the selected treatment site be bulldozed so that the tank can be located on very level ground, i.e., slope less than 5 cm in 8 m (2" in 25'). Problem in supporting the large tank occurs when the slope exceeds 15 cm in 8 m (6" in 8')'

The pillow tanks should also be placed on smooth and level ground to facilitate filling and emptying. The trailer does not have to be on level surface to function properly, but it is preferred. Front-back inclination can be corrected by the landing gear on the trailer. Side to side inclination can be corrected by skimming under the wheels and landing gear with planks. The ground level at the trailer should not be more than 10' above the ground level at the reaction-sedimentation and effluent storage tanks because of suction limitations on the trailer-mounted, self-priming pumps which draw from these tanks.

In assembling the system, care must be taken to allow a minimum of three feet clearance all around the trailer for access to control valves and sampling points. Ready access should be provided to the control platform which is on the right forward portion of the trailer. When laying hardwall hoses in areas where there is vehicle traffic, either the hoses should be buried or straddled with planks which are pinned to the ground. This will prevent crushing of hoses and the resultant flow limitation.

The abovedescribed treatment system provides the necessary response vehicle that can be activated for a wide variety of spill situations in the shortest possible time. Operating experience with the vehicle and predetermined treatment procedures for specific chemicals will insure effective spill response and treatment.

Figure 5.4: Recommended Plan View Layout of System

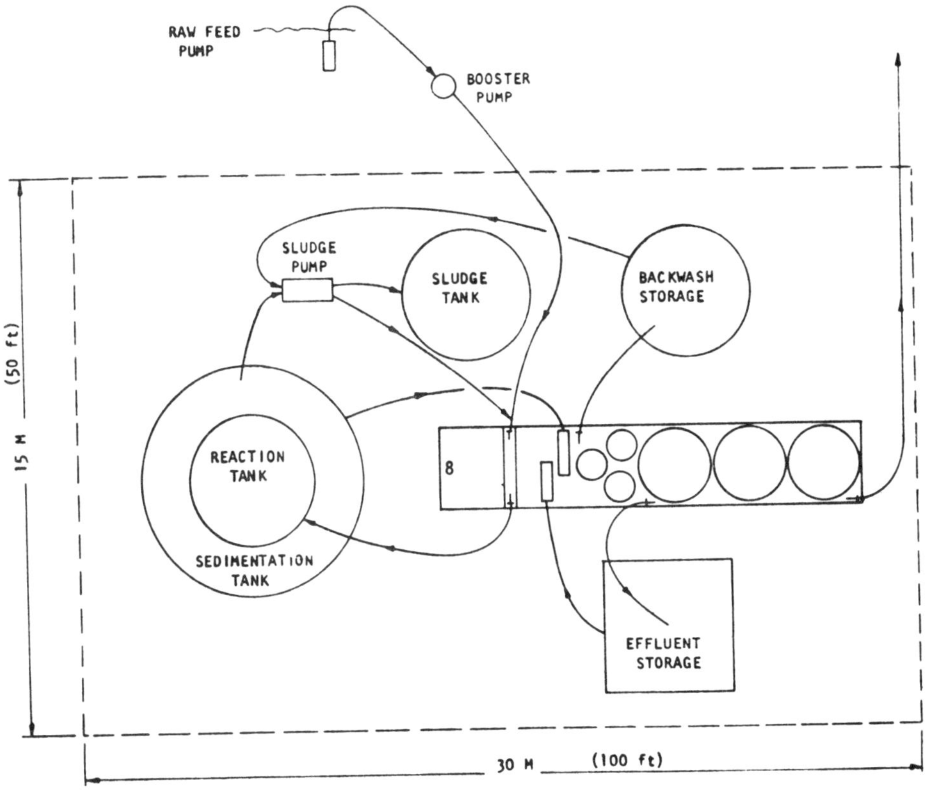

Source: EPA-600/2-76-109

MOBILE DISPENSING SYSTEM FOR MULTIPURPOSE GELLING AGENT

The material in this section is based on *System for Applying Powdered Gelling Agents to Spilled Hazardous Materials,* EPA-600/2-78-145 prepared by J.G. Michalovic, C.K. Akers, R.W. King, and R.J. Pilie of Calspan Corporation for the Environmental Protection Agency.

Introduction

Under sponsorship of the U.S. Environmental Protection Agency (USEPA), Calspan Corporation has developed a mechanical system for delivering Multipurpose Gelling Agent (MGA) to areas endangered by chemical spills. Development included the determination of equipment needed to build a Mobile Dispensing System (MDS), design and construction of a functional MDS prototype, and testing and demonstration of the MDS prototype under field conditions. The completed prototype unit and operational manual were delivered to the Edison Research Laboratory in May, 1977.

Design and Fabrication

Based on preliminary work done under EPA Contract No. 68-03-2093 on *Multipurpose Gelling Agent and Development of Means of Applying to Spilled Hazardous Materials* (14), a powdered form of MGA was chosen for this program. Powdered MGA provides a surface area greater than that for the alternative forms considered, thereby achieving quicker reaction time with spilled liquids.

Dispersing System Selection: In evaluating dispersion systems for powdered MGA, four alternatives were considered. These alternatives included (1) a pressurized tank; (2) a Venturi/compressed air combination; (3) a centrifugal blower; and (4) an auger-fed/pneumatic conveyor.

Because the auger-fed/pneumatic system showed the most promise of dispensing MGA in a safe and effective manner, it was selected as the dispersal component for the Mobile Dispensing System (MDS) (14). In addition to its increased load-delivery capacity and its efficient MGA dispersal volume, the auger-fed/pneumatic system can be powered by either a DC motor or a gasoline-driven engine, making it ideal for portable field application.

Equipment Selection: The auger-fed/pneumatic conveyor system selected for the MDS was a Bantam 400 Rockduster (Mine Safety Appliances). The unit has a hopper which auger-feeds MGA into a moving airstream, fluidizing the agent and transporting it through a 5 cm (2 in.) i.d. delivery hose for distances of approximately 60 m (200 ft). In preliminary testing, using a 30 m (100 ft) section of exit hose, the Rockduster lost less than 20% of the MGA load to wind effects.

In its off-the-shelf configurations, the Bantam 400 Rockduster distributor can be powered by either a 440 volt, 2.7 kW (5 hp), direct current (DC) electric motor or by a hydraulic oil flow of 49 l/min (13 gal/min) at 68 atm (1,000 psi). Since it was required that the MDS design be completely mobile and operable without external power sources, the Rockduster model selected was hydraulically powered using a K-series 31.8 cm^3 (1.94 cu. in.) per revolution displacement, single-stage hydraulic pump (Webster Electric Company, Inc). This hydraulic pump is driven by a 22.4 kW (30 hp) air-cooled gasoline engine (Teledyne Wisconsin Motor). The engine has a four-stroke 1.76 liter (108 cu. in.) piston displacement with 12 VDC electric starter, alternator and 23 liter (6 gal) fuel tank. This power supply incorporates a 151 liter (40 gal) hydraulic oil reservoir (Hydrocraft, Inc.).

Because MGA powder reacts with moisture and clogs the dispersal equipment, the MDS requires sheltering from rain or snow. To provide this weather protection, it was decided to package the MDS in a small enclosed trailer. Based on the dimensions of the Rockduster components and space required for their operation, a 2 x 4 x 2 m (7 x 12 x 6 ft 6 in.), tandem-axle utility trailer (Wells Cargo, Inc.) was selected. This trailer is equipped with a four-wheel, large hydraulic brake system and a jack-ram hoist coupling system. The 2 x 4 m (7 x 12 ft) trailer design allows for easy towing by a three-quarter-ton pickup truck or other vehicle equipped with a 5 cm (2 in.) ball hitch and capable of towing 1,400 kg (3,080 lb).

Trailer Modification: Wells Cargo, Inc. modified the basic utility trailer unit (Model UT-12) according to MDS design specifications to include double rear doors, awnings, roof vents, dome lights, and a double door on the right front corner. The MDS design contains sufficient access doors and openings to operate and maintain MDS components, to provide ventilation for the air-cooled engine, and to minimize MGA dust in the operating area (Figure 5.5).

MDS Assembly: In designing and building the MDS unit, the weight of the Rockduster, engine and hydraulic reservoir were distributed so that the utility trailer hitch did not exceed a 54 kg (120 lb) tongue load. The balance point for a tandem-axle trailer was located between the axles to divide the total volume of compartment space into a 60 to 40 ratio.

To balance weight load and maintain tongue load capacity, the hydraulic oil reservoir and engine/hydraulic pump unit were offset from the area above the tandem axle to the rear of the trailer. With this configuration, both the reservoir and engine were made accessible by the double rear doors and through a 66 cm (26 in.) walkway between the units. As shown in Figure 5.5, the Rockduster was installed across the front of the utility trailer compartment so that the MGA delivery hose can be run through the front corner double doors.

The hydraulic oil reservoir was fitted with a Multiport, dual-element suction line filter (AMBAC Industries, Inc.), a Maxiflow return line filter (UCC Inc.), a filter gauge (UCC Inc.), and a filler breather cap. The reservoir unit was then mounted in the left-rear section of the trailer compartment, 25 cm (10 in.) from the rear doors. The hydraulic system was filled with 151 liters (40 gal) of Mobil 10W30 motor oil. With filters and gauges, the hydraulic reservoir weighs approximately 89 kg (195 lbs).

The 22.4 kW (30 hp) gasoline engine was mounted in the right-rear section of the trailer compartment, 31 cm (12 in.) from the double rear doors, and isolated with a plywood enclosure. The engine exhaust pipe and a noise muffler were then connected to the upper-rear corner of the trailer wall. The engine housing has an outside access door for maintenance work and also opens to the outside at the rear of the trailer to cool the engine and ventilate fumes when in use.

Figure 5.5: Floor Plan and Hydraulic System

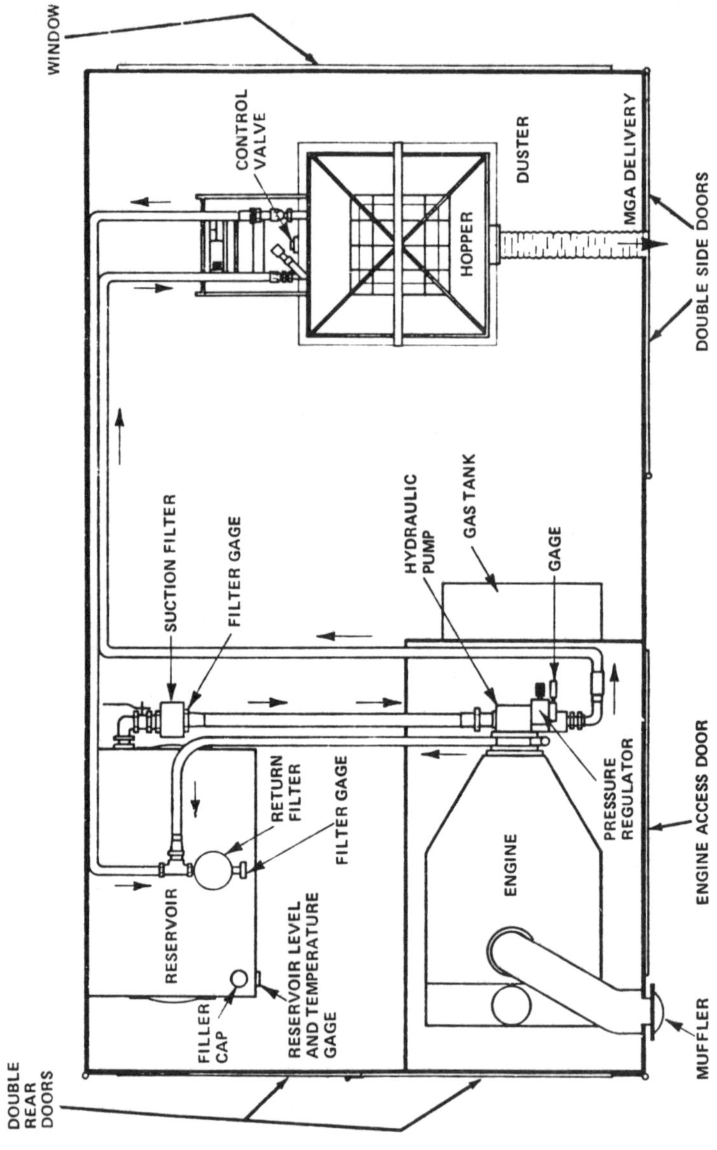

Source: EPA-600/2-78-145

After mounting the engine, the single stage, 31.8 cm^3 (1.94 cu. in.) displacement hydraulic pump was adapted to the engine drive. Total approximate combined weight of the hydraulic pump/engine unit is 145 kg (320 lb).

Buna-N 2.5 cm (1 in.) i.d. tubing was installed under a raised floor in the front of the trailer compartment to connect the hydraulic pump and reservoir with the Rockduster unit.

The hopper was fitted with an overlapping Plexiglas cover to minimize MGA powder from blowing throughout the work area during operation. A control valve near the hopper permits ease of operation of the MGA blower and auger. The work area is ventilated by the front awning opening and corner double doors. The Rockduster unit weighs approximately 143 kg (315 lb).

Testing and Demonstration

Test Site Preparation: Calspan modified a company-owned facility in Bethany, New York, so that chemical test spills could be treated using the MDS prototype and demonstrations could be performed for EPA representatives. The Bethany test site configuration is illustrated in Figure 5.6.

Spill tests were performed in three 23 m (75 ft) ditches that sloped at a 1% slope and terminated in 4 m (12 ft) diameter pools. The trench configuration allowed spill tests to simulate both stream and shallow pool conditions. Test site modification consisted of reworking old trenches to remove eroded soil and digging new shallow pools at the end of each trench.

Field Testing: In developing guidelines for the MGA program, each of the 35 hazardous compounds tested were classified as either an aqueous solution; nonpolar organic material; polar organic material; or alcoholic material. Small-scale spill tests were planned to include one chemical compound from each of these classes. Evaluators also planned to simulate a major chemical spill of approximately 950 liters (250 gal) using dyed water adjusted to pH 3.0 with sulfuric acid.

The MDS was performance-tested by expelling agent over a 6 x 6 m (20 x 20 ft) plastic sheet to determine its dispersion pattern and maximum delivery rate. At maximum throttle, the MDS unit delivered approximately 5.4 kg (12 lb) of agent per minute to the test target area. With the exit nozzle held 1.2 m (4 ft) above the ground, the MDS unit distributes 85% of the MGA over distances 2 to 6 m (6 to 20 ft).

Testing of MDS on small liquid spills began October 21, 1976. Temperatures during testing were near 0°C (32°F). The first spill consisted of pouring 76 liters (20 gal) of water into a linear ditch 76 cm (30 in.) wide. After 15 seconds of spilling, MGA was applied to the head of the spill, containing the flow in 2 m (8 ft) and using 9 kg (20 lb) of agent. The kerosene was tested under similar conditions, MGA contained the spill in 6 m (20 ft) using 41 kg (90 lb) of agent. A flow of trichloroethylene was gelled in 6 m (20 ft) using 18 kg (40 lb) of MGA. The methanol was also further contained in 6 m (20 ft) with a 20 kg (45 lb) application of MGA. Further information on the minor spill tests is presented in Table 5.7.

Figure 5.6: Demonstration Spill Site, Bethany, New York

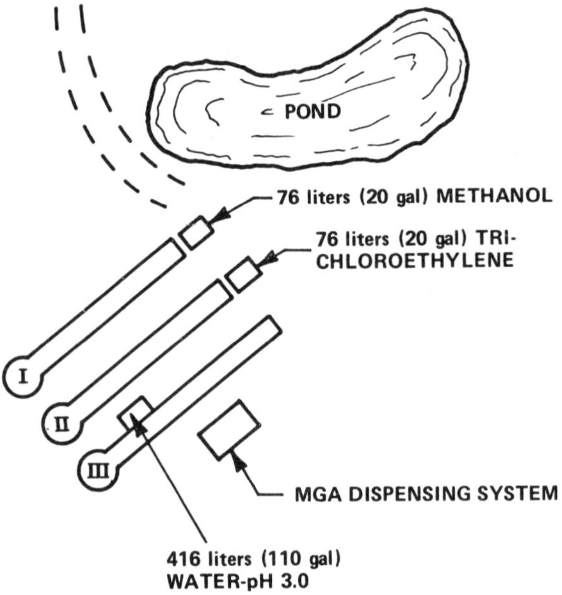

Source: EPA-600/2-78-145

Table 5.7: 76 Liter (20 Gallon) Chemical Spills

Chemical	Class	Spill Rate [liter (gal)/min]	MGA Used [kg (lb)]	MGA Used per Liter Spilled (kg/l)	Distance Spill Stopped [m (ft)]
Water*	Aqueous	151.4 (40)	9.1 (20)	0.12	2.4 (8)
Kerosene**	Hydrocarbon	151.4 (40)	40.8 (90)	0.54	6.1 (20)
Trichloroethylene**	Chlorinated	151.4 (40)	18.1 (40)	0.24	6.1 (20)
Methanol**	Alcohol	151.4 (40)	20.4 (45)	0.27	6.1 (20)

*Air temperature 7°C, wind 8 km/hr.
**Air temperature 0°C, wind 8 to 16 km/hr.

Source: EPA-600/2-78-145

To test MGA performance on a major spill, 946 liters (250 gal) of a simulated toxic substance were poured into a 76 cm (30 in.) wide trench running 9 m (30 ft) into an open pool area. The simulant consisted of water dyed with Rhodamine WT and adjusted with sulfuric acid to pH 3.0. Six 208 liter (55 gal) drums of the toxic simulant were arranged in a pyramid at the end of the trench. The center drum was opened first and treated with MGA until a stiff gel formed on the surface. The other five drums were opened as quickly as possible and within

2 minutes all drums were emptied. Treatment with MGA produced a stiff gel on the surface [5 to 10 cm (2 to 4 in.)] of the spill but failed to immobilize depths greater than 10 cm (4 in.). After the top layer gelled, it became difficult to penetrate with additional MGA, limiting the possibility of treating deep spills, unless external mixing can be provided. Earlier experiments showed that impenetrability can also be partially alleviated by use of the rolled compressed form of the agent (14).

These field experiments showed that the MDS can be effectively operated by a two-man crew, with one crew member operating the mechanical portion of the system and filling the hopper and the other directing the MGA onto the spilled material. For optimum operation, a third crew member is useful for directing the operation, assisting in maintaining a supply of MGA filled drums near the hopper, and recording treatment data.

Demonstration for USEPA: A major spill demonstration was conducted at the Bethany site on November 4, 1976. 76 liters (20 gal) of methanol were spilled into a linear ditch and immobilized in 3.3 min with a 20 kg (45 lb) application of MGA. A second demonstration consisted of spilling and gelling 76 liters (20 gal) of trichloroethylene. Gelling for this second spill was completed in 3 min with 18 kg (40 lb) of agent.

For the major spill demonstration, 416 liters (110 gal) of a simulated toxic substance were poured down 9 m (30 ft) of linear ditch to a 3 m (10 ft) pooling area. The simulant consisted of water which had been dyed red with Rhodamine WT and adjusted to pH 3.0 with sulfuric acid. The demonstration team immobilized this major spill in 8 min with 34 kg (75 lb) of MGA. After treatment, the consistency of the gelled materials was tested.

The ground at the spill site was frozen on the day of the demonstration with the surface thawing gradually under sunny skies. Air temperature was -5.5°C (22°F) and wind speed 8 to 16 km/hr (5 to 10 mph).

Following all tests and demonstrations, team members shoveled the gelled chemicals into 208 liter (55 gal) drums; the material was then sent to an approved landfill for proper disposal.

MDS Modification

Safety modifications were added to the MDS prototype design based on suggestions made during the testing and field demonstration programs. These modifications included the addition of a nonslip floor to the hopper work area and relocation of the gas tank to the opposite side of the partition wall from the engine. The separating wall between the gas tank and engine reduces the effects of engine heat on the fuel supply and makes the tank easily accessible from the front corner door. Two brackets were mounted on the trailer sidewall near the Rockduster for hanging 15 m (50 ft) and 31 m (100 ft) coils of 5 cm (2 in) i.d. MGA delivery hose. A mast was attached to the top of one outside wall to hold optional wind-monitoring equipment.

ACTIVATED CARBON REGENERATION MOBILE SYSTEM

The information in this section is based on a paper presented at Conference 1978 by G.H. Griwatz of MSA Research Corporation and J.E. Brugger of the U.S. Environmental Protection Agency. The title of the paper was "Activated Carbon Regeneration Mobile Field-use System".

Introduction

Water supplies, contaminated with hazardous material spills, have been successfully cleaned up using the EPA mobile physical-chemical treatment trailer (15). This system is based on suitable stream pretreatment followed by final adsorption columns using granular activated carbon beds. Utility of this cleanup system would be improved by the capability for on-site regeneration of spent carbon together with disposal of collected contaminants. To this end, EPA contracted for the development of a complementary mobile carbon regeneration system as further described. This carbon regeneration system is now in the final assembly and testing stage.

System Design

On-site regeneration of activated carbon, with detoxification of the adsorbed hazardous spill materials, is accomplished by a mobile system for field operation. The unit is trailer-mounted for rapid transport to the spill site. It is self-contained for operation, by trained personnel, when deployed near a source of fresh water and provided with fuel. Rating is 100 lb/hr of $\geqslant$40 mesh size granular carbon, regenerated at less than 20% carbon loss and greater than 75% adsorption capacity restoration.

Spent carbon feed is drained of excess water, and transferred to a feed hopper at the rear of the trailer. A screw feeder meters spent carbon into the kiln barrel through the fume breech. The carbon is regenerated as it progresses through the rotating kiln barrel, which is direct fired with controlled gas composition. Regenerated carbon product is discharged from the kiln barrel, through the firing breech, to a slurry quench tank and screening recovery section for reuse. Flue and adsorbate gases are ducted from the fume breech into a direct-fired incinerator to break down the objectionable contaminants. Incinerator gasses are then quenched with water sprays and scrubbed with alkaline solution to neutralize acids and remove particulates before venting to atmosphere. Spent process water is filtered and carbon treated to prevent further land or water contamination.

System design parameters include the size, configuration, weight and stress factors imposed by trailer-van mounting, preferably without exceeding conventional road limitations. Lightweight and resilient fabrication was emphasized throughout, including light structural containment, ceramic fiber thermal insulation and flexible piping sections. Areas subject to abrasion or corrosion with high temperatures are protected using Inconel or stainless steel.

Transport Vehicle: All process equipment is mounted in a special semi-trailer van. The van is fitted for transportation by tractor-truck or piggy-back by rail in accord with all applicable requirements and regulations. It meets conventional width-height-length-weight limitations, to avoid special permit delays. A self-supporting platform base was selected, for rugged off-road operation and for adequate support at minimal bed deflection. Full enclosure is provided by an upper van section to withstand prolonged storage, transit and inclement weather effects. Multiple openings, provided for operation, are of the awning/platform style, to provide rain protection and equipment access walkways.

The basic trailer is a Fruehauf Closed Top Van, special model FB-8-PB-F2-45, measuring 8 ft wide by 45 ft long by 13.4 ft high, when closed for storage or transit. When opened for operation, measurements are 21 ft wide by 53 ft long by 21 ft high with the stack extended. Permissible equipment payload is 42,300 lb, without special road permits based on the 15,980 lb basic van weight when coupled to a 15,000 lb stub tractor for 73,280 lb gross. Capacity trailer loading of 60,000 lb would require road permits.

Equipment is mounted above and below the trailer deck. The primary kiln-incinerator-scrubber section occupies the rear half of the trailer. Spent carbon feed is drained of excess water and loaded into the feed hopper at the tailgate section. Regenerated carbon is water quenched in a slurry tank, mounted below deck at the center of the trailer. The forward closed van section contains the carbon product screening recovery equipment and associated analytical, service and operating accessories.

The system control panel serves as a sound barrier for a diesel-generator mounted in the front portion of the trailer. Below-deck mounting is further utilized for diesel fuel, caustic and water storage tanks together with various system pumps, blowers, and related plumbing. General storage provisions are included throughout the trailer.

Kiln: Thermal regeneration of spent activated carbon is accomplished in the rotary kiln assembly. This is basically a three-step process, occurring progressively in approximately equal lengths of the kiln barrel. In the feed inlet section, lifting flights cascade the wet carbon into the exhausting hot gas stream, to vaporize all water and most low boiling contaminants.

In the mid-section of the barrel, increasing carbon temperature vaporizes higher boiling contaminants and decomposes others. Typical decomposition products have been identified as free carbon and such gases as formaldehyde, acetaldehyde, benzene, acetic acid, SO_2, HCl, NO, and NO_2 depending on the type of spill chemicals adsorbed.

In the final section of the kiln barrel, carbon is gently tumbled to minimize fracture and to promote uniform heating to final temperature with exposure to the activating burner gases for regeneration completion. Burner gases are adjusted for optimum $CO_2/CO/O_2/H_2O$ content. At regeneration temperatures of 1600 to 1800°F, steam reacts rapidly with the free decomposition carbon and at a slower rate with the base carbon product, forming CO and H_2. Some NO and NO_2 may also be formed at these temperatures.

Kiln equipment design includes resilient lightweight thermal insulation which will withstand road stresses at minimal added trailer weight. For this purpose, an alumina-silica ceramic fiber system, rated compatible to 2300°F, was selected. This was B&W Kaowool insulation; vacuum-formed cylinders at 16 lb/ft^3 density were installed to 4.5" in minimum thickness. For abrasion resistance to carbon particles, liners were provided in the kiln barrel and firing breech. Inconel-601 was selected for the hot corrosive sections; stainless steel (316L/304L) affords protection at milder conditions and for transition to carbon steel shells.

Kiln firing is countercurrent to carbon flow for efficient heat utilization. Burner equipment can be operated with either propane or natural gas. Water spray injection is used to moderate the flame for good carbon temperature and activating atmosphere control. Kiln accessories include carbon feed and product recovery systems; fabricated of type 304 stainless steel and suitable trim, for corrosion resistance to wet carbon in air.

The feed system is located at the rear of the trailer. A 20 ft^3 hopper provides a 4 hr operating supply at design feed rate. Wet carbon delivery of 2 to 10 ft^3/hr was measured with the dual pitch screw when operated over its full speed range of 4 to 22 rpm using a variable speed drive. The 2" pitch is used for accurate metering from the hopper; the 3" pitch provides free conveyance through the 4" diameter transfer tube, a minimal resistance without plugging.

A retractable hoist arrangement is provided to unload drums of wet carbon feed delivered to the trailer. Feed is then gravity drained of free water on the tailgate before charging to the feed hopper. A conventional dewatering screw could not be utilized conveniently with trailer limitations.

The product recovery system is located in the forward section of the trailer. A 50 gal slurry tank is agitated with recirculated and makeup water. A 7% slurry at 150°F is pumped through an air cooler and delivered at 130°F to a screening separator (Kason 24" diameter, 2 decks). Carbon product (>40 mesh, 400 microns) and carbon/silica fines (>325 mesh, 44 microns) are separated and the effluent water is further filtered (10 microns) before recirculating. Moyno pumps pump the slurry at 3.5 gpm and 15 psig. Piping is sized for 2 to 3 ft/sec to minimize carbon solids separation with potential plugging. Velocities above 5 to 6 ft/sec would accelerate erosion of containment materials.

Incinerator: Incineration of the gaseous by-products was selected for convenient detoxification and disposal of adsorbed spill contaminants. Vaporized and partially decomposed gaseous products from carbon regeneration are ducted from the fume breech into the incinerator mounted above the kiln at the feed end. Operation is adjusted to convert all gases to CO_2, H_2O, HCl, HNO_3 and SO_3 depending upon adsorbates present.

Fume gases enter the incinerator below the burner for good mixing. The firing rate is controlled by the flue gas temperature and will vary with the fuel content of fumes present. Excess air is adjusted to assure adequate oxygen for complete conversion as determined by detector tube check of stack gas. A 1.0 sec residence time, for complete mixing and combustion, is afforded by the large firebox volume (2 ft wide by 2 ft high by 5 ft long).

Lightweight resilient fabrication was maintained. Original 8 lb/ft^3 Kaowool blanket facing with mineral wool backup was modified to include a 1" thick liner of 16 lb/ft^3 vacuum formed Kaowool board for added mechanical strength in road transit. Overall insulation thickness is 5.6" installed with a type 330 stainless steel stud/washer anchoring system, welded to the ⅛" carbon steel shell. Burner equipment is similar to that for the kiln; operation is in parallel, using a common air and gas supply.

Scrubber: Incinerator flue gases are treated in a scrubber system to neutralize acid gas components and remove entrained particulates, before venting without atmospheric contamination. Hot incinerator gases are adiabatically quenched in a water spray tower, followed by caustic scrubbing and cooling in a packed column. Ideally, stack gases are comprised only of CO_2, N_2 and O_2 saturated with water vapor. Acid gases, which were formed, are converted to sodium salts and removed with a bleed stream from the circulated scrubbing liquor. Particulates are removed using filters in the circulating liquor. Liquor circulation is through an air cooler for heat removal followed by caustic addition for pH control. Vent gases are delivered to a stack using a blower to control induced draft.

Scrubber configuration yields maximum column lengths in the available trailer space for efficient operation. Materials were selected for corrosion and heat resistance at operating conditions. The spray tower is 16" diameter by 6' high Inconel-625 weldment. The inlet end is flanged to the incinerator outlet, provided with six spray nozzles and thermally insulated. The outlet is welded to a common sump [2 ft high by 5 ft long (100 gal)] fabricated from 316L stainless steel, annealed to minimize chloride stress corrosion. The sump is normally operated ⅔ full of spent scrubbing liquor with the upper portion serving to duct quenched incinerator gases to the caustic scrubber.

Should acid mists in the quenched gas stream pose corrosion problems, excess spray volume can be reduced for a higher (~300°F) effluent temperature with reduced condensation. The sump outlet is fitted with a 24" i.d. by 5" high scrubbing column, packed with 4.5' of 1" Pall rings below the liquid distributor, and provided with a flanged head containing an MSA mist eliminator. Materials are all stainless steel (316L/304) except for Teflon fiber portions of the mist eliminator and Viton/EPM gaskets.

The scrubbing liquor cooler (600,000 Btu/hr) is mounted above the kiln at the forward firing breech end. Spent cooling air (11,700 cfm) is discharged through a roof vent arrangement. The caustic treating system includes an ample 7 day supply (260 gal) of 20% NaOH (-15°F MP) solution, stored in two tanks below deck. For prolonged operation, replenishment provisions must be considered.

The water distribution system includes an ambient receiver and a pressurized reservoir for operation from an adjacent lake or stream, when a local pressurized supply is unavailable. All effluent water is piped through portable activated carbon filters, to prevent further land or water contamination.

Generator: A Katolight diesel-electric generator provides 74 kVA-240 V-3 phase power for field operation, when a local supply is unavailable. System consumption is about 50 kVA at steady state operation, leaving a 25 kVA standby reserve for motor starting surges and for additional field applications.

Installation, in the front portion of the trailer van, includes provision for quiet operation. The generator is compartmented from the operating area by the system control panel with added acoustical insulation. Vibration absorption pad mounting isolates the engine from the floor. A snubbing muffler, for silencing exhaust gases, is externally mounted.

Diesel fuel storage (390 gal below deck) is adequate for three days continuous operation. With regularly scheduled refills, the operating crew will be relieved of constant concern over power outage resulting from diesel fuel shortages.

The generator control console is mounted in the common system panelboard, for remote operation. Access to the generator compartment for periodic servicing is provided by an outside door, which serves as a cooling air inlet during operation.

Pilot Regeneration

Mobile system operation will be demonstrated using spent carbon from actual spill cleanup work. Regeneration properties will be checked using the MSAR pilot-plant scale system(16) which served as the design base for the mobile unit. In preparation for this work, MSAR performed pilot runs on an EPA shipment of drummed wet spent activated carbon, which was used in a spill cleanup occurring in Plains, Virginia.

The pollutant in this carbon was primarily toxaphene ($C_{10}H_{10}Cl_8$), with lesser quantities of other chlorinated compounds, such as aldrin ($C_{12}H_8Cl_6$); dieldrin ($C_{12}H_8Cl_6O$); heptachlor ($C_{10}H_5Cl_7$); and chlordane ($C_{10}H_6Cl_8$). The carbon was discolored with a brown material, which appeared to be a thin layer on the outside surface of the granules. On the basis of bulk density measurements, pollutant content was approximately 13% and water content was approximately 52% by weight, based on the dry weight of the regenerated carbon product.

Regeneration gave an 88% bulk volume yield, with the large loss attributable to removal of the brown surface coating. Complete regeneration was indicated by density and iodine number if the unidentified carbon was Pittsburgh (Calgon) type CAL. Vent gas analysis showed on HCl, CO_2 and H_2O with CO and hydrocarbons below detectable limits. Negligible HCl was removed by water spray scrubbing according to a test of the scrubbing liquor with silver nitrate. Caustic scrubbing, as planned for the mobile unit, should effectively remove HCl.

Summary

The mobile carbon regeneration system was expected to be operational and available for field use during the latter part of 1978. Performance was expected to exceed minimum specified requirements based on pilot studies made under conditions which can be duplicated in the mobile unit.

The mobile system is completely self-contained for field operation with normal maintenance. Operating requirements include a suitable parking area (~40 ft by 80 ft); a source of fresh water (ambient or pressurized); propane (or natural) gas for the kiln/incinerator; diesel fuel for the generator (or a local power supply); and replenished caustic for extended operation. Provisions for delivery of spent carbon feed and removal of regenerated carbon product must be implemented. Continuous operation can be effected with three trained operators given bulk carbon handling assistance.

MOBILE OZONE TREATMENT SYSTEM

The information in this section is based on a paper presented at Conference 1978 by J.P. Lafornara of the U.S. Environmental Protection Agency. The title of the paper was "Feasibility of a Mobile Hazardous Materials Spills Water Treatment System Based on Ozonolysis."

Introduction

Treating spills of hazardous substances which have reached watercourses is often a complex and expensive problem. Although wastewater treatment technology has developed to the point where, at a fixed plant site, it is technically, if not economically, feasible to devise a method to remove almost any given substance from an industrial effluent, the fact that a spill of the same substance can occur at any place, at any given time of day or night, and in any weather, introduces logistical problems which may make it impossible to conduct a water treatment operation.

For example, a plating shop's effluent containing waste cyanide ion can be detoxified by several different currently available methods such as ion exchange, electrochemical oxidation, ozonation, etc. However, the treatment of a spill of a cyanide solution from a tank truck or tank car into a remote rural pond or stream will be possible only if the spill can be contained and the proper equipment can be transported to the site, set up and operated before the spill has done its damage or been diluted.

Several containment methods, varying from the deployment of earthen dikes with bulldozers to the spraying of a foam-in-place dam can extend spill response time considerably, but the treatment of water contained behind these improvised barriers is still feasible only if a safe, effective system can be delivered quickly. Recognizing the need, EPA's Oil and Hazardous Materials Spills Branch has initiated a program to develop such systems. The first system, the Mobile Physical/Chemical Treatment Trailer, which is capable of treating water by chemical addition, precipitation, flocculation/sedimentation, mixed-media filtration, and carbon adsorption at the rate of 200 gpm, was delivered to the USEPA in 1972 and has responded to ten toxic materials incidents (17) (18). A second trailer-mounted unit, utilizing a Dynactor, capable of treating 250 gpm of water with a number of gaseous, liquid or solid reagents including ion exchange resins and powdered activated carbon, was discussed previously.

These mobile systems will find their main utility in treating waterborne spills of heavy metal or other inorganic compounds, relatively water-insoluble organic compounds and biodegradable organic compounds. A technology gap still exists, however, for soluble polar organic chemicals such as ketones, aldehydes, glycols, ethers, and others which are not amenable to biological degradation.

The use of ozonation has been shown to be useful in water treatment including drinking water and sewage treatment disinfection. It is a powerful oxidizing agent which forms by-products which are relatively innocuous when compared with those formed by chlorine. It was the purpose of this investigation to determine the feasibility of fabricating a self-contained mobile wastewater treatment system based on ozonation, in order to fill the technology gap for treatment of spills of soluble biorefractory organic compounds. This study was accomplished by conducting a literature search and laboratory and pilot scale experiments, producing a preliminary mobile system design and evaluating the design against logistical, operational and cost criteria.

Pilot Plant Studies

Selection of Analytical Method and Compounds: The purpose of the pilot study was to determine whether ozonation showed sufficient promise in removing a selected group of organics from water. Since complete removal, not partial conversion to unknown oxygenated products, is the desired goal of hazardous spill treatment, the analytical parameter selected to determine effectiveness was Total Organic Carbon. It has been determined that the oxygen demand parameters (chemical, biological, total, etc.) are not accurate indicators of a compound's complete removal from water unless the products of the treatment reaction are characterized and quantified. Specific analytical techniques for each of the tested compounds were not warranted, since these would not necessarily indicate the complete removal of a compound, either. Samples were analyzed on a Beckman 915 Total Organic Carbon Analyzer after liberation of carbon dioxide by acidifying with concentrated HCl and sparging with CO_2-free nitrogen.

The following industrial organic compounds were selected based on their likelihood of spillage; toxicity to aquatic life; ease of handling in the laboratory; resistance to biodegradation; and infeasibility of carbon adsorption treatment: acetone, amyl alcohol, ethylene glycol, methyl isobutyl ketone, and phthalic acid.

Pilot-Plant Apparatus: Although a custom-designed pilot plant is necessary to predict the performance of a specific system, it was the aim of this study to give an indication as to whether ozonation had promise as a technique for removing biorefractory, polar organics from water. A package pilot-plant system, consisting of ozone generator and contactor, was sought for rental or purchase.

The pilot system was based on a model 1101 ozone pilot plant offered by W.R. Grace & Co. The main feature of this apparatus was a 10 ft high, double pass contactor, using a specially designed gas-liquid positive pressure injector head for the initial contact. The contactor was made from PVC with clear sight ports at the appropriate places so that its operation might be viewed. Also included with the pilot plant was a rotameter (calibrated in gpm), flow control and shut-off valves, and an electrically driven centrifugal pump capable of 14 gpm in pilot plant operation.

Ozone was generated by a W.R. Grace & Co. Model 62-2-62 ozone generator. This air-cooled unit was capable of generating up to 2.5 lb O_3/day from an oxygen feed. All gas tubing was passivated aluminum tubing. The liquid tubing was made from PVC pipe or rubber hose with camlock couplers. The samples were drawn from a tap located at the bottom of the contactor.

In addition to the equipment supplied with the pilot plant, a 100 gal polyethylene tank was used to mix and store solutions. This tank was fitted with a cover in which ports for two 14 ft^3/min electrical deozonators were installed. The electrical deozonators decomposed the residual ozone by heating the off-gas above the instantaneous ozone decomposition temperature.

The effluent gas from the contactor was deozonated chemically by reduction with sodium thiosulfate solution to which a potassium iodide/starch indicator had been added.

Testing and Results: 100 gal of tap water were used for each test run. A sample of this water was taken and used for a pH measurement and as a blank in the TOC analysis. The organic contaminant was then added and mixed with a motor-driven stirrer and pumped directly from the tank into the pilot plant.

Test runs were 20 min long at a liquid flow rate of 10 gpm. The effluent from the system was returned to the mixing tank so that the entire 100 gal were recirculated twice. Samples were taken at the beginning of each run and at 5 min intervals thereafter.

Efforts were made to complete the entire test sequence from mixing of solution to analyses of samples in eight hours to minimize the effect of aging.

The optimum settings for 10 gpm liquid flow rate were an oxygen flow of 15 ft^3/hr, a pressure of 14.5 psig, and a power setting of 400 W. Under these conditions, the generator produced 1.4 lb O_3/day at a concentration of 3.2% by weight. Assuming total dissolution, the theoretical dose of ozone to the liquid stream was 11.6 mg/l. The operating conditions were chosen, not on the basis of maximum ozone output, but on the basis of the conditions of pressure and gas flow that optimized the contacting system.

Removal of organics varied from a low of 1.2% for phthalic acid to a high of 20.19% for methylisobutylketone.

Preliminary Design of Mobile Ozone Treatment System

Design Criteria: Since the pilot scale removal experiments were not conclusive, they alone could not be used to determine the feasibility of a mobile ozonation system. A preliminary design was prepared and evaluated. Several criteria necessary for a mobile system that are not normally encountered for fixed site ozone installations are: overall system and component dimensions and weight must be compatible with weight, length, width and vertical clearance restrictions on highway vehicles; equipment must be rugged enough to withstand vibrational and other stresses encountered in over-the-road transport; the system must be self-contained; it must be capable of safe operation without causing secondary air or water pollution; and it must be flexible enough to allow changes in process conditions from one spill to another and as treatment of a given spill progresses. These special conditions are in addition to the general criteria of: commercial availability of components, low capital, operation and maintenance costs.

System Design: The flow diagram developed for the mobile ozone system is shown in Figure 5.7. The contaminated water would be introduced into the system through a submersible pump. It would then flow through a booster pump and into a spray-type deaerator where dissolved gases (mostly nitrogen) will be removed. The water would be pumped from the deaerator directly to the ozone contactor and flow from there to a holding tank.

Air would enter the system through a compressor and would be successively cooled and dried before entering the ozone generator. From the generator, the ozone-enriched air would pass under its own pressure to the gas/liquid contactor. Excess ozone/air would be vented to a deaerator.

Commercially available and where necessary custom-fabricated devices were selected for unit processes corresponding to those in Figure 5.7; the commercial units are listed in Table 5.8. The whole system would be fabricated from 316 stainless steel to provide the necessary mechanical strength and corrosion resistance. The entire system would be mounted on one, 45 ft long by 8 ft wide, low-boy flatbed, triaxle trailer.

Figure 5.7: Flow Diagram for Mobile Ozone Spill Treatment System

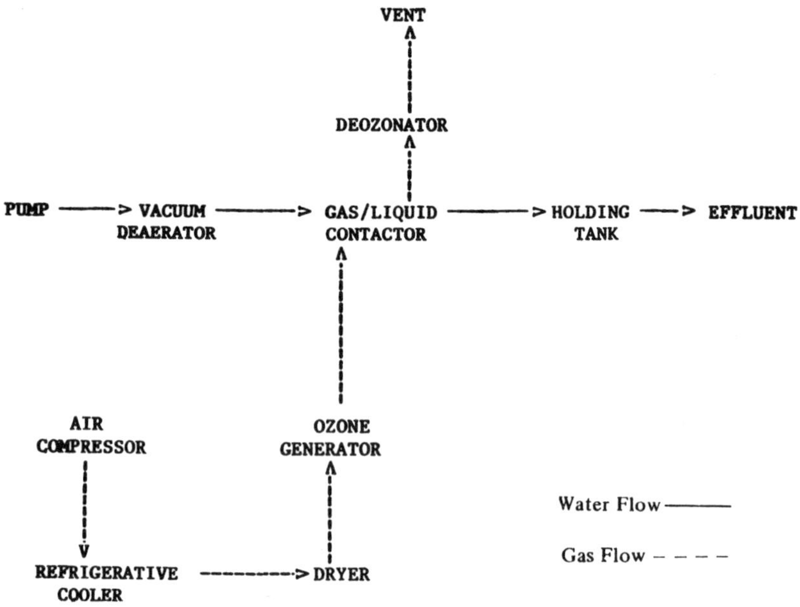

Source: Conference 1978

Table 5.8: Equipment List for Mobile Ozone System

Item	Custom or Commercially Available
Pumps	Available
Valves, piping, hoses	Available
Deaerator	Custom fabrication
Gas/liquid contactor	Custom
Holding tank	Available
Compressor	Available package
Refrigerative cooler	Available package
Dryer	Available package
Ozone generator	Available package
Deozonator	Custom
Electrical and control panel	Custom
Diesel-electric generator	Available
Trailer	Available
Fabrication	Custom
Engineering	Custom

Source: Conference 1978

All components were selected to be compatible with limitations imposed on over-the-road vehicles.

Discussion

Neither the results of the pilot plant studies nor the preliminary design analysis could be used to invalidate the development of a mobile ozone spill response system. However, the data from the pilot plant study did not indicate sufficient contaminant removal to warrant a decision to build a full-scale system, especially because of the unit's high estimated cost.

According to the literature, the state-of-the-art in industrial waste treatment using ozone is moving rapidly. Advances in interfacing ozonation and other treatment methods such as ultraviolet (UV) photolysis, and activated carbon adsorption have indicated that much more is possible in the area of ozone technology than what was thought during this study. UV/ozone systems have shown particular promise with a synergistic enhancement of removal efficiencies for various organic compounds. It is anticipated that, in the near future, demonstrated full-scale equipment for treating industrial wastes by UV/ozone will be commercially available.

Because of the rapidly advancing state-of-the-art and because of the inconclusiveness of the pilot scale tests, it was decided to defer the development of a mobile ozone spill response trailer until UV/ozone or some other interface with ozone is shown to be more effective in detoxifying spills of biorefractory water-soluble organic chemicals. A preliminary investigation into this process is scheduled to begin shortly and, if results are positive, development of the full-scale interface system should follow.

POLYURETHANE FOAM

The information in this and the following section is based on *Control of Hazardous Chemical Spills by Physical Barriers*, EPA-R2-73-185 prepared by J.V. Friel, R.H. Hiltz and M.D. Marshall of MSA Research Corporation for the Environmental Protection Agency.

Introduction

One of the strategies for the control of hazardous chemical land spills depends on the availability of a small, portable system which could be carried on transporting vehicles and utilized by the carrier personnel. Polyurethanes, already commercially available as a portable foam dispensing system, potentially offered a quick response solution. Polyurethanes are resins derived from the catalyzed reaction between polyols and isocyanates. Once formed, they are reasonably inert to most other chemicals. The foamed form, which is currently used in a wide variety of applications, is obtained by blending a low boiling liquid, usually a Freon, into the reaction mass.

A typical portable, man-carried dispenser is packaged as a two tank unit. This size delivers some 10 cu ft of foam with an expansion of 25:1 and an output of approximately 1 cu ft/min. This unit would appear to be quite suitable to fill the requirement for a small, portable system to provide emergency diking. The only deficiency in the unit is the low output rate. A discharge of 4 to 5 cfm would be a more suitable situation for spill control. With the existing formulations and portable packages as a base, a program was undertaken to adapt this device for use in the control of hazardous chemical land spills. Three major items were considered: (1) increasing the discharge rate to an overage of 5 cfm; (2) limited evaluation of the chemical compatibility of the urethane with hazardous chemicals; and (3) determination of the effectiveness of the material as a barrier against a variety of substrates.

Portable Units

The basic objective of the polyurethane development was the adaptation of existing commercial equipment to the spill control application. Two portable units have been evolved. The small unit is hand portable; the larger is available as a back pack or cart mounted unit. Pertinent details are given in Table 5.9.

Table 5.9: Polyurethane Systems

	Model 1	Model 2
Part number	510028	210029
Volume of foam delivered, ft^3	22-25	50-55
Delivery rate (average), cfm	5	5-7
Weight (gross), lb	27	65
Size, inches	20 x 10 x 5	20 x 20 x 10
Storage time (tentative), months	6	6
Storage temperature, °F	>50	>50
Useful temperature range (substrate), °F	15-120	15-120

Source: EPA-R2-73-185

Operating instructions are provided on the package. The polyurethane is dispensed from each unit in a spray cone as a thick liquid. There is a delay time of some 8 to 12 seconds after generation before foaming begins. The foam rigidizes rapidly after expansion. The rigidized foam is inert to a wide variety of chemicals and when bonded to the substrate can effectively contain or divert chemical spills.

Evaluation

The polyurethane formulation bonds well to dry surfaces such as cement, asphalt and packed dirt. The material also bonds to itself, thus new foam can be effectively deposited over old foam. Control on dirt may be limited with failure occurring due to liquid seepage through the substrate at the dirt-urethane interface.

Some control is exercised on gravel, rocky or vegetated ground. Some penetration will ultimately occur below the substrate-urethane interface. In such situations, double or triple barriers can be set to extend the time of control.

Temperature also has an effect. The unit temperature at the time of use should preferably be above 50°F, but at least above 40°F. Substrate temperatures can be as low as 15°F. At the lower temperatures expansion may be reduced at least for that material applied directly to the substrate.

At the present time good adhesion is not obtained on wet surfaces nor can flowing streams be directly blocked. If the foam can be mechanically locked onto the surface the problem of wet surfaces can be overcome. This is the case with sealing storm sewer grates, manhole covers, etc. By generating foam into a plastic or rubber bag, pipe, culverts and the like can be plugged even when liquid is flowing.

Although the chemical nature of polyurethane would indicate resistance to most chemicals, actual evaluation is limited. Control has been demonstrated for water base liquids except strong acids, nonpolar organics, and some selected materials such as chlorine and ammonia. Polar compounds are a question. It has been established that the present formulation is not effective against methyl alcohol.

FOAMED CONCRETE EMERGENCY UNIT

Introduction

Foamed concrete barriers were discussed in the preceding chapter. Foamed concrete of density about 40 pcf can be made to take an extremely fast set (2 to 3 seconds) with the addition of sodium silicate to form a gelled structure with sufficient strength to build dikes 2 ft high or better. The equipment requirements are simple. A unit following the schematic of Figure 5.8 would consist of a mixer for blending a cement-water slurry, a slurry pump, preformed foam generator, sodium silicate solution, storage tank and nozzle. The preformed foam is metered into the slurry stream where it blends to produce foamed concrete. The silicate solution is injected as the foamed concrete exits the nozzle, producing a chemical reaction with the cement to cause the rapid set.

Figure 5.8: Schematic of Foamed Concrete System

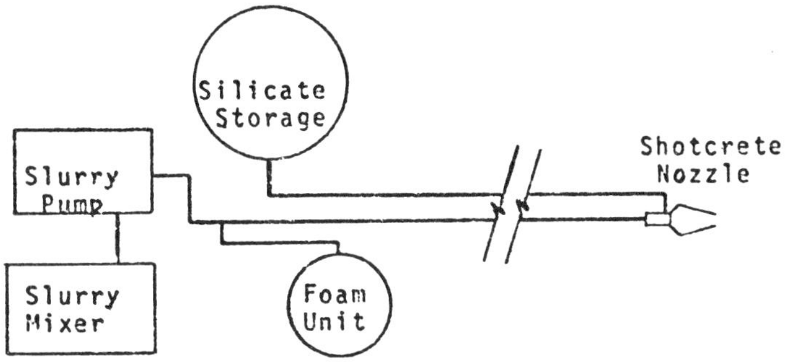

Source: EPA-R2-73-185

A foamed concrete barrier has strength enough to impound liquid immediately after being placed, and flowing water of slow stream velocities has been successfully pooled. The initial gel set has surprising strength, but to impound liquids to any appreciable depth, such as 2 ft, behind the barrier needs additional strength from the hydraulic set of the cement. This comes on slowly after the initial gel set and is a function of many factors, including water temperature, water/cement ratio and type of cement.

Under ideal conditions significant strength from the hydraulic set could be attained in less than 5 min. Thus, the rate of build-up of hydraulic head on a newly poured barrier is probably the single most important factor to successfully impounding a spill. For this reason, broad, flat rather than narrow, high-pitched impounding basins are favored.

The type of substrate is not an important factor. Tests on clay, shale, chipped limestone, grass and weed covered ground have been successful. In addition, such chemicals as methanol, 1,1,1-trichloroethane, phenol, acetone cyanhydrin and acrylonitrile do not appear to affect the gel set action.

As might be expected in a water-based system, extremely cold, sub-freezing working conditions make the equipment operations difficult. The preformed foam and solution lines tend to freeze first. The use of freezing point lowering additives in the solution has only had moderate success.

Emergency Field Unit

A suggested design for an Emergency Field Unit is shown in Figure 5.9. The unit consists of the basic components described earlier, driven with hydraulic motors. A gasoline engine furnishes power to the hydraulic pump and a small air compressor. The components are all commercial off-the-shelf items. The parts list is given in Table 5.10.

The unit is trailer mounted, suitable for a pick-up truck operation. It will produce approximately 50 ft^3 of foamed concrete at 45 pcf per batch within 30 min of delivery to the site. Repeat cycles would take approximately 25 min each.

Three such batches (150 ft^3) could be produced in approximately 80 min to build barriers of such sample dimensions as: 2' x 2' x 38'; 1.5' x 2' x 50'; and 1.5' x 3' x 33'.

Figure 5.9: Proposed Emergency Field Unit

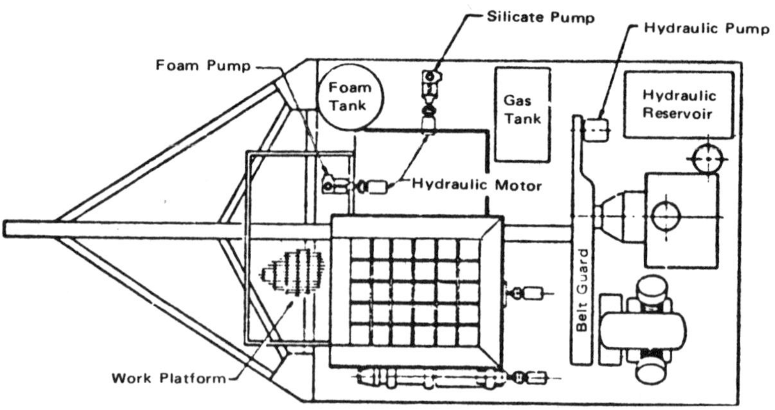

(continued)

Figure 5.9: (continued)

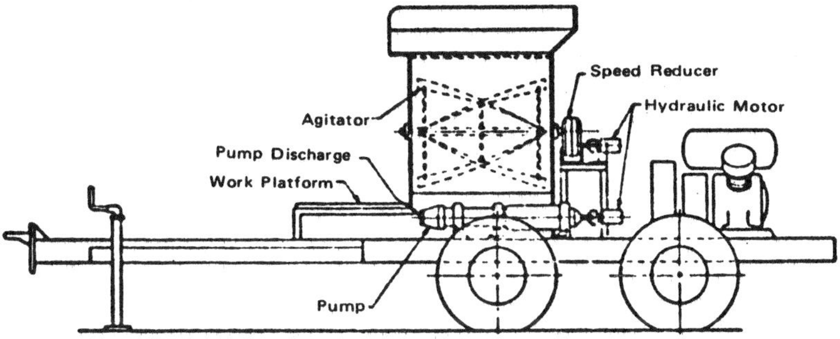

Source: EPA-R2-73-185

Table 5.10: Mobile Foamed Concrete Emergency Unit

Quantity	Description
1	Ingersoll-Rand Air Compressor, Type 30, Model 71T2XGT with gas engine drive, Wisconsin Model V465D engine and V465D power take-off clutch. Gasoline tank included with motor.
1	Hopper similar to MSA P/N A91959, 400#
1	Agitator blade similar to MSA P/N 379689 Agitator B91962
2	Bearings, P/N 66802
1	Slurry pump, Moyno 3L4, P/N 68572
4	Motor, hydraulic, Char-Lynn Size AA
1	Pump, hydraulic, commercial shearing P30 Series, 1½" gear
1	Silicate proportioning pump, Viking Model HL124S, carbon bushing and stuffing box, all iron, built in relief
1	Foam pump, Viking Model HL156, all iron, built in relief
1	Foam tank, similar to pressure tank paint container
1	Foam gun
1	Hydraulic tank, P/N 456313
4	Couplings, T.B. Woods #6, P/N 625462
1	Speed reducer, Dodge #11
1	Valve, hydraulic–4 way 3 position Double A P/N R6-175-FF-N
1	Valve, hydraulic–3 way 2 position Double A P/N R6-175-FZ-N
1	Valve, hydraulic–flow control Double A P/N QXA175 with P15-06 sub plate
1	Filter, hydraulic P/N 625468
1	Strainer, suction P/N 69806
1	Breather and filler cap P/N 69081
1	Water control valve P/N 69412
1	Air on-off valve, 1½" size P/N 69452
1	Gauge, hydraulic press, 0-3000 psi P/N 56414
1	Gauge, air pressure, 0-150 psi P/N 68052
1	Gauge, water pressure, 0-100 psi

(continued)

Table 5.10: (continued)

Quantity	Description
1	Battery, electric, 12V
2	Mounts for hydraulic motor with Viking Pump
1	Mount for hydraulic motor with Moyno pump
1	Mount for hydraulic motor with Dodge speed reducer
	Hydraulic hoses
	Hydraulic fittings
	Compressor piping
	Moyno pump piping
1	Sheave, hydraulic pump P/N 625860
2	V-belt P/N 69753
1	Mount for hydraulic pump
1	Trailer, Model 12001-2 double axle, 12,000 lb capacity
2	Bearings, pillow block, $1\frac{15}{16}$" diameter P/N 66058
1	Stub shaft for mounting drive sheave
1	Coupling, T.B. Woods #7, P/N 63951

Source: EPA-R2-73-185

Raw materials for one batch are: 10 bags of cement (940 lbs); 30 gal of sodium silicate solution; and 1.5 qt of foam concentrate. Material for one run would be transported on the unit. Material for three batches would weigh approximately 4,900 lbs and require about 1,800 lbs of water (215 gal) at the site. Additional material could be brought in if needed. The sodium silicate and foam concentrate would be the only material required to be stored in quantity for multiple batches. Enough Type 1 cement could be stored for 1 to 3 batches, with additional cement picked up as needed. A detailed description of the procedure for placing the mobile unit in operation, with estimated time commitments for 2 men, is as follows:

Operation	Time (min)
Start gasoline engine, slurry and slurry mixer	2
Attach water source, fill tank to level mark (70 gal)	8
Add cement (10 bags)	5
Add 1.5 qt of MSA reg foam concentrate to 10 gal foam tank–fill with water to level	4
Attach silicate drums to pumper	2
Attach foam cement and silicate, quick connect hose to unit	2
Total	23

The first cycle including set up and a run time of 10 min should require approximately 33 min. Additional cycles should be possible in less than 25 min.

ENVIRONMENTAL EMERGENCY RESPONSE UNIT (EERU)

The information in this section is based on a paper presented at Conference 1976 by R.W. Fullner and H.J. Crump-Wiesner of the U.S. Environmental Protection Agency. The title of the paper was

"Use of EPA's Environmental Emergency Response Unit in a Pesticide Spill."

Introduction

Since 1970, the EPA and its predecessor agencies have assumed responsibility for reducing the frequency of spills of hazardous substances and for minimizing environmental damage caused by those spills which do occur. While waiting for the Hazardous Substances Regulations under Section 311 of the 1972 Federal Water Pollution Control Act (P.L. 92-500) to be promulgated in final form, the federal revolving fund has been unavailable for expenditures incurred in the mitigation and removal of hazardous substances. The EPA has been responding to hazardous material spill incidents with in-house resources such as the EERU. The EERU consists of the response personnel and trailer for on-site removal of hazardous substances amenable to treatment by carbon adsorption. Operation of the EERU is supported by the Oil and Special Materials Control Division and the Edison Industrial Environmental Research Laboratory.

The emergency response trailer is particularly suited to the treatment of organochlorine insecticides, which were first introduced in about 1940. These synthetic organic pesticides were so successful in controlling pests that there was extremely rapid and general adoption of them and development of new ones.

In the United States, the annual production of pesticides increased from approximately 300 MM lb in 1954 to over 1,200 MM lb in 1973; approximately 30 MM lb were organochlorine insecticides. Because of the large quantities of pesticides produced, transported and used, there is an increasing potential for accidental spillage and environmental damage. Evidence indicates that pesticides pose the greatest hazards to aquatic organisms, which seem to be much more susceptible to them than terrestrial organisms; aquatic organisms also concentrate pesticides in their tissues more readily.

Design

In general, technology has not developed as rapidly for the removal or mitigation of hazardous substances as it has for oil. Although techniques to counteract hazardous substances have been demonstrated in the laboratory, they have not been tested extensively on actual spills. One technique, however, that has received considerable attention in the treatment of hazardous substances is carbon adsorption. The EERU's primary treatment capability is based on carbon adsorption; additionally other processes can be incorporated in the treatment scheme, depending on the properties of the hazardous substance spilled and what methods of removal are employed.

Prior to the construction of the EERU, various treatment processes for a representative number of hazardous materials were evaluated in the laboratory under an EPA contract. The design of the EERU was based on the findings obtained in the laboratory evaluation phase. The unit incorporates chemical reaction, flocculation, sedimentation, filtration and activated carbon adsorption in a flexible and versatile system for on-site removal of hazardous materials. The treatment trailer itself consists of the power unit, pumping system, three multi-media filters and three activated carbon filtration columns where the final adsorption of the chemical pollutant takes place. Removal of the suspended or particulate matter is achieved by means of the multi-media filters, which can be backflushed and thereby prevent fouling and choking of the carbon columns. The maximum hydraulic capacity of the EERU's pumping system is 200 gpm, while the design process flow rate is 100 gpm.

The EERU has been used successfully six times in EPA's spill response program for hazardous substances. Its use involves close coordination between three EPA program elements; Washington headquarters, regional offices and the Edison Industrial Environmental Research Laboratory. The Oil and Special Materials Control Division in Washington has the ultimate responsibility for determining whether or not to activate the EERU.

In hazardous material spill response, analytical support is absolutely essential. Without it, a successful response is impossible. The main analytical support problem encountered in past spills has been the long interval between the time the samples are collected and the time the results are received at the spill site. Since hazardous substances when spilled into water are generally invisible and soluble, analytical data are crucial in determining the magnitude of the spill area, in deciding on what strategy to be taken in spill cleanup, and in measuring the effectiveness of the treatment process. It is impossible to control the treatment efficiency of the response treatment system without timely analytical data. Therefore, the dedicated services of an analytical laboratory are of utmost importance in hazardous material spill management.

Hazardous material spill response also requires a streamlined procedure for obtaining financial assistance. Although limited emergency contracting authority exists, further financial assistance can only be obtained through more time-consuming, conventional contracts. After the Hazardous Substances Regulations are finally promulgated, the federal revolving fund will become available for hazardous substances spill response as well as for oil spills.

ADAPTS

> The information in this section is based on a paper presented at Conference 1976 by S. Shaw of Seaward International, Inc. and C. Gower of U.S. Coast Guard Headquarters. The title of the paper was "Extending the Use of ADAPTS to Hazardous Materials."

Introduction

The Air Deliverable Antipollution Transfer System, ADAPTS, was developed by the U.S. Coast Guard in 1970, as a self-contained means for removing oil from grounded or disabled tankers. The basic elements of this system consist of an aircooled, explosion-proof, 40 hp diesel-driven, hydraulic power unit; a hydraulic motor-powered, two-stage, 10" vertical turbine pump; a 140,000 gal rubberized fabric, collapsible oil storage container; 300 ft of 6" discharge hose; hydraulic transmission hoses; a rubberized, flexible fuel container; and associated handling equipment. The equipment is packaged in watertight containers and rigged for either parachute delivery from a C-130 aircraft or for delivery by helicopter.

In use, the power unit is located on the deck of the stricken vessel near the cargo tank to be pumped. The submersible pump, with discharge and hydraulic supply and return hoses attached, is lowered into the cargo tank through the standard Butterworth deck openings. The other end of the 6" hose is routed either to the collapsible storage container, to an intact tank on the same vessel, or to a lightering vessel alongside. Each ADAPTS pump has a nominal capacity of 1,000 gpm delivery at a discharge head of approximately 70 ft.

At present, eighteen ADAPTS systems, without the collapsible storage containers, are maintained in readiness for emergency use by the Coast Guard's National Strike Teams. This equipment, along with other pollution control devices, is divided among the Strike Team bases at Elizabeth City, North Carolina; San Francisco, California; and Bay St. Louis, Mississippi.

Although the storage container portion of the system has not yet been fully proven out, the pumping subsystem has been employed with great success on numerous occasions over the past four years. The most famous of these occasions was the offloading and salvage of the 237,000 dwt tanker Showa Maru near Singapore in January, 1975. In both of these grounding accidents, timely pumping of the oil cargo from the disabled tanker prevented thousands of tons of oil from being spilled.

Because of its demonstrated effectiveness, the Coast Guard feels that the ADAPTS pumping equipment can also be an effective tool in dealing with tank vessel incidents involving certain hazardous substances other than oil. In fact, because many chemicals either sink, rapidly dissolve, or chemically react when combined with water, as opposed to oil which tends to float and not mix with water, pumping off prior to spilling may be the only practical way to mitigate the pollution hazard associated with some grounded or damaged chemical vessels.

In order to establish in advance what chemicals could or could not be adequately handled by this system, a research program was undertaken with Seaward International, Inc. as the prime contractor. A list of 266 chemicals which were believed to be transported by barge or tank ship in liquid bulk form was developed. Establishing the applicability of the ADAPTS system to each chemical required the collection of data relating to the properties of each chemical and its mode of transport and on the materials and design of the ADAPTS elements in contact with the pumped liquid.

Data Collection and Evaluation

Information on the materials of the ADAPTS components in contact with the pumped fluid were obtained from the component specifications and drawings and in some cases by direct contact with the component manufacturers. The directly affected portions of the pumping system included the following: submersible pump, interior and exterior; submersible pump hydraulic drive motor, exterior only; pump/motor shaft coupling, housing, and bearing, interior and exterior; 6" cargo discharge hose, interior and exterior; discharge hose couplings, interior and exterior; hydraulic hoses, exterior only; hydraulic hose couplings, exterior only; and pump lifting wire and grip hoist.

The exposed materials in these components contained cast iron, steel, stainless steel, aluminum, zinc and cadmium plating, copper, bronze, neoprene, Buna-N and nylon.

In order to handle the mass of data which was to be collected, a standardized data sheet form was developed and filled out for each chemical, as the data became available. In addition, summary status sheets, showing all of the 266 chemicals by name and sequential number along with the type of data collected and its source, were generated and continually updated.

The individual data sheets contained necessary data on each of the following important characteristics: vapor pressure; viscosity; density; personnel safety hazards and protection; flammability; compatibility with specific metals and elastomers (both short and long term) and conditions of shipment.

Many of the required data were readily available from the Coast Guard's CHRIS (19) manual. CHRIS was used almost exclusively as the source of data on personnel hazards and fire hazards. Although not quite as complete in the area of physical properties and conditions of shipment, CHRIS still provided the bulk of this information. Furthermore, CHRIS provided the names of selected manufacturers for each chemical. When the necessary data could not be obtained by other means, these manufacturers were queried directly.

The least readily available data were the short and long-term effects of the chemicals on the various metallic and elastomeric materials. This information was obtained largely from material suppliers' compatibility summaries and from direct contact with chemical manufacturers. In some cases different sources provided conflicting information with regard to materials compatibility. In these cases engineering judgement and/or testing were used to determine the relative reliability of the sources. In about 10% of the cases, lacking data were obtained by extrapolating properties or behavior from similar chemicals for which data was available.

In order to verify material compatability data and to resolve questionable data, ten chemicals from different chemical families were selected for laboratory testing. In addition to collecting data on the ADAPTS components and on the chemical substances, a number of chemical transportation barges were inspected to determine their characteristics. Fortunately, it was established that almost without exception, all atmospheric pressure, bulk liquid transport barges have Butterworth deck access openings into the cargo tanks. These fittings allow direct, unobstructed access for the ADAPTS pump into the cargo.

Once the data on chemical properties, material compatability, and conditions of shipment had been acquired and the pump performance for each chemical had been determined, the chemicals were easily categorized according to pumpability. The following three major categories were used to group the chemicals: 50 chemicals pumpable; 21 chemicals unpumpable; and 195 chemicals pumpable with appropriate modifications. 33 of the chemicals listed as either pumpable by the existing or modified ADAPTS were designated as conditionally pumpable, because their pumpability is dependent on their exact properties; i.e., temperature, solution concentration, etc., must be within certain specified limits.

The chemicals requiring equipment modifications in order to be pumpable were grouped into subcategories requiring similar modifications.

The unpumpable chemicals were further classified by the reason for their being unpumpable; such as too viscous; too corrosive; vapor pressure too high; not shipped in bulk, etc. In order to extend the applicability of the ADAPTS pumping subsystem to the largest number of the chemicals, a variety of system component modifications were considered.

Modification Analysis

Incompatibilities between chemicals and specific materials used in the ADAPTS components were noted on the data summary sheets. Then, wherever practical, suitable alternate materials were listed for each application. By totaling up the number of chemicals which could be accommodated by each alternate material in a given component, the materials with the broadest application were selected. In order to simplify equipment logistics and to reduce the risk of using the wrong equipment with the wrong chemical, an attempt was made to develop modifications which would apply to the largest number of chemicals. Naturally, this led to the selection of materials with excellent chemical resistence such as type 316 stainless steel and Teflon for most applications. The modifications suggested are in a hierarchical order with each succeeding one requiring the existence of the preceding modifications to be effective.

The simplest and most significant modification in terms of the number of chemicals it added to the pumpable category is the inclusion of a pump shaft face seal. This seal prevents the pumped liquid from contacting the thrust bearing and bearing grease. A total of 78 additional chemicals become pumpable with the addition of this silicone carbide seal.

The next sequential modification is the elimination of the bronze sleeve bearings in the cargo pump. Replacement of these bearings by either Teflon or stainless steel sleeve bearings enables the system to handle an additional seven chemicals.

The present Buna-N cargo hose liner is not compatible with a number of the chemicals considered. Changing this hose to an all 316 stainless steel construction renders the hose interior and exterior compatible with all of the chemicals considered with the exception of hydrofluoric acid. This modification adds an additional 14 to the pumpable list.

When combined with the preceding modifications, a 316 stainless steel exterior on the portion of the hydraulic hose sections immersed in the pumped chemical will provide for the pumping of an additional 59 chemicals, making a total of 208 chemicals or 78%.

Replacing miscellaneous system parts made from aluminum, galvanized or cadmium plated steel, and carbon steel with 316 stainless steel will increase the acceptable chemicals by another 23 for a total of 231. Of the remaining 35 chemicals, 14 can be handled satisfactorily by changing to a 316 stainless steel pump instead of the present cast iron one. This brings the total number of pumpable chemicals to 245 or 92%. Among these 14 chemicals are phosphoric acid, sulfuric acid, and other commonly shipped bulk chemicals.

Conclusions

While only 50 or 19% of the chemicals are amenable to pumping using the original ADAPTS design, the inclusion of a pump shaft seal will raise this number to 48%. Further modifications to the pump internals, cargo hose lining, and to the exterior portions of the hydraulic transmission and cargo hoses which are immersed in the chemical will enable 231 or 87% of the chemicals to be handled. Finally, the use of an all stainless steel pump and hoses will allow all but 21 of the chemicals to be pumped.

Because the replacement of present ADAPTS hoses and pumps with stainless steel components would constitute a major system redesign and would be extremely costly, it was recommended that only the first two modifications, i.e., the addition of the pump shaft seal and the elimination of the bronze bearings, be made on all existing ADAPTS pumps. A smaller capacity, special chemical pumping system more appropriately sized to the chemical vessel capacities was recommended for pumping the 114 chemicals for which ADAPTS would not be suitable. This system should have a pump and hoses made of stainless steel and/or Teflon.

REMOTELY CONTROLLED COUNTERMEASURE DEVICES

> The information in this section is based on a paper presented at Conference 1974 by L.A. Witzeman of the Rural/Metro Fire Department, Inc., Scottsdale, Arizona. The title of the paper was "Snail–A Remotely Controlled Countermeasure Device."

Introduction

The automation and remote control technology that has been common and essential in handling hazardous materials for years is now just beginning to appear in fire fighting technology.

Today, we have the technology to produce hazardous materials in bulk in plants in which automatic valving, computer control and remote-operation technology not only reduce the manpower required in production but make it unnecessary, in many cases, for manpower to even work in an area in which hazardous materials are used. However, when bulk materials are shipped, because of collision or other mischance, these materials present a danger to the public. To date, the problem most typically has been solved with what at best might be described as hand technology.

While there is much concern with what happens in the production plant when a spill occurs, there should be even greater concern with what happens on the highways, railroads, and other transportation channels of the world when a mischance scatters these same chemicals, liquid or solid, volatile or stable, in the heart of a populous area, for modern technology will be of far less assistance in the field than in the factory. But a beginning technology offers help, a little now, and almost certainly a great deal more in the future.

This technology is, in fact, only an application to the field of mobile fire protection, much of which is already routinely dealt with in plant safety.

If a butane tank car develops a leak in the United States today, the leak will probably be sprayed with water to achieve dilution and control of a spreading blanket of potentially explosive, flammable fumes. A generation ago, these incidents were handled exactly the same way, by putting three men on the end of a length of 2½" hose and pouring 250 gal/min of water towards the leak. In this day of advanced technology, men are still placed 50 to 75 ft from the leaking tank car, exposing them personally to the grave effects of the rampaging, uncontrolled chemical.

Firemen are among the group that must respond when the efforts of the chemical industry fail to provide safe manufacture, safe storage, and/or safe transportation. In 1973, when a blazing butane tank car exploded in Kingman, Arizona, members of the Rural/Metro Fire Department, Inc. temporarily restored a minimum level of protection as a third of the entire Kingman Fire Department's manpower and equipment was eliminated in a few fearful seconds.

When a potentially similar incident happened some months later in southern Arizona, in an area for which the Rural/Metro has a responsibility, control was achieved by one man safely behind a railroad embankment operating what is called the snail. This incident was not a fire, but was a butane tank car leak of the type that apparently preceded the catastrophic Kingman blast. The tank car was on a siding east of Tucson. The tank car was cooled with a hose stream from the snail while men were flown in from El Paso, about 100 miles away, to repair the leak in the relief valve mechanism of the railroad tank car.

Snail

The snail is typical of a new generation of remote control fire fighting devices, frequently dubbed robots, which are being developed for the fire fighters of today.

It is a tracked, battery-operated device controlled by an umbilical cable on which is mounted a similarly remote controlled nozzle. The nozzle can operate through a full range of patterns from on and off to a straight stream or any fog pattern. It can be elevated or depressed or turned through a range of approximately 240°.

The device is about 26" wide, 46" long and 45" high. It weighs approximately 600 lb. Its present capability includes the ability to drag 400 ft of charged 2½" hose. Its operator controls it from a belt-carryable console strapped to his waist through an umbilical cable; he can stand about 150 ft away from the snail itself, far enough to be shielded from the usual effects of heat, but not far enough for some blast effects unless he has further shielding.

The equipment development is part of a project not only to evolve safer fire fighting conditions, but to lessen the need for brute manpower. The snail, for instance, when operating at a fire at 500 gal/min, its theoretical maximum gallonage, is doing the work of 6 hosemen and nozzlemen. The snail has the advantage of being lightweight and flexible. It would also be an economical unit to construct, being potentially available at a price many fire departments could afford.

Fire Cat

Another unit, however, is already on the market. The fire cat is commercially available today. It can drag 1,500 ft of hose and can be operated by radio from a similar distance. It is thus a far safer unit for the operator. Maneuverable up a 60° grade according to its manufacturer, having a 3,500 lb draw bar pull and able to travel 5 mph, it can achieve a gal/min discharge of 1,200 gal. Dimensionally, the unit is 4' high, 4' 4" long, 3 ft 10¾" wide and is driven by two 36V reversible motors rated at 2 hp each.

By contrast, the smaller, lighter snail travels approximately 3 mph on metal treads, is powered by two 6 V and two 8 V batteries supplying two small electric dc motors. It is deliberately sized to go through a standard residential doorway opening and built for transportation on existing trucks.

The usefulness of these two devices to those who handle hazardous materials is obvious. The first objective in the handling of hazardous materials spills is the same as the first objective of those in the fire service, life safety. Now there is available a first generation of remote control devices to do much of this dangerous work.

In contrasting these two devices, one finds the fire cat to be considerably more powerful in terms of rate of water application, safe distance between operator and machine, and speed. The snail on the other hand, has an enhanced portability due to its lighter weight and is vastly less costly to build, but is not yet on the market, although it can be leased through its builder. The higher cost of the fire cat has been a factor which has not enhanced its sales because, to the average relatively small fire department, it represents an investment too close to the cost of a fire truck itself. To be readily marketable, a remote device must be made cheaply.

The significant point to be made is simply that these devices are available and in the future they are going to be the vital tools of hazardous response technology. Obviously, they need more development such as the addition of such devices as television viewing cameras, infra-red sensing, etc; these developments, however, are occurring. The development of the snail and the fire cat went on simultaneously, apparently unknown to each other. While this was occurring, developers in Japan and Germany were at work on, and have experimentally produced, remote control devices to achieve the same goal.

Foreign Devices

The Yokohama Fire Department has experimented with a device which, unlike its American counterparts, is wheeled instead of tracked and has a generally vertical rather than horizontal configuration. Television viewing capability has already been installed and was a primary initial part of the development instead of being something on the future development list as in the case of the U.S. counterparts. It is being used experimentally by the Yokohama Fire Department, but has been developed under the aegis of Professor Saburo Takada of Kanagawa University, Kanagawa-Ken, Japan. It is the speed demon of the units under development to date, achieving approximately 10 mph, but due to its vertical configuration has displayed an alarming tendency to roll over when operating on difficult terrain. Standing approximately 6 ft tall, it weighs approximately 3,100 pounds.

Another unit under construction in Frankfurt, Germany under the overall direction of Chief Achilles of that department. Being developed for the Frankfurt airport, the most obvious feature of the German vehicle was its resemblance to a tanker with a cab at both ends, literally one cab facing one way and the other facing the opposite way. The Chief explained that a great deal of the Department's off-airport crash work involves negotiating narrow roads in wooded areas where it is impossible to turn around as rapidly as may be required under fire fighting conditions, and with the large water-carrying-capacity vehicles required. Consequently, the two cabs have immediate capability to reverse directions instantly with full control and speed.

Underneath this truck in the center were two compartments in which were to be remotely controlled devices. Chief Achilles' plan was to have the devices constructed in such a way that they could be driven out of these compartments by the drivers, who would not need to leave the heat-shielded cabs. The remote devices are intended to be primarily foam handling units, but could also handle water. Any of the devices discussed herein could alternatively handle either water or some types of foam.

Impending Developments

The robots of the United States, Germany and Japan are but one phase of the automated control devices that are being developed and will be of great assistance in handling of hazardous materials spills. Other impending technology developments which will be of assistance include:

> Construction of a number of completely automated fire trucks. Operating by remote radio or imbilical control, these units offer further safety because fewer men are required and they may be farther from the hazard. These units are being placed on the market.
>
> Programmable deluge guns available through Akron Brass Company. They may be set up at the scene manually and then programmed to sweep a given area with a pre-set water pattern. At this point, manpower may back up to the water source 1,000 ft away, and let nature take its course in relative safety. Completely remote controlled stationary deluge guns are also available from three manufacturers.
>
> New chemical approaches, ranging from so-called light water to various powders, are available to the fire fighter for handling hazardous materials.
>
> Infra-red heat detecting devices, specifically made for the fire service, expected on the market in an advanced form. These should give response teams the capability to detect heat buildup in a truckload of chemicals prior to ignition. For information on this device and other automation data, consult particularly with Public Technology, Inc., Washington, D. C.
>
> A number of items which make it possible for water to be supplied from a longer distance and still effectively handle fires and other incidents requiring large volumes of water. These might be categorized as items of water-moving technology and include rapid water of Union Carbide Company, which by reducing friction in flowing water increases the volumetric rate of flow. Large diameter hoses, 4" to 5",

are beginning to replace conventional 2½" hoses. This development has the effect of replacing the old-fashioned hose line standardized at about the turn of the century with new portable pipeline techniques.

Conclusion

The importance of the development to hazardous materials handling specialists is that the fire protection field will be better prepared as a back-up. Communications between the hazardous materials handling field and the fire protection field have vastly improved. In the alarm rooms in Arizona, there are publications describing the emergency handling of many items, publications which are more complete than they were a few years ago. Firemen can instantly obtain information from a 24 hr switchboard operated by the chemical manufacturers. The Rural/Metro Fire Department is in turn ready to fly the snail on short notice throughout a wide area and to provide standard complements of equipment on short notice for categories including typical railroad, mining, highway, and wildfire problems.

It is certainly time the hazardous materials engineer and the fire chief work closer together than ever before. The technology both need is coming fast; by working together, it can be used to its fullest.

REFERENCES

(1) Hiltz, R., Roehlich, F., and Brugger, J., "Emergency Collection System for Spilled Hazardous Materials," p 208-211, *Proceedings of National Conference on Control of Hazardous Material Spills,* AICHE/EPA, San Francisco (August 1974).

(2) American Public Health Association, *Standard Methods for the Examination of Water and Wastewater*, 13th Edition, Washington D.C. (1971).

(3) Sanders, R.G., and Rich, S.R., "A Short Contact Time Physical-Chemical Treatment System for Hazardous Material Contaminated Waters," *Proceedings of the National Conference on Control of Hazardous Material Spills*, Houston, Texas (March 1972).

(4) Dawson, G.W., Shuckrow, A.J., and Swift, W.H., *Control of Spillage of Hazardous Pollution Substances*, Battelle Memorial Institute, FWQA Report 15090 FOZ 10/70 (Contract 14-12-866), (November 1, 1970).

(5) Wilder, I., and LaFornara, J., *Control of Hazardous Material Spills in the Water Environment–An Overview,* paper presented before the Division of Water, Air and Waste Chemistry, American Chemical Society, Washington, D.C., (September 1971).

(6) King, P.H., Yeh, H.H., Warren, P.S., and Randall, C.W., "Distribution of Pesticides in Surface Waters," *Journal AWWA*, Vol. 61, No. 483, (September 1969).

(7) Lotse, E.G., Graetz, D.A., Chesters, G., Les, G.B., and Newland, L.W., "Lindane Adsorption by Lake Sediments," *Environmental Science Technology,* Vol. 2, No. 353, (May 1968).

(8) Schwartz, H.G., "Adsorption of Selected Pesticides on Activated Carbon and Mineral Surfaces", *Environmental Science and Technology*, Vol. 2, No. 353, (May 1968).

(9) Huang, J.C. and Liao, C.S., "Adsorption of Pesticides by Clay Minerals," *Journal Sanitary Engineering Division, Proceedings of the Chemical Society of Civil Engineers,* 96:SA5, (October 1970).

(10) Morris, J.C., and Weber, W.J., *Adsorption of Biochemically Resistant Materials*

from Solution, 2, USPHS Environmental Health Series 999-WP-33, U.S. Government Printing Office, Washington, D.C., (1966).
(11) Aly, O.H., and Faust, S.D., "Studies on the Removal of 2,4-D and 2,4-DCP from Surface Waters," *Proceedings of the 18th Industrial Waste Conference,* Purdue University, p 6, (1963).
(12) Robeck, G.G., Dostal, K.A., Cohen, J.M., and Kreissl, J.F., "Effectiveness of Water Treatment Processes in Pesticide Removal," *Journal AWWA*, Vol. 57, No. 181, (February 1965).
(13) Shapiro, H., and Frey, F.W., *The Organic Compounds of Lead*, Inter-Science Publisher, New York, N.Y., (1968).
(14) Michalovic, J.G., et al, *Multipurpose Gelling Agent and Development of Means of Applying to Spilled Hazardous Materials,* EPA Report No. 600/2-77-151, U.S. Environmental Protection Agency, Cincinnati, Ohio, (January 1977).
(15) Freestone, F.J., "Shakedown and Demonstration of R&D Prototype Devices and Techniques at Hazardous Material Spills," *Proceedings of the 1976 National Conference on Control of Hazardous Material Spills,* New Orleans, La., (April 1976).
(16) Juhola, A.J., and Brugger, J.E., "Pilot Study of Activated Carbon Regeneration." *Proceedings of the 1976 National Conference on Control of Hazardous Material Spills*, p 219, New Orleans, La., (April 1976).
(17) Gupta, M.K., *Development of a Mobile Treatment System for Handling Spilled Hazardous Materials*, EPA Report 600/2-76-109 (July 1976).
(18) Lafornara, J.P., "Clean-up After Spills of Toxic Substances," *Journal Water Pollution Control Federation* (in press)
(19) *Chemical Hazards Response Information System*, Vol. 2, U.S. Coast Guard Publication CG-446-2, (January 1974).

SOURCES UTILIZED

Conference 1972
Proceedings of the 1972 National Conference on Control of Hazardous Material Spills, PB-228-736, sponsored by Environmental Protection Agency, Houston, TX, March 21-23, 1972.

Conference 1974
Proceedings of the 1974 National Conference on Control of Hazardous Material Spills, G.F. Bennett, editor; sponsored by Environmental Protection Agency and American Institute of Chemical Engineers, San Francisco, CA, August 25-28, 1974.

Conference 1976
Proceedings of the 1976 National Conference on Control of Hazardous Material Spills, G.F. Bennett, editor; sponsored by Environmental Protection Agency and Oil Spill Control Association of America, New Orleans, LA, April 25-28, 1976.

Conference 1978
Proceedings of the 1978 National Conference on Control of Hazardous Material Spills, G.F. Bennett, editor; sponsored by Environmental Protection Agency, United States Coast Guard, and Hazardous Materials Control Research Institute, in participation with Oil Spill Control Association of America, Miami Beach, FL, April 11-13, 1978.

CG-D-16-77
A Feasibility Study of Response Techniques for Discharges of Hazardous Chemicals that Disperse through the Water Column, by E. Drake, D. Shooter, W. Lyman, and L. Davidson of Arthur D. Little, Inc., prepared for the U.S. Coast Guard, July 1976.

CG-D-38-76
Agents, Methods and Devices for Amelioration of Discharges of Hazardous

Chemicals on Water, by W.H. Bauer, D.N. Borton, and J.J. Bulloff of Rensselaer Polytechnic Institute, prepared for the U.S. Coast Guard, August 1975.

CG-D-56-77
Feasibility Study of Response Techniques for Discharges of Hazardous Chemicals that Float on Water, by J.S. Greer of MSA Research Corporation, prepared for the U.S. Coast Guard, October 1976.

CG-D-56-78
A Feasibility Study of Response Techniques for Discharges of Hazardous Chemicals that Sink, by T.D. Hand, A.W. Ford, P.G. Malone, D.W. Thompson, and R.B. Mercer of the U.S. Army Engineer Waterways Experiment Station, prepared for the U.S. Coast Guard, June 1978.

EPA-R2-73-185
Control of Hazardous Chemical Spills by Physical Barriers, by J.V. Friel, R.H. Hiltz, and M.D. Marshall of MSA Research Corporation, prepared for the Environmental Protection Agency, March 1973.

EPA-600/2-76-109
Development of a Mobile Treatment System for Handling Spilled Hazardous Materials, by M.K. Gupta of Envirex, prepared for the Environmental Protection Agency, July 1976.

EPA-600/2-76-245
Removal and Separation of Spilled Hazardous Materials from Impoundment Bottoms, by M.A. Nawrocki of Hittman Associates, Inc., prepared for Environmental Protection Agency, September 1976.

EPA-600/2-77-151
Multipurpose Gelling Agent and Its Application to Spilled Hazardous Materials, by J.G. Michalovic, C.K. Akers, R.E. Baier, and R.J. Pilie of Calspan Corporation, prepared for the Environmental Protection Agency, August 1977.

EPA-600/2-77-162
Emergency Collection System for Spilled Hazardous Materials, by R.H. Hiltz and F.R. Roehlich, Jr. of MSA Research Corporation, prepared for the Environmental Protection Agency, August 1977.

EPA-600/2-77-164
In Situ Treatment of Hazardous Material Spills in Flowing Streams, by G.W. Dawson, B.W. Mercer, and R.G. Parkhurst of Battelle-Northwest, prepared for the Environmental Protection Agency, October 1977.

EPA-600/2-77-227
Manual for the Control of Hazardous Material Spills. Vol. 1: Spill Assessment and Water Treatment Techniques, by K.R. Huibregtse, R.C. Scholz, R.E. Wullschleger, J.H. Moser, E.R. Bollinger, and C.A. Hansen of Envirex, prepared for the Environmental Protection Agency, November 1977.

EPA-600/2-78-145
System for Applying Powdered Gelling Agents to Spilled Hazardous Materials, by J.G. Michalovic, C.K. Akers, R.W. King and R.J. Pilie of Calspan Corporation, prepared for the Environmental Protection Agency, July 1978.

EPA-670/2-73-078
Treatment of Hazardous Material Spills with Floating Mass Transfer Media, by B.W. Mercer, A.J. Shuckrow, and G.W. Dawson of Battelle-Northwest, prepared for the Environmental Protection Agency, September 1973.

EPA-670/2-74-073
Evaluation of MTF for Testing Hazardous Material Spill Control Equipment, by C.R. Thomas and G.M.L. Robinson of Environment Control Corporation and E.J. Martin of Environmental Quality Systems, Inc., subcontracted by Hancock County Port and Harbor Commission for the Environmental Protection Agency, December 1974.

EPA-670/2-75-042
Methods to Treat, Control and Monitor Spilled Hazardous Materials, by R.J. Pilie, R.E. Baier, R.C. Ziegler, R.P. Leonard, J.G. Michalovic, S.L. Peck, and D.H. Bock of Calspan Corporation, prepared for the Environmental Protection Agency, June 1975.